The "African" Honey Bee

Westview Studies in Insect Biology

Michael D. Breed, Series Editor

The "African" Honey Bee, edited by Marla Spivak, David J.C. Fletcher, and Michael D. Breed

Biocontrol of Arthropods Affecting Livestock and Poultry, edited by Donald A. Rutz and Richard S. Patterson

Applied Myrmecology: A World Perspective, edited by Robert K. Vander Meer, Klaus Jaffe, and Aragua Cedeno

The Genetics of Social Evolution, edited by Michael D. Breed and Robert E. Page, Jr.

The Entomology of Indigenous and Naturalized Systems in Agriculture, edited by Marvin K. Harris and Charlie E. Rogers

Interindividual Behavioral Variability in Social Insects, edited by Robert L. Jeanne

Integrated Pest Management on Rangeland: A Shortgrass Prairie Perspective, edited by John L. Capinera

Management of Pests and Pesticides: Farmers' Perceptions and Practices, edited by Joyce Tait and Banpot Napompeth

Fire Ants and Leaf-Cutting Ants: Biology and Management, edited by Clifford S. Lofgren and Robert K. Vander Meer

The "African" Honey Bee

EDITED BY
Marla Spivak, David J.C. Fletcher, and Michael D. Breed

Routledge
Taylor & Francis Group
LONDON AND NEW YORK

First published 1981 by Westview Press, Inc.

Published 2019 by Routledge
52 Vanderbilt Avenue, New York, NY 10017
2 Park Square, Milton Park, Abingdon, Oxon OX14 4RN

Routledge is an imprint of the Taylor & Francis Group, an informa business

Library of Congress Cataloging-in-Publication Data
The "African" honey bee / edited by Marla Spivak, David J.C. Fletcher, and Michael D. Breed
p. cm. — (Westview studies in insect biology)
Includes index.
ISBN 0-8133-7209-7
1. Africanized honeybee. I. Spivak, Marla. II. Fletcher, David J.C. III. Breed, Michael D. IV. Series.
QL568.A6A35 1991
595.79'9—dc19

86-32606
CIP

ISBN 13: 978-0-367-28999-7 (hbk)

Contents

PART THREE
POPULATION BIOLOGY, ECOLOGY, AND DISEASES

PART FOUR
DEFENSIVE BEHAVIOR

PART FIVE
BEEKEEPING IN SOUTH AMERICA

1

INTRODUCTION

Marla Spivak,[1] David J.C. Fletcher,[2] and Michael D. Breed[3]

This book is the first review of the scientific literature on the Africanized honey bee. The African subspecies *Apis mellifera scutellata* (formerly *adansonii*) was introduced into South America in 1956 with the intent of cross-breeding it with other subspecies of bees already present in Brazil to obtain a honey bee better adapted to tropical conditions. Shortly after its introduction, some of the African stock became established in the feral population around São Paulo, Brazil, and spread rapidly through Brazil. It has since migrated through most of the neotropics, displacing and/or hybridizing with the previously imported subspecies of honey bees. Africanized bees have been stereotyped as having high rates of swarming and absconding, rapid colony growth, and fierce defensive behavior. As they have spread through the neotropics they have interacted with the human population, disrupting apiculture and urban activities when high levels of defensive behavior are expressed.

Our goal as editors was to bring together the large body of information that has become available concerning the Africanized bee and its spread and impact through the New World. Accordingly, we present chapters from a diversity of authors, include important research not previously reviewed in English, and cover a wide range of scientific methods including both basic and applied research. A couple important investigators were unwilling or unable to deliver promised chapters; nevertheless, we feel that this book is a success in that it brings some much needed objective balance and clarity to a subject that has often been clouded by emotions.

---

[1]Dr. Spivak is a Research Associate in the Center for Insect Science, University of Arizona, Tuscon, AZ, 85719, USA. Her address for correspondence is the USDA, Carl Hayden Bee Research Laboratory, 2000 E. Allen Rd., Tuscon, AZ 85719, USA.

[2]Professor Fletcher is in the Department of Biology and Program in Animal Behavior, Bucknell University, Lewisburg, PA, 17837, USA.

[3]Professor Breed is in the Department of Environmental, Population and Organismic Biology, The University of Colorado, Boulder, CO, 80309-0334, USA.

While we as editors do not have a consensus even among ourselves as to the best solution to the "Africanized bee problem," we agree that there is sufficient variation in behavior within the honey bee population in the neotropics to provide the basis for future selection programs to yield manageable and productive honey bee populations that can coexist with humans. Others believe that the best way of ameliorating whatever undesirable traits these bees may display in the United States will be to modify them genetically by controlled hybridization with the honey bees already present in this country.

Our intention is to provide reviews of the major topics concerning Africanized honey bees. Recently, Needham *et al.* (1988) published a compendium of current research on Africanized honey bees and bee mites. Various other symposia and congresses are cited in Needham *et al.* (1988) or in this work. There is a large body of unpublished Master's theses and Doctoral dissertations from Brazilian universities on Africanized bees which provide a substantial base of knowledge concerning the biology of these bees. These have, unfortunately, been neglected in most English language reviews on Africanized bees; a list of these works is available from Dr. Lionel Gonçalves in Brazil.[4] Our hope is that readers will be able to identify common themes and principles concerning Africanized bees among the diverse views presented in this book.

WHAT SHOULD WE CALL THIS BEE?

The introduced bees from which our Africanized bees are derived were from a race distributed in southern and eastern Africa, *Apis mellifera scutellata* Lepeletier. Most of the other bees in the New World are derived from European races, *A. m. mellifera* L. (Germany), *A. m. ligustica* Spin. (Italy), *A. m. caucasica* Gorb. (Caucasus mountains, northern Europe and west central Russia), and *A. m. carnica* Pollman (southern Austrian alps, northern Yugoslavia, Danube Valley) (Winston, 1987).

Our opinion is that in general there has been too great of a tendency to pigeon-hole bees found in the New World as either European or African in origin. While the genetic origins of New World bees can be ascertained, the distinctions among the Old World subspecies, have been blurred by processes of hybridization, artificial selection, and natural selection in their current ecological context. Thus, knowledge concerning honey bees will probably be better advanced by considering New World bees as representing, to greater or lesser extents, new subspecies based on novel gene combinations and adaptations to local conditions.

---

[4]Depto. de Biologia, Faculdade de Filosifia, Ciências e Letras de Riberão Preto, Universidade de São Paulo, Riberão Preto, 14.049, São Paulo, Brazil.

We have allowed the authors of the individual chapters to choose their own terminology when referring to the populations of bees discussed in this book. This book deals with a highly variable set of populations, some of which fit the stereotype of swarming, absconding and defense mentioned above and others which do not. As readers assimilate the information in this book, some statements may seem contradictory because they refer to different populations in different life zones that come under the general grouping of Africanized bees. We feel that, given the genuine differences of opinion among experts, this is the proper way to handle this difficult problem of systematics and nomenclature in this rapidly expanding and evolving population.

A BRIEF HISTORY OF THE INTRODUCTION OF AFRICAN BEES TO THE NEW WORLD

The most widely disseminated narrative of the introduction of African bees is presented by Kerr (1957). In this account, Professor W. E. Kerr brought 170 queens from the savannah of eastern and South Africa to Brazil. Forty-seven or 48 survived importation and were successfully introduced into colonies at Piracicaba (São Paulo state); one of these queens was from Tanzania (called Tanganyika at the time) and the others were from South Africa. These were moved in November of 1956 to Camapuã, Rio Claro, São Paulo state, Brazil, in a *Eucalyptus* forest of the Paulista Railroad. In early 1957, an ill-informed technician of the Railroad removed queen excluders from the entrances of 26 of these colonies, which soon swarmed (Nogueira-Neto, 1964; Kerr, pers. comm.; Kerr, 1966/67; Kerr, 1967; Gonçalves, 1974; Gonçalves and Stort, 1978). Coincidentally, Kerr (1957) characterized 26 of these colonies as "the most prolific, productive and industrious bees that we have seen up to now." It is not clear from the accounts whether this refers to the same 26 colonies which escaped.

There are several studies which support an argument that not all of the original colonies escaped, and which are not consistent with the commonly held view that the original spread was due to just 26 colonies. Kerr and Mendoça-Fava (1972) refer to experiments performed in 1957 using "hybrids" between African queens and *ligustica* drones, which indicates early rearing of African queens. Widely cited articles (Kerr, 1966/67, 1967; Portugal-Araújo, 1971) present tables in which honey production is compared among three races of bees, African, German, and Italian. Ten colonies of African bees were used in both 1958 and 1959 in these experiments, after the escape of the 26 colonies. In fact, African bees were propagated through queen rearing and artificial insemination during or after the 1958 and 1959 experiments and distributed widely among beekeepers in southern Brazil (Kerr, pers. comm.). By the time of the first International Congress of Apiculture held in Florianopolis in 1970, there were published reports of African queens that were reared and inseminated with African drones (Kerr *et al.*, 1972).

Kerr (1966/67) presented a detailed discussion of the problems associated with the African bees and proposed a series of solutions, including enhancement of drone production in manageable stock, use of better protective equipment, placement of apiaries away from people and livestock, and use of Italian queens. The solutions presented in this article were of considerable interest, but unfortunately it was never translated and circulated in Spanish-speaking Latin America or in North America. Many of the solutions now being proposed stem from his original ideas, although his work is sometimes not cited.

In a survey of beekeepers Kerr (1966/67) found a slight preference for Africanized bees although some beekeepers reported quitting the business because of difficulties managing these bees. De Jong (1984) also reports a strong preference among the remaining beekeepers for African bees because of the bees' high productivity. Another reason beekeepers in some regions may prefer Africanized bees is that their highly defensive nature deters theft of equipment and honey. A preference for African bees is also probably dependent on the availability and affordability of both feral swarms and beekeeping technology that allows management of highly defensive bees. Many beekeepers in developing countries do not have access to movable frame hives, smokers, veils and other protective clothing.

Although problems with defensive behavior were quickly recognized in Brazil and Argentina, the first English language discussion in the scientific literature of a serious problem with defensiveness came from Nogueira-Neto in 1964. In 1969, Kerr published a map of the distribution of Africanized bees in Brazil (Kerr, 1969). Soon after, the National Research Council (USA) commissioned a report (Anonymous, 1972) on the developing problem. Professor C. D. Michener participated in the NRC study and published his assessment separately (Michener, 1975). Shortly thereafter, Taylor (1977) published maps which illustrated the predicted rate of spread of Africanized bees, their climatic limits in North America, and discussed the potential impact on beekeepers. Professor Taylor's predictions have been remarkably accurate, although the hypotheses about northern climatic limits have not yet been tested. The balance of the story concerning the spread, ecology, and demography of Africanized bees is dealt with in various chapters of this book.

HAS HYBRIDIZATION OCCURRED AMONG SUBSPECIES?

This critical and controversial issue is addressed in the first section of the book. Because of rapid advances in research, it cannot be conclusively reviewed in this book. Three major techniques have been applied to the question. Daly (Chapter 2) discusses the use of morphometrics and allozymes. Hall (Chapter 3) focuses on the use of DNA techniques. Some morphometric analyses have supported the argument that intermediate phenotypes occur in Brazil and southern South America (Buco *et al.*, 1987; Rinderer, unpubl. data) while another study (Boreham and Roubik, 1987) has argued that the bee population in Panama has, over time, reverted to an African phenotype. Allozyme studies of the malate dehydrogenase locus have shown evidence of hybridization (Nunamaker and Wilson, 1981; Lobo *et al.*, 1989), however after initial hybridization in a given region, the alleles acquired from European bees may decline in frequency (O. R. Taylor, pers. comm.). Studies of mitochondrial DNA by Hall and Muralidharan (1989) and Smith *et al.* (1989) show similar restriction fragment patterns between the bees they sampled in the neotropics and African (*A. m. scutellata*) bees. These results are consistent with a study of nuclear DNA in Mexico in which the bees show a lack of persistent hybridization with European bees (Hall, 1990). However, these DNA patterns may be different in Argentina where the feral African population approaches its climatic limits in northern latitudes and European bees persist further south (Sheppard, unpubl. data). We hope these issues will be resolved in the near future.

THE SPREAD OF AFRICANIZED BEES AND THE AFRICANIZATION PROCESS

In the second section of the book, the Africanization process is considered. This consists of two distinct processes, migration of Africanized or African bees into new areas and gene flow among populations. To a considerable extent, workers in this area have been unable to distinguish between the migration of the bees into an area and subsequent evolutionary events. Some of these issues are covered in the chapters on identification, discussed above; in this section the chapters focus on ecological and behavioral mechanisms of the spread of Africanized bees. There are fundamentally differing views on how Africanization proceeds. For example, Rinderer and Hellmich, Chapter 5, argue that hybridization is a key factor, while one of us, Fletcher in Chapter 4, sees this population as African, rather than Africanized in nature. Fletcher (Chapter 4), proposes that a new introduction of African bees be undertaken, with pre- and post-introduction selection programs to insure that manageable and productive phenotypes are obtained. These differences may in fact be reconciled in part by Ratnieks (Chapter 6) approach which considers both that hybridization and selection processes may have taken place at and behind the migratory front. Another of the editors, Spivak, presents a case study of Africanization in Costa Rica (Chapter 7). This section of the book closes with a general review of honey bee genetics and breeding by Page and Kerr; this includes a review by Kerr of what measures have been taken in Brazil. Hellmich, Chapter 9, gives practical advice on preparing for Africanization.

POPULATION BIOLOGY AND ECOLOGY

Knowledge is somewhat better based concerning population biology and ecology than systematics and the Africanization process. Authors here provide the important basic data on colony dynamics (Winston, Chapter 10), population biology (Otis, Chapter 11), foraging (Rinderer and Collins, Chapter 12), ecology (Roubik, Chapter 13), and diseases (Shimanuki *et al.*, Chapter 14). These data have formed the fundamental characterization of Africanized bees and will provide the basis for further studies involving their ecology.

DEFENSIVE BEHAVIOR

The greatest cause of public concern over the Africanized bee has been its defensive behavior and particularly its stinging behavior. One of us, Breed, gives an overview of honey bee defensive behavior (Chapter 15). Collins and Rinderer (Chapter 16) and Stort and Gonçalves (Chapter 17) give detailed accounts of the genetic work on defensive behavior in European and Africanized bees. In our opinion there are two major gaps in knowledge of this area. First, we do not understand the range of defensive phenotypes in Africanized bees. It is difficult or even impossible to design rational selection programs without this knowledge. Second, there has been virtually no work on the ethology of colony defense of Africanized bees. We do not know if defense is organized in the same fashion as in European bees or if there are completely different modes of communication and division of labor in Africanized bees. The ethology of defense of European bees is reviewed in Breed's chapter. It is remarkable, given the span of time since the introduction of African bees and the public, agricultural, and scientific concern over defensiveness that more studies have not been done.

BEEKEEPING IN SOUTH AMERICA

To date the greatest experience with keeping Africanized bees has been in South America. We present three case studies (Brazil, Chapter 18; Peru, Chapter 19; and Venezuela, Chapter 20). In addition Spivak (Chapter 7) discusses issues related to beekeeping in Costa Rica. We view the Brazilian experience as being particularly important because they have had Africanized bees for 30 years and claim success in their selection programs. The Peruvian study also holds particular interest because the investigator brings a social scientist's viewpoint to his study. Our intent with these chapters was to bring together enough information so that research directions might become apparent. We also hope that the chapters will be useful to beekeepers in the United States.

CLIMATIC LIMITS OF AFRICANIZED BEES

The climatic limits of Africanized bees has been a controversial issue and there is no clear conclusion about survivorship of Africanized bees in temperate areas. Readers interested in this topic are referred to studies in Argentina (Kerr et al., 1982; Dietz *et al.*, 1985, 1988, 1990; Krell *et al.*, 1985), in Germany (Villa *et al.*, 1990) and predictions by Taylor (1977, 1985) and Taylor and Spivak (1984). The rapid spread of Africanized bees into North America may soon provide an empirical answer to this question. It is difficult to translate the experiences of South and Central American beekeepers into projections for North America because of differences in climate and beekeeping technology. Nevertheless, it is clear that Africanized bees have disrupted beekeeping and public activities wherever they have migrated and established permanent populations. The extent of the continued disruption appears to depend largely upon beekeepers and the implementation of selection programs.

CONCLUSIONS AND FUTURE DIRECTIONS

We have drawn together information on the identification, spread, ecology, defensive behavior, and practical implications of the Africanized honey bee in the New World. The most rapidly growing field of inquiry deals with identification and the extent of hybridization. How and why are European bees being hybridized and displaced while Africanized bees appear to have maintained a high degree of phenotypic and genetic similarity to bees from South Africa? More information should be available on these critical issues soon. We hope that this book stimulates further work on defensive behavior, particularly on phenotypic variation and the ethology of defense in Africanized bees. Many studies have assumed that morphometric or other identification techniques predict defensiveness; this linkage has not been firmly established and when studies of variability in defensiveness are conducted it will be important to pursue correlates between defensive traits and traits used in identification. Despite our use of the term "Africanized" we do not believe that this is a completely appropriate characterization of a population that has been subject to both natural and artificial selection since 1956 and which occurs in many different ecological zones. One of the challenges facing researchers on Africanized bees is to acknowledge that phenotypic differences among bees in different areas may explain apparent contradictions among studies. We hope that the focus of research will shift to unifying principles with an underlying comprehension of phenotypic variation.

LITERATURE CITED

Anonymous. 1972. *Final report of the Committee on the African honey bee*, Washington, D. C.: Natl. Res. Counc. Natl. Acad. Sci. 95 pp.

Boreham, M. M., Roubik, D. W. 1987. Population change and control of Africanized honey bees in the Panama canal area. *Bull. Entomol. Soc. Am.* 33:34-38.

Buco, S. M., Rinderer, T. E., Sylvester, H. A., Collins, A. M., Lancaster, V. A., Crewe, R. M. 1987. Morphometric differences between South American Africanized and South African (*Apis mellifera scutellata*) honey bees. *Apidologie* 18:217-222.

De Jong, D. 1984. Africanized bees now preferred by Brazilian beekeepers. *Am. Bee J.* 124:116-118.

Dietz, A., Krell, R., Eischen, F. A. 1985. Preliminary investigation on the distribution of Africanized honey bees in Argentina. *Apidologie* 16:99-108.

Dietz, A., Krell, R., Pettis, J. 1988. Survival of Africanized and European honey bee colonies confined in a refrigeration chamber. In *Africanized Honey Bees and Bee Mites*, ed. G. R. Needham, R. E. Page, M. Delfinado-Baker, C. E. Bowman, pp. 237-242. Chichester, England: Ellis Horwood Limited.

Dietz, A., Krell, R., Pettis, J. 1990. Study on winter survival of Africanized and European honey bees in San Juan, Argentina. *Apidologie* In Press.

Gonçalves, L. S. 1974. The introduction of the African bees (*Apis mellifera adansonii*) into Brazil and some comments on their spread in South America. *Am. Bee J.* 114:414-415, 419.

Gonçalves, L. S., Stort, A. C. 1978. Honey bee improvement through behavioral genetics. *Ann. Rev. Entomol.* 31:197-213.

Hall, H. G. 1990. Parental analysis of introgressive hybridization between African and European honey bees using nuclear DNA RFLP's. *Genetics* 125:611-621.

Hall, H. G., Muralidharan, K. 1989. Evidence from mitochondrial DNA that African honey bees spread as continuous maternal lineages. *Nature* 339:211-213.

Kerr, W. E. 1957. Introdução de abelhas africanas no Brasil. *Brasil Apicola* 3:211-213.

Kerr, W. E. 1966/67. Solução e criar uma raca nova. *Guia Rural* 67:20-22.

Kerr, W. E. 1967. The history of the introduction of Africanized bees to Brazil. *S. Afr. Bee J.* 39:3-5.

Kerr, W. E. 1969. Some aspects of the evolution of social bees (Apidae). In *Evolutionary Biology, Vol. 3*, ed. T. Dobzhansky, M. K. Hecht, W. C. Steere, pp. 119-175. New York: Meredith Corporation.

Kerr, W. E., Del Rio, S. Barrionuevo, M. D. 1982. Distribuicão de abelha africanizida em seus limites ao sul. *Ciencia e Cultura* 34:1439-1442.

Kerr, W. E., Del Rio, S., De Barrionuevo, M. D. 1982. The southern limits of the distribution of the Africanized bee in South America. *Am. Bee J.* 122:196-198.

Kerr, W. E., Gonçalves, L. S., Blotta, L. F., Maciel, J. B. 1972. Biología comparada entre as abelhas italianas (*Apis mellifera ligustica*) Africana (*Apis mellifera adansonii*) e suas híbridas. *1° Cong. Bras. Apic. (Florianópolis, 1970)*, ed. H. Wiese pp. 151-185. Santa Catarina, Brazil.

Kerr, W. E., Mendoça-Fava, J. F. de 1972. Contribução para a apicultura migratoria racional no estado de São Paulo. *1° Cong. Bras. Apic. (Florianópolis, 1970)*, ed. H. Wiese pp. 80-87. Santa Catarina, Brazil.

Krell, R., Dietz, A., Eischen, F. A. 1985. A preliminary study on winter survival of Africanized and European honey bees in Cordoba, Argentina. *Apidologie* 16:109-118.

Lobo, J. A, Del Lama, M. A., Mestriner, M. A. 1989. Population differentiation and racial admixture in the Africanized honey bee (*Apis mellifera* L.). *Evolution* 43:794-802.

Michener, C. D. 1975. The Brazilian bee problem. *Ann. Rev. Entomol.* 20:399-416.

Nogueira-Neto, P. 1964. The spread of a fierce African bee in Brazil. *Bee World* 45:119-121.

Nunamaker, R. A., Wilson, W. T. 1981. Comparison of MDH allozyme patterns in the African honey bee (*Apis mellifera adansonii* L.) and the Africanized population in Brazil. *J. Kans. Entomol. Soc.* 54:704-710.

Needham, G. R., Page, R. E., Delfinado-Baker, M., Bowman, C. E. eds. 1988. *Africanized Honey Bees and Bee Mites.* Chichester, England: Ellis Horwood Limited.

Portugal-Araújo, V. de 1971. The central African bee in South America. *Bee World* 52:116-121.

Severson, D. W., Aiken, J. M., Marsh, R. F. 1988. Molecular analysis of North American and Africanized honey bees. In Africanized Honey Bees and Bee Mites, ed. G. R. Needham, R. E. Page, M. Delfinado-Baker, C. E. Bowman. pp. 294-302. Chichester, England: Ellis Horwood Limited.

Smith, D. R., Taylor, O. R., Brown, W. W. 1989. Neotropical Africanized honey bees have African mitochondrial DNA. *Nature* 339:213-215.

Taylor, O. R. 1977. The past and possible future spread of Africanized honey bees in the Americas. *Bee World* 58:19-30.

Taylor, O. R. 1985. African bees: Potential impact in the United States. *Bull. Entomol. Soc. Am.* 31:15-24.

Taylor, O. R., Spivak, M. 1984. Climatic limits of tropical African honey bees in the Americas. *Bee World* 65:38-47.

Villa, J. D., Koeniger, N., Rinderer, T. E. 1990. Overwintering of Africanized, European, and hybrid honey bees in Germany. *Environ. Entomol. Physiol. & Chem. Ecol.* In Press.

Winston, M. L. 1987. *The Biology of the Honey Bee.* Cambridge, Mass: Harvard University Press.

PART ONE

Systematics and Identification

2

SYSTEMATICS AND IDENTIFICATION OF AFRICANIZED HONEY BEES

Howell V. Daly[1]

The purpose of this chapter is to evaluate the current status of systematics of Africanized bees and of several methods of identification based on the phenotype. Both subjects are more complex than often appreciated, hence the need to discuss the history and background in some depth. The name "African bees" will be used for native bees of Africa and "Africanized bees" for their relatives in the Western Hemisphere.

Identification is the first step for all research and efforts to mitigate the problems created by Africanized bees. Furthermore, reports about Africanized bees are accurate only to the extent that the identifications are trustworthy. For these reasons, reliable methods are currently being developed for the purposes of scientific investigation, regulation (Stibick, 1984), and for future breeding and certification (Page and Erickson, 1985). For reports of recent progress in identification, see Needham *et al.* (1988).

SYSTEMATICS

Apis mellifera is well known for its remarkable communication and environmental control in the hive. Less familiar is its extraordinary biogeography. In contrast to other Apoidea where congeneric sympatry and limited distributions are common, *A. mellifera* occupies an immense and varied geographic area. Except for a narrow overlap with *Apis florea* in the east, *A. mellifera* coexists naturally with no other member of its genus. The distribution extends from southern Scandinavia south to the Cape of Good Hope and from Senegal east to about 60° E longitude (Ural Mountains; Mashhad, Iran; and coast of Oman) (Ruttner *et al.*, 1978). Colonies are found from sea level to about

[1]Professor Daly is in the Department of Entomology, The University of California, Berkeley, Berkeley, CA, 94720, USA.

1000 m in the Alps in the temperate zone (F. Ruttner, pers. comm.) and, in the tropics, from sea level to 3100 m on Mt. Kilimanjaro in Africa. They are missing from extreme deserts, but survive as wild colonies in hot, arid Oman at 200-1500 m (Dutton *et al*., 1981).

Populations throughout this vast distribution are believed to be largely interfertile. They are similar in morphology, have the same number of chromosomes, and exhibit low protein polymorphism. They probably share the same gene loci, but differ in allelic frequencies at some loci. Ruttner (1988a) has argued that the present distribution might be no older than the late Pliocene. Adaptation to local environments has created geographic races of greater or lesser distinction, depending mainly on physical barriers. Differences among races are found in morphometry, behavior, and physiology (Cornuet and Louveaux, 1981).

Beginning in 1956 in Brazil, African bees (one colony from Tanzania and 46 from Pretoria, South Africa) were said to be crossed with European bees (primarily *A. m. ligustica* and *A. m. mellifera*), to produce Africanized bees (Filho *et al*., 1964; Kerr, 1967, 1969). This was a cross between distantly related races: bees of Europe and Southern Africa had evolved under different physical and biotic ecology and biogeographic history; they were separated by over 70° latitude; and genetic exchange had been further restricted for at least the last 2,000 years by the Sahara desert (Ruttner, pers. comm.). The relative contributions of European and African ancestry in Africanized bees at the outset is unknown. Nor do we know the genetic consequences of subsequent hybridization with European bees and of natural selection in new habitats of the Western Hemisphere. During the period 1982 to 1985 in Panama, Boreham and Roubik (1987) found morphometric measurements of Africanized bees to become smaller or more African-like. It is clear that the entity we now call Africanized is not a singular population, but rather a series of variable populations.

Partial reproductive isolation is known to exist between European and Africanized bees (Kerr and Bueno, 1970). As Africanized bees spread through South and Central America, their reproductive biology apparently has operated to perpetuate their African ancestry and give them sufficient advantage to replace European bees (Taylor, 1985; Rinderer, 1986). Africanized bees resemble their African parents more than their European parents in mitochondrial DNA, morphometry, hemolymph proteins, biochemistry of cuticular hydrocarbons, and behavioral characteristics. In view of this similarity, Taylor (1985) has speculated that Africanized bees are essentially African bees. The new DNA technology described in the next Chapter and elsewhere (Hall, 1988; Hall and Mulralidharan, 1989; Severson *et al*., 1988; D. Smith, 1988; D. Smith *et al*., 1989) will provide a method for assessing the genomes of these variable populations.

In the meantime, I provide here a graphic representation (FIGURE 1) of the morphometric relationships among African, Africanized, and European bees by

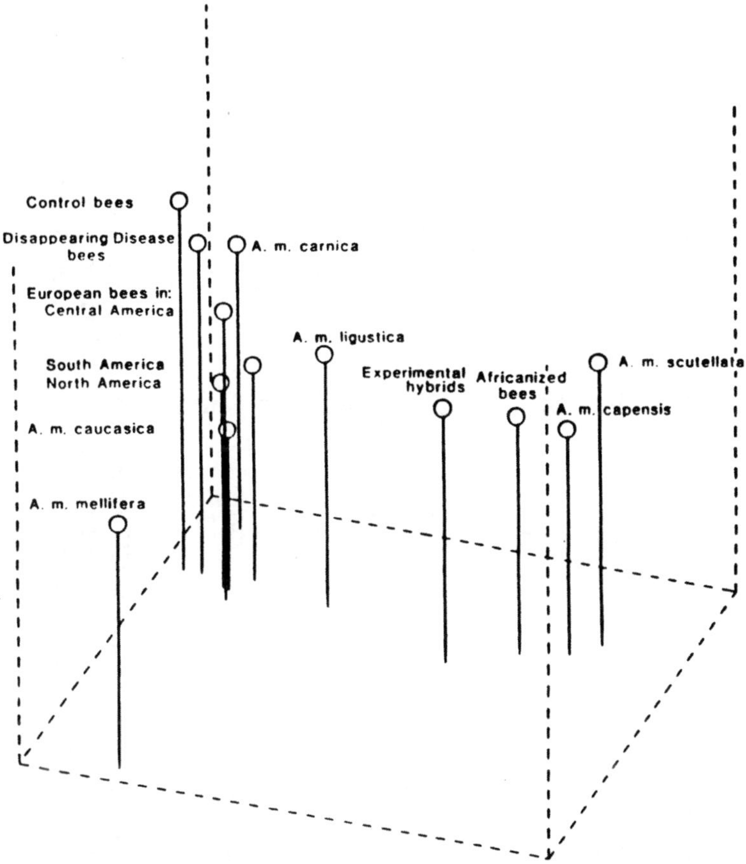

FIGURE 1. Morphometric relations among 13 groups of honey bees as shown in the first 3 dimensions of a discriminant analysis. Circles show positions of group centroids. The diameters of circles are arbitrary. See text and TABLE 1 for further explanation.

using a discriminant analysis of 25 characters (method discussed further on in this chapter). The characters were measured on ten bees from each of 820 colonies. Average measurements from each colony were used in the analysis. Colonies were divided into 13 groups on the basis of geography and known identity (TABLE 1). The "hybrids" were experimental colonies of miscellaneous origin in South America that were crosses of Africanized and European parents for one or more generations. Colonies designated as having "disappearing disease" and the "controls" were provided from the United States by W. C. Rothenbuhler (see Kulincevic *et al.*, 1984).

TABLE 1. Discriminant analysis of African, Africanized, and European honey bees.

No.	Name of groups and geographic origin	NC	CC	MC	(No.)
1.	*A. m. scutellata* (South Africa)	51	80.4	11.8	(3)
2.	*A. m. capensis* (South Africa)	12	91.7	8.3	(1)
3.	Africanized bees (South America)	191	68.6	16.2	(4)
4.	Hybrids of Africanized & European bees	10	90.0	10.0	(3)
5.	*A. m. carnica* (inbred lines, Germany)	8	87.5	12.5	(10)
6.	*A. m. liguistica* (Kangaroo Island, Austr.)	11	100.0		
7.	*A. m. mellifera* (Tasmania & S. Amer.)	36	100.0		
8.	*A. m. caucasia* (lab colonies, U.S.A.)	7	85.7	14.3	(9)
9.	European bees: North America	182	67.6	11.5	(10)
10.	European bees: Central America	135	69.6	8.9	(9,11)
11.	European bees: South America	118	66.1	15.3	(10)
12.	European bees: Disappearing Disease	36	73.9	13.0	(13)
13.	European bees: Control	23	86.1	11.1	(10)

Key: NC = number of colonies (10 bees measured in each colony). CC = per cent of colonies classified in the correct group out of total number of colonies in that group. MC = highest per cent of colonies misclassified in another group and the code number (No.) for the other group in parenthesis. See text for further explanation.

The graph shows positions of the centroids for the groups in the first three dimensions of the 12 dimensional space required for the 13 groups. The first three functions explain 62.8%, 15.76%, and 5.8% of the total variance, respectively, making a total of 84.4%. The graph is therefore an approximate representation of the morphometric relationships among the groups. Of interest here are the relations among African (*A. m. scutellata*), Africanized, and European bees (using the North American group as an example). For an exact measure of these relations, I computed Euclidean distances in the 12 dimensional space. The distance between African and Africanized bees is 2.56; between Africanized and European bees, 5.35; and between European and African bees, 6.65. From the approximate positions in the graph and the relative distances it can be seen that Africanized bees are closer morphometrically to African bees than to European bees in North America. Africanized bees, however, differ from African bees and are situated in the graph between African and European bees in a position that could indicate some hybridization has taken place or a "founder effect" has occurred (Mayr, 1963) or both.

After the analysis was completed, the original colonies were reclassified into the 13 groups (TABLE 1). All colonies of *A. m. mellifera* and *A. m. ligustica* were correctly classified into their respective groups, indicating that they formed discrete clusters. The clusters of the other groups overlapped to various degrees such that some colonies of a given group were misclassified as members of one or more other groups. The 191 Africanized colonies were reclassified as Africanized (131 colonies or 68.6%), hybrid (31 or 16.2%), *A. m. scutellata* (17 or 8.9%), *A. m. capensis* (10 or 5.2%), European bees: Central America (1 or 0.5%) and South America (1 or 0.5%). In this respect also, the Africanized bees show a closer morphometric relation to African bees and Africanized hybrids than to European bees. Buco *et al.* (1987) compared 24 of the same measurements among Africanized, African, and European bees. They report 18 measurements of Africanized bees are more similar to African bees than to European bees.

Disappearing disease among European bees in North America was hypothesized to be a genetic trait derived from Africanized bees (Roberge, 1978). The 13-group analysis indicates that such bees, like their controls, show no morphometric evidence of Africanization.

SCIENTIFIC NAMES FOR AFRICAN AND AFRICANIZED BEES

The concept of a widespread race of honey bees in subsaharan Africa that is closely related to the European races can be found in most reviews of bee classification. Before 1958, however, opinions differed as to whether subsaharan bees were distinct species (Smith, 1865; Ashmead, 1904; Friese, 1909; Skorikow, 1929a; Goetze, 1930, 1940; Maa, 1953) or an infraspecific form of *A. mellifera* (Buttel-Reepen, 1906; Enderlein, 1906; Ruttner and Mackensen, 1952; Kerr and Laidlaw, 1956). The modern era of bee classification started with

Kerr and Portugal-Araújo (1958) who applied the biological species concept to the problem. They cite genetic crosses among European and African races as justification that all belong to one species. All evidence to date supports their conclusion, but partial reproductive isolation may exist between some races (Kerr and Bueno, 1970).

The species name for the African bee, therefore, should be *Apis mellifera* Linnaeus (1758). Even this name has been subject to controversy. In 1761 Linnaeus changed the name to *Apis mellifica* because the original name means "honey carrier" rather than "honey maker" which he preferred. According to the Principle of Priority (Art. 23, International Code of Zoological Nomenclature, hereafter abbreviated as ICZN; see Ride *et al.*, 1985) the oldest available name is the valid name of a taxon. In spite of this long standing rule, the junior name still appears in some European literature.

The name *Apis mellifera*, however, does not distinguish African bees from all other forms of the species. We obviously need a name that is simple, stable, and communicates just which honey bees we intend. Zoological nomenclature is a system of scientific names for animals known to occur in nature. The system has been remarkably successful in providing unique names for species of animals. In the past, systematists attempted to extend the hierarchic classification to populations within species by naming subspecies, varieties, races, phases, forms, etc. In the twentieth century attention focused on biological variation among populations within a species. This led to the "New Systematics" of Huxley (1942) and Mayr (1942). Of several infraspecific categories, the subspecies traditionally has been favored as the category deserving formal recognition in nomenclature. All infrasubspecific categories are excluded from our formal nomenclature (ICZN Art. 1(b)(5)).

Earlier in this century, naming of subspecies became a major preoccupation of systematists. The practice drew criticism, of which the most influential was the critique of Wilson and Brown (1953). They argued that while criteria for species had proved to be objective and practicable, the delimitation of subspecies was not only subjective and arbitrary but also inefficient for reference purposes.

The intent of naming subspecies was to recognize genetically distinct, geographic segregates of a species that were capable of interbreeding along lines of contact. Yet the number of subspecies appeared to vary directly with the number of characters used to distinguish them. Wilson and Brown pointed out that popular quantitative methods still required an arbitrary decision on the degree of difference. There was essentially no lower limit to the definition of subspecies. More importantly, careful analyses of geographic variation of characters in various species of animals often revealed a lack of concordance among characters. Two populations having the same subspecies name have an implied similarity in characteristics. Likewise, two populations of the same species having different subspecies names have an implied discontinuity in character variation. Yet these implications may not have a basis in fact. Wilson and Brown recommended the formal trinomial be replaced by the species name plus a simple vernacular name based on geographic origin. Following the paper of Wilson and Brown, interest in analysis of geographic variation continued to increase, but among professional systematists enthusiasm for naming subspecies markedly declined.

The difficulty in delimiting races of honey bees was already apparent in the last century. After comparing variation in size and color of specimens from diverse localities, Gerstaecker stated (in translation, 1863:342): "The variability of the coloration... gives transitions from one form to another; and thus it becomes impossible to define clearly limited varieties. Latreille and Lepeletier made eight species out of the Honey-Bee; with equal justice we might now, from the existing materials, make 20-30." He proposed six "main varieties" of *A. mellifera* of which one was in Egypt, one widespread in Africa and one in Madagascar.

Prior to 1958, 14 species or subspecies names had been proposed for honey bees in Africa, Madagascar and neighboring islands. In addition to a widespread race, many authors acknowledged variation among bees in the subsaharan region by including names for local varieties or distinct species. The small, scattered collections available for study undoubtedly had the effect of accentuating real differences among populations. In one instance, incorrect identification led to errors in biogeography. In 1906, both Buttel-Reepen and Enderlein stated that forms of *A. indica* (now under the name *A. cerana*) existed in West Africa. This Asian species is now known not to occur naturally in Africa.

Kerr and Portugal-Araújo (1958) recognized a single, widespread subsaharan subspecies and correctly named it *A. m. adansonii* Latreille (1804) according to the oldest available name for honey bees in this region of the African continent. For this reason, when African bees were introduced to Brazil, the name *A. m. adansonii* was applied to the bees in Brazil. They also recognized *A. m. capensis* Eschscholtz (1822) at the Cape of Good Hope and *A. m. unicolor* Latreille (1804) for bees on Madagascar and neighboring islands. Smith (1961) followed their classification and added *A. m. monticola* (Mt. Kilimanjaro and Mt. Meru) and *A. m. litorea* (coast of Tanzania). These were described as new "varieties," but are to be treated now as subspecies (ICZN Art. 45(g)(ii)(1)). Goetze (1964) recognized two subspecies: *A. m. capensis* for the Cape region and *A. m. adansonii* for the rest of Africa, Madagascar, and neighboring islands.

At this point, we have 19 names for species or subspecies in the Ethiopian Zoogeographic Region, of which 16 might apply in subsaharan Africa, Madagascar and neighboring islands. All are available names under the terms of ICZN, but much nomenclatorial housekeeping is needed in the future because most of the descriptions fail to include the designation and deposition of types, and types are missing.

The main issue still remains: How many geographic segregates deserve formal subspecies names? If we recognize one subspecies for subsaharan Africa (excluding Madagascar and neighboring islands), then the name should be *A. m. adansonii.* Ruttner (1975a), however, presented evidence for six subspecies in subsaharan Africa: *A. m. adansonii* in West Africa from Senegal at least to the Republic of Congo, *A. m. scutellata* in Savanna of East and South Africa, *A. m. litorea* in East coast of Africa from Somalia to Mozambique, *A. m. monticola* in mountains of Ethiopia, Kenya, and Tanzania, *A. m. capensis* in Cape of Good Hope, and *A. m. unicolor* in Madagascar and neighboring islands.

Ruttner's subspecific classification is based on multivariate statistical analysis of 40 morphometric characters, plus distribution and behavior. In the statistical analysis, each subspecies forms a cluster that is partly or entirely separated from other such clusters. Overlapping clusters can be separated by detailed analysis of the subspecies involved (Ruttner, 1986, 1988a,b). Quantitative differences among subsaharan subspecies are similar in magnitude to those among European subspecies that have long been accepted by taxonomists and apiculturists. Future studies on intervening populations in Africa may show that these clusters intergrade to some extent. This is to be expected because no major physical barriers exist between the populations studied. Ruttner and Kauhausen (1985) explain the geographic diversification as adaptation to local environments.

Following their arguments, I agree that it is reasonable to recognize more than one subspecies in subsaharan Africa. The names can be justified both on grounds of convenience in communication as well as in recognition of distinctive populations for which the exact boundaries are still uncertain. If a separate subspecies in the savanna of East and South Africa is recognized, the name should be *A. m. scutellata* Lepeletier (1836). The name *scutellata* is based on specimens collected in "Caffrerie," a region in South Africa extending from the Great Kei River on the south to Natal Province on the north and between the Drakensberg Mountains and the coast. Populations near Pretoria are probably similar. According to Ruttner, *A. m. scutellata* extends north to East Africa.

In summary, we now have two choices for the scientific name of the African bees that were introduced to Brazil from Tanzania and Pretoria: If the classification of Kerr and Portugal-Araújo (1958) is followed, then the name should be *A. m. adansonii.* If we follow Ruttner (1975a) as I recommend and recognize six subspecies, then the name of the introduced bees is *A. m. scutellata.*

The issue of the correct name is further complicated by questions regarding the possible hybrid origin of Africanized bees. Mitochondrial DNA of Africanized bees is African in origin (Hall and Muralidharan, 1989; D. Smith *et al.*, 1989). This indicates that the spread has been by maternal migration and not by paternal gene flow into European populations. Additional research is needed to determine whether their nuclear DNA is similarly African in origin or is partly European through hybridization. If Africanized bees that exist now are not hybrids and still have essentially an African genome, then their scientific name is *A. m. scutellata*. On the other hand, if Africanized bees are a genetic hybrid swarm (Buco *et al.*, 1987), then they do not have a convenient scientific name in zoological nomenclature.

Hybrids are sufficiently common among plants to receive special treatment in the International Code of Botanical Nomenclature (Voss *et al.*, 1983; Appendix I; see also Wagner, 1969). Several options are available to provide unique scientific names for hybrid plants at different taxonomic levels. The collective prefix "notho-" or hybrid form, as in nothosubspecies, can be used for any thriving population of hybrids between natural subspecies, whether F1, segregate or backcross.

In zoology, hybrids as such are explicitly excluded from the provisions of ICZN and do not receive separate scientific names (Art. 1 (b) (3)). Thus, if the Africanized bee is a genetic hybrid swarm between races originating in the Old World it cannot have a formal name that conveys that fact. Informally, animal hybrids are often designated by a formula indicating the parental taxa. In our case, the formula for the Africanized population would presumably be: *A. m. carnica* x *A. m. caucasica* x *A. m. ligustica* x *A. m. mellifera* x *A. m. scutellata*. This is obviously too cumbersome a name for practical use and fails to indicate the close similarity of Africanized and African bees. Unless and until the International Commission on Zoological Nomenclature comes to grips with the nomenclatural problems of hybrids, we will have to use common names such as "European" and "Africanized" even though these bees are permanent, identifiable additions to the fauna of the Western Hemisphere.

IDENTIFICATION

Identifications are made with various degrees of assurance. When specimens of a species have unique and clearly defined structural or other characters, then the identifications are irrefutable within the context of the current classification. For example, *A. mellifera* is distinguished structurally from its nearest relative, *A. cerana*. The latter species has two veinlets extending distad from the large basal cell of the hind wing rather than one as in *A. mellifera*. This and other key characters have proven to be consistent and species specific. Specimens of *A. mellifera*, therefore, can be conclusively identified (Daly, 1988).

The geographic races and other distinctive populations of *A. mellifera*, however, usually can not be conclusively identified. They exhibit characters that may overlap to some degree or may grade imperceptibly into adjacent populations. Based on comparison of samples known to be typical of two or more populations, one can estimate how often an identification based on certain characters is likely to be correct. If quantitative characters are used, statistical analysis can provide a statement about the probability that a new sample is correctly identified. In this case, the identification is probable rather than conclusive. Identifications of subspecies, geographic races, genetic hybrid swarms, ecotypes, or biotypes are usually of this nature.

The accuracy of probable identifications depends entirely on how representative the initial samples are with respect to the total populations to be identified. Both Africanized and European bees in the Western Hemisphere appear to be genetically heterogeneous. European bees are a mixture of races, including minor introductions from Africa even before the advent of Africanized bees (Morse *et al.*, 1973). Any procedure for making probable identifications should be based on a broad sampling of this heterogeneity.

The sample unit is usually a collection of bees from a colony and identification is based on pooled extracts or averages of characters of the collection. Some procedures can identify individual bees. A complication for all procedures is the possible mixture of Africanized and European workers in a single colony. This could occur by drift, or when an Africanized colony is in the process of taking over a European colony, or the queen may produce a mixture of daughters because she was inseminated by both kinds of drones.

Probability statements of identification must be interpreted within the context of the procedure. For example, with current methods in morphometrics, the statement that a colony collection is Africanized at 0.7 or 70% probability also indicates the sample is European at 0.3 or 30% probability. The sample could be of normal Africanized bees or normal European bees, but it is more likely to be the former based on previous analysis of known Africanized and European bees. The statement does not mean that the colony is composed of 70% Africanized bees and 30% European bees or that workers have 70% Africanized genes and 30% European genes. To make such statements, the procedures must be able to distinguish individuals or be based on genetic analyses, respectively. Furthermore, the statement that a new sample is Africanized at 1.0 or 100% probability is not a conclusive identification; it is still a probable identification based on the initial analysis of known Africanized and European bees.

All probable identifications carry the risk of actual misidentification. Any method (morphometric, biochemical, behavioral, genetic) that yields a probable rather than conclusive identification carries this risk. When large numbers of samples are being identified, even a small risk becomes an important consideration in terms of the numbers of samples that may be misidentified.

The problem of identifying Africanized bees can be considered at two extremes: "new introductions" or detection of the first Africanized bees to arrive in areas previously occupied by European bees; and "hybrids" or detection of genetic crosses and backcrosses between Africanized and European bees in areas where they have interbred. In the first situation, Africanized and European bees are relatively distinct and phenotypic methods are effective. However, special care must be exercised when one or a few Africanized samples are suspected in the midst of a large population of European bees. Because identifications are based on probability statements, the suspected Africanized bees may, in theory, be indistinguishable statistically from the "tail" of a very large distribution of European samples. In this case, the best action would be to combine evidence from several methods of identification (Spivak *et al.*, 1988). In the second situation a spectrum of genotypes or hybrid swarm may exist in an area together with one or both parental types. Current phenotypic methods were not intended to discriminate among a series of genotypes and, even with further development, will never be as precise as genetic methods.

When Africanized and European bees are carefully compared, statistical differences in morphometry, behavior and physiology are not difficult to find. In behavior, for example, differences have been shown in defense of the nest (Collins *et al.*, 1982), weight of swarms and nest cavity selection (Rinderer, Tucker *et al.*, 1982), nectar foraging (Rinderer *et al.*, 1984), and hoarding (Rinderer, Bolten *et al.*, 1982).

Some of the differences between Africanized and European bees provide a practical basis for identification. Three approaches that are now used or might be used with further development will be discussed here: morphometrics, protein electrophoretic variants, and biochemistry of cuticular hydrocarbons. Average size of worker brood comb cells provides a useful character if natural comb can be obtained (Fletcher, 1978; Rinderer, Sylvester, Brown *et al.*, 1986; Spivak *et al.*, 1988). The collection of comb, however, from feral colonies is time-consuming and sometimes impossible because the host tree or building cannot be damaged. The comb may have been produced by the progeny of a previous queen and not the current queen. Furthermore, comb samples are often sticky and difficult to keep for later measurement without crushing. Unless fumigated or frozen, wax moths often infest comb samples. Other approaches that are still being explored range from the chemistry of venom and alarm pheromones (Mello, 1970; Shipman, 1975; Shipman and Vick, 1977; Blum *et al.*, 1978) to wing-beat frequency (Anonymous, 1986).

MORPHOMETRICS

Morphometrics is the measurement and analysis of form. In biology the forms measured are morphological structures of organisms and analysis is usually by statistics. Morphometrics is widely applied to problems in insect life history, physiology, ecology, and systematics (Daly, 1985). Because it is the phenotype that is measured, an insect's morphometrics includes both genetic and environmentally induced variation. To be useful in identification, the genetically determined racial differences between taxa must be large enough to provide distinguishing characters in spite of environmentally induced and local genetic variation within each taxon.

Historical background

Morphometrics of bees have been extensively analyzed, especially during the first third of this century when apiculturists sought bees with longer tongues that could reach nectar in flowers with deep corolla tubes. Without controlled matings, attention turned to natural variation of bees with the hope of finding useful stock for breeding. Although the objectives were not realized, these early studies gave us much information on geographic variation and inheritance of morphometrics and environmental influences on morphometrics (Merrill, 1922).

In 1929, Alpatov reviewed the pioneering studies of Russian scientists on the roles of genetics and environment in geographic variation in bees. By transplanting colonies to new localities in Russia and observing European races in the United States, he concluded the races and geographic variants within races had specific characters, including morphometrics, that were genetically determined. Unless artificially selected or crossed with other races, the characters were stable in new habitats.

Statistical analysis of individual morphometric characters (univariate analysis), therefore, formed the basis for early studies on bee races by Alpatov (1948), Goetze (1940, 1964) and Skorikow (1929a,b, 1936). The genetic basis for seven morphometric characters in bees was first established by Roberts (1961) who estimated heritabilities at 0.28 (number of hamuli) to 0.85 (wing width and tongue length). Morphometric characters are regularly used for breeding and certification in Europe (Ruttner, 1988a).

Careful measurements and experiments by Alpatov's contemporary, A. S. Michailov, also revealed environmentally induced variation within and among colonies. As summarized by Alpatov (1929), "the following conditions have a pronounced effect on the body size of worker bees: (1) the season of the development, (2) the temperature of the surroundings during the pupal stage, (3) the size of the cell, (4) feeding by nurse bees of different age, and (5) individuality of the colony." Alpatov noted that absolute body size and changes in some proportions could be related to reduced larval feeding.

During the same early period in the United States, Kellogg and Bell (1904), Casteel and Phillips (1903), and Phillips (1929) produced major papers on bee biometry and showed that drones were more variable than workers. This feature of drones was later explained by Brueckner (1976) to be the consequence of reduced developmental homeostasis that arises from their hemizygous genome. Grout (1937) demonstrated that workers reared from enlarged brood cells were significantly larger than workers from normal brood cells. Recently, Eischen *et al*. (1982, 1983) reared worker larvae with different numbers of nurse bees, finding positive correlations between the number of nurse bees and dry weight and life span of the progeny.

The use of morphometrics in bee classification was accelerated by DuPraw (1965a,b) who introduced the use of multivariate analysis. DuPraw's purpose was to create a multidimensional framework based on discriminant analysis to show relationships of bee races. In this multivariate technique measurements of two or more characters are weighted and combined linearly to give maximal separation of two or more groups. For explanations of the method see Pimentel (1979) or a guide to mainframe computer packages such as for SPSS by Norusis (1985).

The multivariate approach, including principal component analysis, has since been applied to discriminate between genetic lines (Louis *et al.*, 1968), ecotypes or strains within a race (Louis and Lefebvre, 1968; Tomassone and Fresnaye, 1971; Cornuet *et al.*, 1978, 1982; Leporati *et al.*, 1983, 1984) and geographic races or subspecies (Louis and Lefebvre, 1971; Cornuet *et al.*, 1975; Gadbin *et al.*, 1979; Santis *et al.*, 1983). Ruttner *et al.* (1978) describe and illustrate 41 characters that are the basis for continuing analysis of all the geographic races of *A. mellifera* in the Old World. The results have appeared in a series of papers by Ruttner (1968, 1969, 1973, 1975a, b, 1981, 1986, 1988a,b) and Ruttner and Kauhausen (1985).

Univariate analysis of Africanized bees

The first efforts to distinguish Africanized from European bees with morphometrics were by univariate analysis. Kerr *et al.* (1967) and Kerr (1969) reported that Africanized bees are smaller than Italian bees except for number of hamuli, width of the basitarsus, and diameter of ocelli in which Africanized bees were said to be larger. Rinaldi *et al.* (1971) computed indices for various measurements of wings, mouthparts, and hind legs of Africanized, Italian, and Caucasian bees. Sarmiento *et al.* (1974) measured widths of abdominal (metasomal) sterna 3, 5, and 6 for Africanized and Italian bees. Authors of these papers do not provide sufficient statistical information to test differences between means.

In the context of a larger study, Woyke (1977) examined four colonies from South Africa, three colonies of Africanized bees, and nine colonies of Italian bees. He found an overlap in more than 30 characters, but counts of bristles were separated. Bristles on the upper surface of the wing were counted within a standard 0.8 mm x 0.5 mm area in the discoidal cell (2nd M cell). Bees with more than 80 bristles were classified as Africanized and those less than 80 were Italian bees. However, counts from one to three colonies in East and West Africa, and in northern and eastern Europe gave partial overlaps.

Multivariate analysis of Africanized and European bees

When Africanized and European bees are compared on the basis of single characters, the variation in characters usually overlaps between the groups. An intermediate specimen or sample from a colony, therefore, can not be identified at a high level of probability by a single character. Multivariate discriminant analysis has features that are useful in identifying Africanized bees. When the same groups are compared in a discriminant analysis of many characters, the combination of characters often gives a clear separation of groups. In the simplest case where two groups are distinguished, new specimens can be identified by multiplying each of the measurements by a corresponding coefficient, summing the products, correcting with a constant, and comparing the resulting discriminant score to the expected values of the scores for the known groups. A probability of membership in each of the groups can be computed. A specimen or collection is usually assigned to the group with which it has a probability of membership greater than 50%.

The first multivariate analyses used 25 characters to demonstrate the feasibility of identifying Africanized bees by morphometrics (Daly, 1975, 1978). The characters were selected from those previously employed by Alpatov, Goetze, DuPraw, and Ruttner. Included were four linear measurements and ten angles between veins of the fore wing, number of hamuli and two linear measurements of the hind wing, four linear measurements of the hind leg, and four of the third sternum. Structures were dissected, cleaned, the sternum stained, and all mounted on a microscope slide. Images of the parts were projected on a table with an overhead projection microscope. Measurements were taken with ruler and protractor. Analyses gave good separation even though overlaps existed for each character. At the outset, it was anticipated that starved European bees might be misidentified if they were quite small in body size (Daly, 1975).

Daly and Balling (1978) describe the further analysis of samples of usually ten bees from each of 101 collections of Africanized bees and 297 collections of European bees. The collections came from diverse geographic areas and were from swarms, feral colonies, well managed and poorly managed colonies, and some were from flowers and other food sources. Analyses were made based on: (1) means of measurements for ten bees in each collection, and (2) on measurements of each individual bee. With all 25 characters in the analysis, all collections and 95.6% of individual bees were correctly identified. The expected rate of misidentification was computed at 0.5% for collections and 4% for individuals.

The 25 character analysis, though successful in identification, was tedious and time consuming to perform by hand, even with a computer for the analyses. The procedure was substantially improved by the addition of a digitizer to form a semi-automatic system for measurement and identification. Daly *et al.* (1982) describe the equipment and computer program for the system.

Blind tests of field versus analytical identifications gave agreement in 95.6% of 135 collections. Five other collections had intermediate scores which might have been genetic in origin (mixtures of Africanized and European bees within a colony or recent hybridization) or the result of environmental influences such as those reported by Michailov that cause a reduction the size of European bees. In the latter case, a score based on atypical measurements that fell within the range of Africanized bees could lead to an intermediate score or misidentification (Daly *et al.*, 1982).

To test some environmental effects on morphometrics, Rinderer, Sylvester, Collins, and Pesante (1986) reared workers of Africanized and European bees under different combinations of nurse-bee genotype and comb cell size. Nurse-bee genotype had small and nonsignificant effects, but cell size had significant effects. The larger European comb resulted in larger bees and the smaller Africanized comb resulted in smaller bees. Despite these influences, the progeny could be correctly identified by the 25 character discriminant analysis. Similarly, Herbert *et al.* (1988) fed larvae various diets to induce nutritional stress with the result that the adult worker bees still were correctly identified by the morphometric procedure.

In summary, the 25 character analysis has been tested under several circumstances and found to give the most reliable identifications to date. The method requires about five hours for one person to process a sample of ten bees. A laboratory is required with stereomicroscope, light, slide making materials and a small chemical hood to remove solvent vapors, plus a projector with high quality optics, computer, digitizer, and computer program. Difficulties may be encountered in interfacing the digitizer and computer. The method requires skilled persons who must exercise care in making the slides and measuring bees. Applications of the method on abnormal bees and hybrids are continuing (Daly *et al.*, 1988; Rinderer *et al.*, 1989)

FABIS methods

Rapid techniques for identifying large numbers of collections have been developed by Rinderer, Sylvester, Brown *et al.* (1986). These are called FABIS for "Fast Africanized Bee Identification System." FABIS is a stepwise procedure leading to identification. To be "identified" in FABIS, the probability of membership must be 90% or greater; if less, then the collection is "unidentified" and is subjected to further analysis. Identifications are based on the means of measurements for 10 bees from a colony. In the latest version (Rinderer, Sylvester *et al.*, 1987), fore wing length, wet weight of freshly killed degastered bees, or dry weight of degastered bees is used as the first step. Bees that remain unidentified at the first step are then measured for one or two additional characters so that a combination can be used in a bivariate or trivariate analysis. Options are given to compute the scores and probabilities for various combinations of fore wing length, wet or dry weight, and length of hind femur.

FABIS has minimal requirements for equipment and can be performed in temporary quarters by unskilled persons. Linear measurements are made by mounting a wing and, if required, a hind leg on microslide coverslips with tape.

This in turn is mounted in a 35 mm slide mount and projected by a slide projector onto a wall at a standardized distance. The image is measured with a meter stick and converted by a magnification factor to the metric system. Weights are taken with a metric balance accurate to 0.01 gm (Sylvester and Rinderer, 1986).

Morphometrics by image analysis

Televised images can be converted to digital information by a suitably equipped computer and the images measured automatically. Of the various structures of bees that are often measured, the wings offer the best images with the current technology. In anticipation of the need to measure bees rapidly by image analysis, Daly and Hoelmer (unpubl. data) used a digitizer to measure 22 lengths of vein segments and 25 angles between veins on the forewings of 100 samples (ten bees each) of Africanized bees and an equal number of European bees. Discriminant analysis of the 47 measurements gave an unbiased estimate of correct classification of 99% for collections of ten worker bees/colony and 91% for single worker bees. This study was used by Batra (1988) as the basis for automatic image analysis of forewings, using an integrated system of optical, television, and computer instruments. The procedure is convenient and provides greater speed than hand-operated digitizers.

As with other phenotypic methods, the reliability of FABIS and the image analysis procedure depends on how similar the initial data sets are to the populations to be discriminated. To prepare for the identification of Africanized bees in a given area, a survey of European bees should be made in advance and compared with the identification standards.

PROTEIN ELECTROPHORETIC VARIANTS

The technique of electrophoresis makes possible the sorting of proteins taken from tissues of organisms. A homogenate of tissue is placed in a gel and an electrical current applied. Within the electrical field, different proteins migrate different distances from the point of origin depending largely on their net charge. The rate of migration may also be influenced by the sizes and shapes of the protein molecules, the properties of the sieve-like gel matrix, and other physicochemical conditions. The protein bands can be made visible in the gel by the addition of suitable stains.

The proteins of interest here are enzymes that can be identified by the use of histochemical stains that are specific for an enzyme. These stains generally couple a specific substrate or another aspect of an enzymatic reaction to a reaction that produces a visible dye. In this way, the localization of a specific enzyme can be determined despite the fact that hundreds or thousands of proteins may be present in a single organism.

Enzymes detected by this technique are called isoenzymes or isozymes. The application of a stain sometimes discloses multiple bands. These are alternate forms of the isozyme. Breeding experiments usually show the isozyme variants are inherited in a Mendelian pattern. Therefore, each variant is considered to be the direct product of an allele of the same gene coding for the enzyme. Such variants are called alloenzymes or allozymes. Although early workers often used terms such as slow, medium, and fast or alphabetic characters to designate allozymes, difficulties were encountered when new allozymes were found or when numerous allozymes were present. The use of "relative mobility" descriptions is much more likely to give unique names to particular allozymes and also to allow results from different studies to be more easily compared. Relative mobilities are calculated by simply measuring the distance a particular allozyme travels in the gel, relative to that of a reference allozyme. The reference allozyme is usually the most common allozyme in the population where the polymorphism was first described (Berlocher, 1980). In some cases, the reference allozyme is based on the slowest or fastest migrating allozyme. Thus, malate dehydrogenase in the honey bee is polymorphic and has five described allozymes: Mdh^{55}, Mdh^{65}, Mdh^{80}, Mdh^{87}, and Mdh^{100} (decimal points are often omitted). The gene responsible for the allozyme is designated similarly and italicized.

Electrophoresis permits analysis of the genetics of natural populations and even single insects in a manner never before possible and provides a valuable tool in systematics (Avise, 1974; Berlocher, 1984). The number of isozymes detected depends on the method employed. The most common are starch gel electrophoresis, polyacrylamide gel electrophoresis, and isoelectric focusing. Various modifications are possible within each method that may improve detection of isozymes. In a comparison of the common methods, Coyne *et al.* (1979) found each method to detect some variation not detectable by the other two.

Studies on proteins of honey bees span almost two decades. Considering the normal roles of enzymes in development and physiology and the potential variation of the genes responsible within and among populations, it is not surprising to learn that isozymes: (1) vary qualitatively and quantitatively over the life of the bee (Gilliam and Jackson, 1972a; Contel *et al.*, 1977; Bitondi and Mestriner, 1983; (2) are present in eggs (Nunamaker and Wilson, 1981a), larvae (Tripathi and Dixon, 1968, 1969; Nunamaker and Wilson, 1982), pupae (Mestriner, 1969; Mestriner and Contel, 1972) and adults (Gilliam and Jackson, 1972b); (3) vary among castes and sexes at the same level of development (Tripathi and Dixon, 1968, 1969; Kubicz and Galuszka, 1971); (4) vary among populations of the same geographic race (Cornuet, 1979; Badino *et al.* 1983a, 1985; Sheppard and Berlocher, 1984, 1985; Sheppard and McPheron, 1986); (5) vary among geographic races of *A. mellifera* (Mestriner and Contel, 1972; Martins *et al.*, 1977; Gartside, 1980; Sylvester, 1982; Badino *et al.*,1983b,

1984; Nunamaker *et al.*, 1984a; Sheppard and Huettel, 1988; (6) vary among species of *Apis* (Tanabe *et al.*, 1970; Nunamaker *et al.*, 1984b; Sheppard, 1985; and (7) vary among genera of the Apoidea (Contel and Mestriner, 1974; Snyder, 1975, 1977).

In comparison with other insects, honey bees and Hymenoptera generally have been reported to have a low level of isozyme polymorphism. For example, Sylvester (1976) used starch gel and 30 stains in his study of adult bees. Thirty-nine bands or loci were found of which only one, malate dehydrogenase (Mdh-1), was polymorphic. Nunamaker and Wilson (1980) used isoelectric focusing and 30 stains to reveal 28 isozymes of which Mdh and non-specific esterase were polymorphic. This unusual feature has stimulated numerous theoretical explanations of which a recent review is by Graur (1985). In contrast, a recent study of sawflies reveals levels of enzyme polymorphism consistent with diploid insects (Sheppard and Heydon, 1986).

The polymorphic isozymes of *A. mellifera* known to date are listed below. Because different techniques are used and different numbers of allozymes are reported, it is difficult to judge if authors are reporting the same allozymes. Furthermore, negative findings such as those of Brueckner (1974) may also be a result of the technique used (Hung and Vinson, 1977). The inheritance of allozymes based on breeding experiments has been determined for alcohol dehydrogenase (Martins *et al.*, 1977), esterase (Mestriner and Contel, 1972; Bitondi and Mestriner, 1983), malate dehydrogenase (Contel *et al.*, 1977), and P-3 protein (Mestriner and Contel, 1972).

•*Aconitase.* Acon-2: 100, 120 in adult worker bees in Czechoslovakia (Sheppard and McPheron, 1986).

•*Alcohol dehydrogenase.* Adh-1: 1, 2, 3 in drone and worker pupae, but absent in young larvae and adults of Italian and Africanized bees (Martins *et al.* 1977); F, S in worker larvae in Australia (Gartside, 1980).

•*Esterase.* Includes a series of loci, each of which has a suite of allozymes, e. g., the esterase loci 1, 3, 5, 6 in larvae and pupae of worker and drone Africanized bees (Bitondi and Mestriner, 1983). The most commonly reported locus is probably Est-3 (Sheppard, pers. comm.). Est: F, S in pupae of workers and drones of Africanized and Italian bees (Mestriner, 1969; Mestriner and Contel, 1972); F, S in worker larvae in Australia (Gartside, 1980); S, M, F in adult worker Italian bees (Badino *et al.*, 1984) and in Sicily (Badino *et al.*, 1985); 100, 130 in adult worker Italian bees (Sheppard and Berlocher, 1985); 70, 100, 130 in adult worker bees in Czechoslovakia (Sheppard and McPheron, 1986).

•*Hexokinase.* HK-1, the fastest allele, in higher frequency among European bees and lower in Africanized bees; other alleles not individually distinguishable found in higher frequency among Africanized bees and lower among European bees (Del Lama and Figueiredo, 1986; Del Lama *et al.*, 1988; Spivak *et al.*, 1988).

•*Malate dehydrogenase.* Mdh-1: A, B, C in larvae, pupae, and adult worker and drone Africanized bees and pupae of Italian worker bees (Contel *et al.*, 1977); a, b, c in adult worker bees in Guadaloupe (Cornuet, 1979); 0.50, 0.63, 1.00 in adult worker European, Italian, and Africanized bees (Sylvester, 1976, 1982); F, M, S in worker larvae in Australia (Gartside, 1980); 0.50, 0.63, 1.00 in adult worker bees in Guatemala and Mexico (Nunamaker *et al.*, 1984a), in Africa and Brazil (Nunamaker and Wilson, 1981b); 65, 80, 100 in adult worker bees in Norway (Sheppard and Berlocher, 1984); S, M, F in adult worker Italian bees (Badino *et al.*, 1983a); S, M, F, F1 in adult worker Italian bees (Badino *et al.*, 1983b, 1984) and in Sicily (Badino *et al.*, 1985); 65, 87, 100 in adult worker Italian bees (Sheppard and Berlocher, 1985); 55, 65, 80, 100 in adult worker bees in Czechoslovakia (Sheppard and McPheron, 1986).

•*Malic enzyme.* Me: 79, 100 in adult worker bees in Norway (Sheppard and Berlocher, 1984); 100, 106 in adult worker Italian bees (Sheppard and Berlocher, 1985); 79, 100 in adult worker bees in Czechoslovakia (Sheppard and McPheron, 1986).

•*Phosphoglucomutase.* Pgm: 75, 100 in adult worker bees in Czechoslovakia (Sheppard and McPheron, 1986; see also Del Lama *et al.*, 1985).

•*Protein.* P-3: F, S in pupae of workers and drones of Africanized and Italian bees (Mestriner, 1969; Mestriner and Contel, 1972).

Five isozymes have been compared between Africanized and European bees: alcohol dehydrogenase, esterase, hexokinase, malate dehydrogenase, and protein P-3. No isozyme has been found that gives complete separation of Africanized and European bees. In other words, we do not find the one allozyme exclusively in one type and another allozyme exclusively in the other type. However, except for the common esterase which is the same in both types, the other four isozymes exhibit partial separation because the frequencies of the alleles differ in each type. Of these, alcohol dehydrogenase was studied in larvae and protein P-3 in pupae, leaving hexokinase and malate dehydrogenase as the only isozymes currently available for use with adult worker bees.

Attention has concentrated on adult worker bees for the purposes of identifying Africanized bees because they are readily collected and have useful isozyme patterns that do not change with the bee's age (Gilliam and Jackson, 1972b). Bees intended for electrophoresis must be fresh or quickly frozen and stored frozen at -60°C. Homogenates of whole bees are prepared or, to avoid contamination from gut contents, the abdomens are discarded or only hemolymph is withdrawn.

Ayala and Powell (1972) propose a method by which partial differences in allozymes can be used as diagnostic characters to distinguish species of *Drosophila*. In brief, the first step is to compute the allelic frequencies from baseline data on the species to be distinguished. Then the expected frequencies of the genotypes in each species are computed by assuming the Hardy-Weinberg equilibrium. For each genotype, the frequency in one of the species will usually be smaller than the other. The overlap of the two species is the sum of the smaller frequencies for each genotype. To identify new specimens by their genotypes, the two species are assumed to be equally common. An individual of a given genotype is assigned to the species with that genotype in the higher frequency. The probability of misidentification is half the computed overlap in the distribution of genotypic frequencies between the two species. Ayala and Powell considered a locus diagnostic if it had a probability of correct assignment at one of two levels: 99% or, more stringently, 99.9%. High probabilities of correct identification are possible only when the differences in genotype frequencies between the species are large.

The method of Ayala and Powell has been considered by Sylvester (1982), Nunamaker *et al.* (1984a), and Page and Erickson (1985) for the purpose of identifying Africanized bees. Sylvester computed the expected genotype frequencies for three alleles of malate dehydrogenase based on samples of (1) his own data for 34 colonies of Africanized bees from Brazil and 24 colonies of European bees from California, and (2) the data of Contel *et al.* (1977) for 78 colonies of Africanized and 34 colonies of Italian bees from Brazil. His evaluations of the probability of correct identification were 92.8 or 94.9%, respectively, depending on the baseline data. Sylvester (1982) and Rinderer and Sylvester (1981) considered the risk of misidentification with Mdh alone was too great for practical use. They propose that if new allozyme systems are discovered in adult bees, these could be combined in the identification procedure to give a joint probability of correct classification and thus reduce the uncertainty.

Nunamaker and Wilson (1981b) compared Mdh in bees from ten colonies of pure African bees from South Africa and 12 colonies of Africanized bees from Brazil. They found the African bees to be monomorphic for Mdh 100 and the Africanized bees polymorphic with the Mdh 100 very high at 93%. From this they conclude the homozygous Mdh 100 genotype is characteristic of African bees. In a subsequent paper, Nunamaker *et al.* (1984a) compared European bees from Australia (4 colonies), Denmark (3), Finland (11), France (2), New Zealand (2), Norway (3), Sweden (3), and Tasmania (3) versus African bees from South Africa (16 colonies). The computed probabilities for correct classification were 99.2-100% This is sufficient to qualify Mdh as a diagnostic locus in the sense of Ayala and Powell (1972). Their choice of samples, however, largely omitted native bees from central and southern Europe.

In Europe, several studies now report Mdh 100 in much higher genic frequency than previously known: northern Italy, 0.073- 0.446, southern Italy 0.521-0.962, Sicily 0.690-1.00 (Badino *et al.*, 1983a,b; 1984); Emilia region of northern Italy, 0.19- 0.46 (Sheppard and Berlocher, 1985); Norway 0.00-0.41 (Sheppard and Berlocher, 1984); Czechoslovakia 0.00-0.53 (Sheppard and McPheron, 1986). Furthermore, Badino *et al.* (1984) showed an inverse relationship between Mdh F (=100) and S in Italy, with F increasing toward the south and warmer winter climates.

It is now clear that homozygous Mdh 100 genotypes can be expected to occur widely in Europe, especially in southern Italy, and are not restricted to southern Africa. The presence of such genotypes in the Western Hemisphere could be from bees imported from either Europe or Africa. They are not exclusively "African" genes. The Mdh 100 allele may confer higher fitness in warmer regions as suggested by Rinderer and Sylvester (1981) and Badino *et al.* (1984). Populations of bees in warmer regions that were established by introducing colonies from cooler regions might be expected to exhibit shifts over time in the relative frequencies of the Mdh allozymes.

The use of allozymes to identify Africanized bees remains a viable option in need of improvement. A distinct advantage is that allozymes are immediate products of structural genes and independent of environmental influences. A disadvantage is the need to kill bees directly by freezing and keep the samples frozen at ultralow temperature before the analysis. Under field conditions, this may be difficult. To be of practical use in the future, one or more additional allozyme systems must be found and added to the Mdh system to meet a stringent joint probability of correct classification. Most important is the development of an adequate baseline on the expected genotype frequencies in critical geographic areas. The method requires special electrophoretic equipment, wet laboratory with chemical hood (some stains are highly toxic) and skilled personnel. If starch gels are made the previous day, run time and incubation time are about four hours each. A number of individual bees can be analyzed at the same time.

BIOCHEMISTRY OF CUTICULAR HYDROCARBONS

Insects contain lipids that are derived partly from their diet and partly from synthesis (Lockey, 1980). Lipids can be extracted from an insect's body by immersion in a fat solvent such as hexane or methylene chloride. Lipids extracted by relatively short immersion are probably derived mainly from the epicuticle and the underlying exocuticle that is also rich in lipids (Hendricks and Hadley, 1983). Longer immersion extracts lipids from cuticular glands and from tissues inside the insect's body. Hemolymph, for example, is rich in hydrocarbon (Chino and Kitazawa, 1981).

In addition to the cuticular lipids, worker bees secreting comb wax may have wax scales on the abdominal sterna that contribute to the extract. The extract also could conceivably include contaminants from the bee's environment such as lipids in honey (Smith and McCaughey, 1966), lipids that have rubbed off from other bees or the combs (Tulloch, 1980), as well as plant lipids from plant cuticular waxes, propolis (Ghisalberti, 1979) and pollen (Stanley and Linskens, 1974).

The extractable lipids from insect cuticle are a complex mixture of compounds in which hydrocarbons often predominate (Hadley, 1986). Analysis of the extract usually involves partitioning the extract followed by gas chromatography and often gas chromatography-mass spectrometry. The hydrocarbons of interest are mostly the long, unbranched chains of saturated (alkanes), mono-unsaturated (alkenes), and di-unsaturated (alkadienes) carbon molecules in an odd-numbered series from C15 to C43. The degree of unsaturation is indicated by "carbon number:degree of unsaturation," e.g., the mono-unsaturated C35:1.

Blomquist *et al.* (1980) determined the cuticular wax of European bees to be 58% hydrocarbon in contrast to comb wax which is predominantly monoester at 31-35%, with hydrocarbon at 13-17%. They further demonstrated that the composition of wax synthesized varies with the age of the bee and with season. In winter, hydrocarbon is the major fraction extracted from the cuticle. In summer months when comb is constructed, hydrocarbon was the major component in younger and older bees, while monoester was the major component in bees 11-18 days following emergence as adults.

Tulloch (1980) pointed out that although comb wax of African and European bees is similar in composition, African wax has proportionately less unsaturated C31 and more C35 hydrocarbon than European wax.

Carlson and Bolten (1984) first reported qualitative and quantitative differences between extracted cuticular hydrocarbons of Africanized and European bees. The samples of bees were of unknown age and varied in method of preservation. Differences in proportions between the two kinds of bees were found in C35:1, C35:2, C37:1, C37:2, C39:1, C39:2, C41:1, C41:2, and C43:2. These totalled 22.4% of hydrocarbons from Africanized bees, but only 1.1-3.1% of hydrocarbons from European bees. Africanized bees had much more C35:1 than European bees. The latter had small, trace, or undetected amounts of the C35 to C43 series. Subsequently, Lavine and Carlson (1987) used multivariate statistics to improve the separation of the two types of bees by the hydrocarbons. Carlson (1988) reported that individual drones of African, Africanized, and European bees could be identified by hydrocarbon patterns.

McDaniel *et al.* (1984) identified and quantified the hydrocarbons extracted from the whole sting apparatus, sting shaft, and general body cuticle of European bees. The samples were of foragers (workers more than 21 days old) collected in plastic bags and frozen. They conclude the sting apparatus is a sufficient source of hydrocarbons for identification of single bees and is relatively free of contamination from extraneous sources. Extracts of the apparatus can be readily made and yield the series C15 to C38 components in contrast to the cuticle with only C23 to C36 components. They did not compare Africanized bees in their analysis.

Francis *et al.* (1985) compared extractable hydrocarbons from workers and drones of four *Apis* species, including samples from native bees in Africa and

Africanized bees from South America. My review is confined to their study of *A. mellifera*. Random aged and newly emerged bees were frozen. Dehydration during storage did not affect the quantity of hydrocarbon extracted. In both random aged and newly emerged bees, they confirmed the higher proportion of unsaturated C35:1 and C35:2 in native African and Africanized bees versus the European subspecies. They also found chain lengths longer than C35 only in random aged African bees. In contrast to previous studies, newly emerged European bees had unsaturated components with chain lengths longer than C35 in amounts sometimes greater than in Africanized bees.

Analyses published to date have used packed column gas chromatography. The most recent analyses by R. K. Smith (1988) utilize the increased resolution of capillary column gas chromatography-mass spectrometry. With this instrument, isomers can be distinguished among molecules of the same chain length and degree of unsaturation. The position of the double bond is counted from the nearest end of the chain and indicated by a numerical prefix, e. g., 14-C35:1.

R. K. Smith (pers. comm.) examined random aged bees that had been collected and stored in isopropanol. Among the mono-unsaturated components, he found the proportions of the following isomers to differ between Africanized and European bees: 9-, 8-, and 7-C29:1; 10-, 9-, and 8-C31:1; and 14-, 12-, and 10-C35:1. Africanized bees exhibited predominantly 9-C29:1 (7-C29:1 also present; 8-C29:1 not observed), only 9-C31:1 (8- and 10-C31:1 not observed), and predominantly 10-C35:1 (12-C35:1 also present; 14-C35:1 not observed). European bees had predominantly 8-C29:1 (9- and 7- C29:1 also present), equally frequent 10- and 8-C31:1 (9-C31:1 not observed), and predominantly 12-C35:1 (14- and 10-C35:1 also present).

In summary, analyses of extractable hydrocarbons have demonstrated a number of differences in composition that are of potential use in identification. The analysis requires a wet laboratory with a chemical hood, appropriate instruments, and skilled personnel. Identification of single bees is possible. About one hour is needed to prepare the sample and one hour for the analysis, but the analytical work can be automated (R. K. Smith, pers. comm.).

Important questions remain to be answered. Do useful hydrocarbons exist that are independent of the bee's age? Will comb secreted by bees of a different geographic type contaminate resident bees with hydrocarbons such that they will be misidentified? Can hybrids and backcrosses between Africanized and European bees be identified?

Procedures for collection, storage, and extraction should be standardized. Precautions must be taken to avoid contamination or loss of lipids. R. K. Smith (pers. comm.) recommends live capture, killing by freezing or cyanide, and storage in dry air either loose or mounted on insect pins. Contact of the bees with fat solvents, petroleum distillates, halocarbons, ethyl acetate, acetone, benzene, ether, etc. will probably render the specimens useless for analysis.

Acknowledgements

I am indebted to S. W. T. Batra, K. Hoelmer, T. E. Rinderer, W. S. Sheppard, R. K. Smith, and H. A. Sylvester for their careful reading of the manuscript and helpful comments. Thanks go to Mark Brown for his patient explanation of the terminology of organic compounds.

LITERATURE CITED

Alpatov, W. W. 1929. Biometrical studies on variation and races of the honey bee (*Apis mellifera* L.). *Q. Rev. Biol.* 4:1-58.

Alpatov, W. W. 1948. The races of honey bees and their use in agriculture. (In Russian). *Sredi prirody,* vol. 4. Moscow Sec. Res. Nat., Moscow.

Anonymous. 1986. $8 million Africanized honey bee barrier proposed. *Am. Bee J.* 126:729-734.

Ashmead, W. H. 1904. Remarks on honey bees. *Proc. Entomol. Soc. Wash.* 6:120-123.

Avise, J. C. 1974. Systematic value of electrophoretic data. *Syst. Zool.* 23:465-481.

Ayala, F. J., Powell, J. R. 1972. Allozymes as diagnostic characters of sibling species of *Drosophila. Proc. Nat. Acad. Sci. (USA)* 69:1094-1096.

Badino, G., Celebrano, G., Manino, A. 1983a. Population structure and Mdh-1 locus variation in *Apis mellifera ligustica. J. Hered.* 74:443-446.

Badino, G., Celebrano, G., Manino, A. 1983b. Identificazione de *Apis mellifera ligustica* Spinola sulla base di sistemi geneenzima. *Boll. Mus. Reg. Sci. Nat. Torino* 1:451-460.

Badino, G., Celebrano, G., Manino, A. 1984. Population genetics of the Italian honey bee (*Apis mellifera ligustica* Spin.) and its relationships with neighboring subspecies. *Boll. Mus. Reg. Sci. Nat. Torino* 2:571-584.

Badino, G., Celebrano, G., Manino, A., Longo, S. 1985. Enzyme polymorphism in the Sicilian honey bee. *Experientia* 41:752-754.

Batra, S. W. T. 1988. Automatic image analysis for rapid identification of Africanized honey bees. 1988. In *Africanized Honey Bees and Bee Mites,* ed. G. R. Needham, R. E. Page, Jr., M. Delfinado-Baker, C. E. Bowman, pp. 260-263. Chichester, England: Ellis Horwood Limited.

Berlocher, S. H. 1980. An electrophoretic key for distinguishing species of the genus *Rhagoletis* (Diptera: Tephritidae) as larvae, pupae, or adults. *Ann. Entomol. Soc. Am.* 73:131-137.

Berlocher, S. H. 1984. Insect molecular systematics. *Ann. Rev. Entomol.* 29:403-433.

Bitondi, M. M. G., Mestriner, M. A. 1983. Esterase isozymes of *Apis mellifera*: substrate and inhibition characteristics, developmental ontogeny, and electrophoretic variability. *Biochem. Genet.* 21:985-1001.

Blomquist, G. J., Chu, A. J., Remaley, S. 1980. Biosynthesis of wax in the honey bee, *Apis mellifera* L. *Insect Biochem.* 10:313-321.

Blum, M. S., Fales, H. M., Tucker, K. W., Collins, A. M. 1978. Chemistry of the sting apparatus of the worker honey bee. *J. Apic. Res.* 17:218-221.

Boreham, M. M., Roubik, D. W. 1987. Population change and control of Africanized honey bees (Hymenoptera: Apidae) in the Panama Canal area. *Bull. Entomol. Soc. Am.* 33:34-39.

Brueckner, D. 1974. Reduction of biochemical polymorphisms in honey bees *(Apis mellifera)*. *Experientia* 30:618-619.

Brueckner, D. 1976. The influence of genetic variability on wing symmetry in honey bees (*Apis mellifera). Evolution* 30:100-108.

Buco, S. M., Rinderer, T. E., Sylvester, H. A., Collins, A. M., Lancaster, V. A., Crewe, R. M. 1987. Morphometric differences between South American Africanized and South African *(Apis mellifera scutellata*) honey bees. *Apidologie* 18:217-222.

Buttel-Reepen, H. V. 1906. Apiacta. *Mitt. Zool. Mus., Berlin.* 3:117-201.

Carlson, D. A. 1988. Africanized and European honey-bee drones and comb waxes: analysis of hydrocarbon components for identification. In *Africanized Honey Bees and Bee Mites,* ed. G. R. Needham, R. E. Page, Jr., M. Delfinado-Baker, C. E. Bowman, pp. 264-274. Chichester, England: Ellis Horwood Limited.

Carlson, D. A., Bolten, A. B. 1984. Identification of Africanized and European honey bees, using extracted hydrocarbons. *Bull. Entomol. Soc. Am.* 30:32-35.

Casteel, D. B., Phillips, E. F. 1903. Comparative variability of drones and workers of the honey bee. *Biol. Bull.* 6:18-37.

Chino, H., Kitazawa, K. 1981. Diacylglycerol-carrying lipoprotein of hemolymph of the locust and some insects. *J. Lipid Res.* 22:1042-1052.

Collins, A. M., Rinderer, T. E., Harbo, J. R., Bolten, A. B. 1982. Colony defense by Africanized and European honey bees. *Science* 218:72-74.

Contel, E. P. B., Mestriner, M. A. 1974. Esterase polymorphisms at two loci in the social bee. *J. Hered.* 65:349-352.

Contel, E. P. B., Mestriner, M. A., Martins, E. 1977. Genetic control and developmental expression of malate dehydrogenase in *Apis mellifera. Biochem. Genet.* 15:859-876.

Cornuet, J. M. 1979. The MDH system in honey bees (*Apis mellifera*) of Guadaloupe. *J. Hered.* 70:223-224.

Cornuet, J. M., Albisetti, J., Mallet, N., Fresnaye, J. 1982. Etude biometrique d'une population d'abeilles landaises. *Apidologie* 13:3-13.

Cornuet, J. M., Fresnaye, J., Lavie, P. 1978. Etude biometrique de deux populations d'abeilles cevenoles. *Apidologie* 9:41-55.

Cornuet, J. M., Fresnaye, J., Tassencourt, L. 1975. Discrimination et classification de populations d'abeilles a partir de caracters biometrique. *Apidologie* 6:145-187.

Cornuet, J. M., Louveaux, J. 1981. Aspects of genetic variability in *Apis mellifera* L. In *Biosystematics of Social Insects*, ed. P. E. Howse, J. L. Clement, pp. 85-93. New York: Academic Press.

Coyne, J. A., Eanes, W. F., Ramshaw, J. A. M., Koehn, R. K. 1979. Electrophoretic heterogeneity of a-glycerophosphate dehydrogenase among many species of *Drosophila. Syst. Zool.* 28:164-175.

Daly, H. V. 1975. Identification of Africanized bees by multivariate morphometrics. *Proc. 25th Int. Apic. Congr., Grenoble, France,* pp. 356-58. Bucharest: Apimondia.

Daly, H. V. 1978. Discriminant analysis of Africanized and European honey bees. In *Simp. Int. Apimondia sobre apiculture en clime quente, Florianópolis, Brazil,* pp. 93-95. Bucharest: Apimondia.

Daly, H. V. 1985. Insect morphometrics. *Ann. Rev. Entomol.* 30:415-438.

Daly, H. V. 1988. Overview of the identification of Africanized honey bees. In *Africanized Honey Bees and Bee Mites,* ed. G. R. Needham, R. E. Page, Jr., M. Delfinado-Baker, C. E. Bowman, pp. 245-249. Chichester, England: Ellis Horwood Limited.

Daly, H. V., Balling, S. S. 1978. Identification of Africanized honey bees in the Western Hemisphere by discriminant analysis. *J. Kans. Entomol. Soc.* 51:857-869.

Daly, H. V., De Jong, D., Stone, N. D. 1988. Effects of parasitism by *Varroa jacobsoni* on morphometrics of Africanized worker honey bees (Acarina, Mesostigmata, Varroidae; Hymenoptera, Apidae). *J. Apic. Res.* 27:126-130.

Daly, H. V., Hoelmer, K., Norman, P., Allen, T. 1982. Computer-assisted measurement and identification of honey bees (Hymenoptera: Apidae). *Ann. Entomol. Soc. Am.* 75:591-594.

Del Lama, M. A., Figueiredo, R. A. 1986. Hexoquinase: um sistema polimorfico em *Apis mellifera. Ciênc. Cult.* 38:923-924.

Del Lama, M. A., Figueiredo, R. A., Soares, A. E. E., Del Lama, S. N. 1988. Hexokinase polymorphisms in *Apis mellifera* and its use for Africanized honey bee identification. *Rev. Bras. Genet.* 11:287-297.

Del Lama, M. A., Mestriner, M. A., Paiva, J. C. A. 1985. EST-5 and PGM: New polymorphisms in *Apis mellifera. Rev. Bras. Genet.* 8:17-27.

DuPraw, E. J. 1965a. The recognition and handling of honey bee specimens in non-Linnean taxonomy. *J. Apic. Res.* 4:71-84.

DuPraw, E. J. 1965b. Non-Linnean taxonomy and the systematics of honey bees. *Syst. Zool.* 14:1-24.

Dutton, R. W., Ruttner, F., Berkeley, A., Manley, M. J. D. 1981. Observations on the morphology, relationships and ecology of *Apis mellifera* of Oman. *J. Apic. Res.* 20:201-214.

Eischen, F. A., Rothenbuhler, W. C., Kulincevic, J. M. 1982. Length of life and dry weight of worker honey bees reared in colonies with different worker-larva ratios. *J. Apic. Res.* 21:19-25.

Eischen, F. A., Rothenbuhler, W. C., Kulincevic, J. M. 1983. Brood rearing associated with a range of worker-larva ratios in the honey bee. *J. Apic. Res.* 22:163-168.

Enderlein, G. 1906. Neue honigbienen und beitrage zur kenntnis der verbreitung der gattung *Apis. Stett. Entomol. Zeit.* 67:331-344.

Eschscholtz, J. F. 1822. *Entomographien.* Vol. 1. Berlin: Reimer.

Filho, C., Francisco, C., Da Silva, R. M. B. 1964. Notas preliminares sobre a *Apis mellifera adansonii. Zootecnia (São Paulo, Brazil)* 11:9-18.

Fletcher, D. J. C. 1978. The Africanized honey bee, *Apis mellifera adansonii,* in Africa. *Ann. Rev. Entomol.* 23:151-171.

Francis, B. R., Blanton, W. E., Nunamaker, R. A. 1985. Extractable surface hydrocarbons of workers and drones of the genus *Apis. J. Apic. Res.* 24:13-26.

Friese, H. 1909. *Die Bienen Afrikas.* Jena: Fischer.

Gadbin, C., Cornuet, J. M., Fresnaye, F. 1979. Approche biometrique de la variete locale d'*Apis mellifera* L. dans la sud tchadien. *Apidologie* 10:137-148.

Gartside, D. F. 1980. Similar allozyme polymorphism in honey bees (*Apis mellifera*) from different continents. *Experientia* 36:649-650.

Gerstaecker, A. 1863. On the geographical distribution and varieties of the honey-bee, with remarks upon the exotic honey-bees of the Old World. *Ann. Mag. Nat. Hist.* 3:270-283, 333-347.

Ghisalberti, E. L. 1979. Propolis: a review. *Bee World* 60:59-84.

Gilliam, M., Jackson, K. K. 1972a. Proteins of developing worker honey bees, *Apis mellifera*. *Ann. Entomol. Soc. Am.* 65:516-517.
Gilliam, M., Jackson, K. K. 1972b. Enzymes in honey bee (*Apis mellifera* L.) haemolymph. *Comp. Biochem. Physiol.* 42B:423-427.
Goetze, G. K. L. 1930. Variabilitats - und zuchtungsstudien an der honigbiene mit beesonderes berucksichtung der langrusseligkeil. *Arch. Bienenkunde* 11:185-236.
Goetze, G. K. L. 1940. *Die beste Biene*. Leipzig: Liedloff, Loth und Michaelis.
Goetze, G. K. L. 1964. Die Honigbiene in naturlicher und kunstlicher Zuchtauslese. Teil I, II. *Monogr. angew. Entomol.* 19:1-120, 20:1-92.
Graur, D. 1985. Gene diversity in Hymenoptera. *Evolution* 39:190-199.
Grout, R. A. 1937. The influence of size of brood cell upon the size and variability of the honeybee (*Apis mellifera* L.). *Res. Bull., Agr. Expt. Sta., Iowa State College, Ames, Iowa.* 218:257-280.
Hadley, N. F. 1986. The arthropod cuticle. *Sci. Am.* 255:104-113.
Hall, H. G. 1988. Characterization of the African honey-bee genotype by DNA restriction fragments. In *Africanized Honey Bees and Bee Mites,* ed. G. R. Needham, R. E. Page, Jr., M. Delfinado-Baker, C. E. Bowman, pp. 287-293. Chichester, England: Ellis Horwood Limited.
Hall, H. G., Muralidharan, K. 1989. Evidence from mitochondrial DNA that African honey bees spread as continuous maternal lineages. *Nature* 339:211-213.
Hendricks, G. M., Hadley, N. F. 1983. Structure of the cuticle of the common house cricket with reference to the location of lipids. *Tissue & Cell* 15:761-779.
Herbert, E. W., Sylvester, H. A., Vandenburg, J. D., Shimanuki, H. 1988. Influence of nutritional stress and the age of adults on the morphometrics of honey bees (*Apis mellifera* L.). *Apidologie* 19:221-229.
Hung, A. C. F., Vinson, S. B. 1977. Electrophoretic techniques and genetic variability in Hymenoptera. *Heredity* 38:409-11.
Huxley, J. 1942. *Evolution, The Modern Synthesis*. London: Allen and Unwin.
Kellogg, V. L., Bell, R. G. 1904. Studies of variation in insects. *Proc. Wash. Acad. Sci.* 6:203-232.
Kerr, W. E. 1967. The history of the introduction of African bees to Brazil. *S. Afr. Bee J.* 2:3-5.
Kerr, W. E. 1969. Some aspects of the evolution of social bees (Apidae). *Evol. Biol.* 3:119-175.
Kerr, W. E., Bueno, E. 1970. Natural crossing between *Apis mellifera adansonii* and *Apis mellifera ligustica*. *Evolution* 24:145-155.
Kerr, W. E., Gonçalves, L., Stort, C., Bueno, D. 1967. Biological and genetical information on *Apis mellifera adansonii*. *Proc. 21st Int. Apic. Congr., College Park, Maryland.* pp. 495-496. Bucharest: Apimondia.
Kerr, W. E., Laidlaw, H. H. 1956. General genetics of bees. *Adv. Genet.* 8:109-153.
Kerr, W. E., Portugal-Araújo, V. de 1958. Raças de abelhas de Africa. *Garcia de Orta* 6:53-59.
Kubicz, A., Galuszka, H. 1971. Polyacrylamide gel electrophoresis of proteins and the acid phosphatase isoenzymes from hemolymphs of the honey bees, *Apis mellifera* L. *Zoologica Pol.* 21:51-58.
Kulincevic, J. M., Rothenbuhler, W. C., Rinderer, T. E. 1984. Disappearing disease: III. A comparison of seven different stocks of the honey bee (*Apis mellifera*). *Ohio State Univ., Columbus, Res. Bull.* 1160:1-21.
Latreille, P. A. 1804. Des especes d'Abeilles vivant en grande societe, et formant des cellules hexagones, ou des Abeilles proprement dites. *Ann. Mus. Nat., Paris* 4:161-178.

Lavine, B., Carlson, D. 1987. European bee or Africanized bee? Species identification through chemical analysis. *Analyt. Chem.* 59:468A-470A.

Lepeletier, A. L. M. 1836. *Histoire naturelle des Insectes. Hymenopteres.* Vol. 1. Paris: Roret.

Leporati, M., Valli, M., Cavicchi, S. 1983. Variazioni ambientali in popolazioni de *Apis mellifera ligustica*: analisi del potere discriminatorio di alcuni caratteri biometrici. *Quad. Doc. F. A. I.* 4, 20pp.

Leporati, M., Valli, M., Cavicchi, S. 1984. Etude biometrique de la variabilite geographique des populations d'*Apis mellifera* en Italie septentrionale. *Apidologie* 15:285-302.

Linnaeus, C. 1758. *Systema Naturae*. tomus I. Holmiae: L. Salvii. 10th ed.

Linnaeus, C. 1761. *Fauna Suecica*. Stockholmiae: L. Salvii.

Lockey, K. H. 1980. Insect cuticular hydrocarbons. *Comp. Biochem. Physiol.* 65B:457-462.

Louis, J., Lefebvre, J. 1968. Etude quantitative de la divergence dans l'evolution morphologique de certaines entites infraspecifiques d'abeilles domestiques (*Apis mellifera* L.). *C. R. Acad. Sci. Paris* 266:1131-1133.

Louis, J., Lefebvre, J. 1971. Les Races d'Abeilles (*Apis mellifera* L.) I. Determination par l'analyse canonique, (etude preliminaire). *Biometrie-Praximedtrie* 12:19-60.

Louis, J., Lefebvre, J., Moratille, R., Fresnaye, J. 1968. Essai de discrimination de lignees consanguines d'abeilles domestiques (*A. m. mellifica* L.) obtenus par insemination artificielle. *C. R. Acad. Sci. Paris* 267:526-528.

Maa, T. 1953. An inquiry into the systematics of the tribus Apidini or honey bees (Hym.). *Treubia* 21:525-640.

Martins, E., Mestriner, M. A., Contel, E. P. B. 1977. Alcohol dehydrogenase polymorphism in *Apis mellifera*. *Biochem. Genet.* 15:357-366.

Mayr, E. 1942. *Systematics and the Origin of Species.* New York: Columbia University Press.

Mayr, E. 1963. *Animal Species and Evolution.* Cambridge, Mass: Belknap Press.

McDaniel, C. A., Howard, R. W., Blomquist, G. J., Collins, A. M. 1984. Hydrocarbons of the cuticle, sting apparatus, and sting shaft of *Apis mellifera* L. Identification and preliminary evaluation as chemotaxonomic characters. *Sociobiology* 8:287-298.

Mello, M. L. S. 1970. A quantitative analysis of the proteins in venom from *Apis mellifera* (including *A. m. adansonii*) and *Bombus atratus*. *J. Apic. Res.* 9:113-120.

Merrill, J. H. 1922. The correlation between some physical characters of the bee and its honey-storing abilities. *J. Econ. Entomol.* 15:125-129.

Mestriner, M. A. 1969. Biochemical polymorphisms in bees (*Apis mellifera ligustica*). *Nature* 223:118-189.

Mestriner, M. A., Contel, E. P. B. 1972. The P-3 and Est loci in the honey bee *Apis mellifera*. *Genetics* 72:733-738.

Morse, R. A., Burgett, D. M., Ambrose, J. T., Conner, W. E., Fell, R. O. 1973. Early introductions of African bees into Europe and the New World. *Bee World* 54:57-60.

Needham, G. R., Page, R. E., Jr., Delfinado-Baker, M., Bowman, C. E. ed. 1988. *Africanized Honey Bees and Bee Mites.* Chichester, England: Ellis Horwood Limited.

Norusis, M. J. 1985. *SPSSX: Advanced Statistics Guide.* New York: McGraw-Hill Book Company.

Nunamaker, R. A., Wilson, W. T. 1980. Some isozymes of the honey bee (*Apis mellifera*). *Isozyme Bull.* 13:111-112.

Nunamaker, R. A., Wilson, W. T. 1981a. Malate dehydrogenase and nonspecific esterase isoenzymes of eggs of the honey bee (*Apis mellifera*). *Comp. Biochem. Physiol.* 70B:607-609.

Nunamaker, R. A., Wilson, W. T. 1981b. Comparison of MDH allozyme patterns in the Africanized honey bee (*Apis mellifera adansonii* L.) and the Africanized populations of Brazil. *J. Kans. Entomol. Soc.* 54:704-710.

Nunamaker, R. A., Wilson, W. T. 1982. Isozyme changes in the honey bee, *Apis mellifera* L., during larval morphogenesis. *Insect. Biochem.* 12:99-104.

Nunamaker, R. A., Wilson, W. T., Haley, B. E. 1984a. Electrophoretic detection of Africanized honey bees (*Apis mellifera scutellata*) in Guatemala and Mexico based on malate dehydrogenase allozyme patterns. *J. Kans. Entomol. Soc.* 57:622-631.

Nunamaker, R. A., Wilson, W. T., Ahmad, R. 1984b. Malate dehydrogenase and non-specific esterase isoenzymes of *Apis florea, A. dorsata,* and *A. cerana* as detected by isoelectric focusing. *J. Kans. Entomol. Soc.* 57:591-595.

Page, R. E., Erickson, E. H. 1985. Identification and certification of Africanized honey bees. *Ann. Entomol. Soc. Am.* 78:149-158.

Phillips, E. F. 1929. Variation and correlation in the appendages of the honey bee. *Cornell Univ. Agr. Expt. Sta. Memoir 121*, 52pp.

Pimentel, R. A. 1979. *Morphometrics*. Dubuque, Iowa: Kendall/Hunt Publishing Company.

Ride, W. D., Sabrosky, C. W., Bernardi, G., Melville, R. V. 1985. *International Code of Zoological Nomenclature*. London: Int. Trust Zool. Nomencl. 3rd ed.

Rinaldi, A. J. M., Popolizio, E. R., Pailhe, L. A. 1971. Indices alares, tarsales y glosales en tres razas de abejas. *Miscelanea* No. 41, pp. 1-10. Universidad Nacional de Tucuman, Argentina.

Rinderer, T. E. 1986. Africanized bees: The Africanization process and potential range in the United States. *Bull. Entomol. Soc. Am.* 32:222-227.

Rinderer, T. E., Bolten, A. B., Harbo, H. R., Collins, A. M. 1982. Hoarding behavior of European and Africanized honey bees (Hymenoptera: Apidae). *J. Econ. Entomol.* 75:714-715.

Rinderer, T. E., Bolten, A. B., Harbo, H. R., Collins, A. M. 1984. Nectar-foraging characteristics of Africanized and European honey bees in the neotropics. *J. Apic. Res.* 23:70-79.

Rinderer, T. E., Daly, H. V., Sylvester, H. A., Lancaster, V. A., Collins, A. M., Buco, S. M., Hellmich, R. L., II, Danka, R. G. 1990. Morphometric differences among Africanized and European honey bees and their F1 hybrids (Hymenoptera: Apidae). *Ann. Entomol. Soc. Am.* 83:346-351.

Rinderer, T. E., Sylvester, H. A. 1981. Identification of Africanized bees. *Am. Bee J.* 121:512-516.

Rinderer, T. E., Sylvester, H. A., Brown, M. A., Villa, J. D., Pesante, D., Collins, A. M. 1986. Field and simplified techniques for identifying Africanized European honey bees. *Apidologie* 17:33-48.

Rinderer, T. E., Sylvester, H. A., Buco, S. M., Lancaster, V. A., Herbert, E. W., Collins, A. M., Hellmich, R. L. II, 1987. Improved simple techniques for identifying Africanized and European honey bees. *Apidologie* 18:179-197.

Rinderer, T. E., Sylvester, H. A., Collins, A. M., Pesante, D. 1986. Identification of Africanized and European honey bees: effects of nurse-bee genotype and comb size. *Bull. Entomol. Soc. Am.* 32:150-152.

Rinderer, T. E., Tucker, K. W., and Collins, A. M. 1982. Nest cavity selection by swarms of European and Africanized honey bees *J. Apic. Res.* 21:98-103.

Roberge, F. 1978. The case of the disappearing honey bees. *Nat. Wildlife* 16:34-35.

Roberts, W. C. 1961. Heterosis in the honey bee as shown by morphological characters in inbred and hybrid bees. *Ann. Entomol. Soc. Am.* 54:878-882.

Ruttner, F. 1968. Les races d'abeilles. in *Traite de Biologie de l'Abeille*, vol 1. ed. R. Chauvin, pp. 27-44. Paris: Masson et Cie.

Ruttner, F. 1969. Biometrische Charakterisierung der osterreichischen Carnica-Biene. *Zietschr. Bienenforsch.* 9:469-503.

Ruttner, F. 1973. Die Bienenrassen des mediterranen Beckens. *Apidologie* 4:171-172.

Ruttner, F. 1975a. African races of honey bees. *Proc. 25th Int. Apic. Congr.*, pp. 325-344. Bucharest: Apimondia.

Ruttner, F. 1975b. Biological and statistic methods as a means of an ingenious intraspecific classification in the honey bee. *Proc. 25th Int. Apic. Congr., Grenoble, France*, pp. 360-64. Bucharest: Apimondia.

Ruttner, F. 1981. Taxonomy of honey bees of tropical Africa. *Proc. 28th Int. Apic. Congr.* pp. 271-274. Bucharest: Apimondia.

Ruttner, F. 1986. Geographical variability and classification. In *Bee Genetics and Breeding*, ed. T. E. Rinderer, pp. 23-56. Orlando, Fla: Academic Press.

Ruttner, F. 1988a. *Biogeography and Taxonomy of Honey bees*. Berlin: Springer-Verlag.

Ruttner, F. 1988b. Principles of geographic variation in honey bees. In *Africanized Honey Bees and Bee Mites*, ed. G. R. Needham, R. E. Page, Jr., M. Delfinado-Baker, C. E. Bowman, pp. 250-259. Chichester, England: Ellis Horwood Limited.

Ruttner, F., Kauhausen, D. 1985. Honey bees of tropical Africa: Ecological diversification and isolation. *Proc. 3d Int. Conf. Apic. Trop. Clim., Nairobi, 1984*. pp. 45-51. London: Int. Bee Res. Assoc.

Ruttner, F., Mackensen, O. 1952. The genetics of the honey bee. *Bee World* 33:53-62, 71-79.

Ruttner, F., Tassencourt, L., Louveaux, J. 1978. Biometrical-statistical analysis of the geographic variability of *Apis mellifera* L. I. Material and methods. *Apidologie* 9:363-381.

Santis, L. de, Bolognese, A., Cornejo, L. G., Crisci, J. V., Diaz, N. B., Lanteri, A. A., Regalia, J. V. S. 1983. Estudio taxonomico de dos subespecies de *Apis mellifera* (*A. m. mellifera* y *A. m. ligustica* Spinola) en proceso de hibridacion, mediante el empleo de tecnicas numericas. *Revista del Museo de la Plata (nueva serie), seccion zoologia*. 8:45-63.

Sarmiento, J. A. V., Santis, L. de, Cornejo, L. G. 1974. La identification de *Apis mellifera adansonii*, segunda contribucion. *Ciencia y Abejas* 2:37-43, 45.

Severson, D. W., Aiken, J. M., March, R. F. 1988. Molecular analyses of North American and Africanized honey bees. In *Africanized Honey Bees and Bee Mites*, ed. G. R. Needham, R. E. Page, Jr., M. Delfinado-Baker, C. E. Bowman, pp. 294-302. Chichester, England: Ellis Horwood Limited.

Sheppard, W. S. 1985. Electrophoretic variation in the honey bees, *Apis*. *Isozyme Bull.* 18:16.

Sheppard, W. S., Berlocher, S. H. 1984. Enzyme polymorphism in *Apis mellifera* from Norway. *J. Apic. Res.* 23:64-69.

Sheppard, W. S., Berlocher, S. H. 1985. New allozyme variability in Italian honey bees. *J. Hered.* 76:45-48.

Sheppard, W. S., Heydon, S. L. 1986. High levels of genetic variability in three male-haploid species (Hymenoptera: Argidae, Tenthredinidae). *Evolution* 40:1350-1353.

Sheppard, W. S., Huettel, M. D. 1988. Biochemical genetic markers, intraspecific variation, and population genetics of the honey bee, *Apis mellifera*. In *Africanized Honey Bees and Bee Mites,* ed. G. R. Needham, R. E. Page, Jr., M. Delfinado-Baker, C. E. Bowman, pp. 281-86. Chichester, England: Ellis Horwood Limited.

Sheppard, W. S., McPheron, B. A. 1986. Genetic variation in honey bees from an area of racial hybridization in Western Czechoslovakia. *Apidologie* 17:21-32.

Shipman, W. H. 1975. Separation of the components of Brazilian bee venom. *Am. Bee J.* 115:56, 59.

Shipman, W. H., Vick, J. A. 1977. Studies of Brazilian bee venom. *Cutis* 19:802-804.

Skorikow, A. S. 1929a. Eine neue basis fur eine revision der gattung *Apis* L. (In Russian). *Repts. Appl. Entomol., Leningrad* 4:249-270.

Skorikow, A. S. 1929b. Beitrage zur kenntnis der kaukasischen honigbienenrassen. (In Russian). *Repts. Appl. Entomol., Leningrad* 4:1-60.

Skorikow, A. S. 1936. Variabilite exterieure de l'abeille (*Apis*) en Eurasie. *Travaux Zoologiques de l'Acad. Sci. URSS* 4:183-243.

Smith, D. R. 1988. Mitochondrial DNA polymorphisms in five Old World subspecies of honey bees and in New World hybrids. In *Africanized Honey Bees and Bee Mites,* ed. G. R. Needham, R. E. Page, Jr., M. Delfinado-Baker, C. E. Bowman, pp. 303-312. Chichester, England: Ellis Horwood Limited.

Smith, D. R., Taylor, O. R., Brown, W. M. 1989. Neotropical Africanized honey bees have African mitochondrial DNA. *Nature* 339:213-215.

Smith, Fredrick. 1865. On the species and varieties of the honey bees belonging to the genus *Apis*. *Ann. Mag. Nat. Hist.* 3:372-380.

Smith, F. G. 1961. The races of honey bees in Africa. *Bee World* 42:255-260.

Smith, M. R., McCaughey, J. F. 1966. Identification of some trace lipids on honey. *J. Food Res.* 31:902-905.

Smith, R.-K. 1988. Identification of Africanization in honey bees based on extracted hydrocarbons assay. In *Africanized Honey Bees and Bee Mites,* ed. G. R. Needham, R. E. Page, Jr., M. Delfinado-Baker, C. E. Bowman, pp. 275-280. Chichester, England: Ellis Horwood Limited.

Snyder, T. P. 1975. Lack of allozymic variability in three bee species. *Evolution* 28:687-689.

Snyder, T. P. 1977. A new electrophoretic approach to biochemical systematics of bees. *Biochem. Syst. Ecol.* 5:113-150.

Spivak, M., Ranker, T., Taylor, O. Jr., Taylor, W., Davis, L. 1988. Discrimination of Africanized honey bees using behavior, cell size, morphometrics, and a newly discovered isozyme polymorphism. In *Africanized Honey Bees and Bee Mites,* ed. G. R. Needham, R. E. Page, Jr., M. Delfinado-Baker, C. E. Bowman, pp. 313-324. Chichester, England: Ellis Horwood Limited.

Stanley, R. G., Linskens, H. F. 1974. *Pollen: Biology, Biochemistry, Management*. Berlin: Springer-Verlag.

Stibick, J. N. L. 1984. Animal and plant health inspection service strategy and the African honey bee. *Bull. Entomol. Soc. Am.* 30:22-26.

Sylvester, H. A. 1976. Allozyme variation in honey bees (*Apis mellifera L.*). PhD Dissertation. Univ. Calif. 56pp.

Sylvester, H. A. 1982. Electrophoretic identification of Africanized honey bees. *J. Apic. Res.* 21:93-97.

Sylvester, H. A., Rinderer, T. E. 1986. Africanized bees: progress in identification procedures. *Am. Bee J.* 126:330-333.

Tanabe, Y., Tamaki, Y., Nakano, S. 1970. Variation of esterase isozymes in seven species of bees and wasps. *Jap. J. Genet.* 45:425-428.

Taylor, O. R. 1985. African bees: potential impact in the United States. *Bull. Entomol. Soc. Am.* 31:15-24.

Tomassone, R., Fresnaye, J. 1971. Etude d'une methode biometrique et statistique permettant la discrimination et la classification de populations d'abeilles (*Apis mellifera* L.). *Apidologie* 2:49-65.

Tripathi, R. K., Dixon, S. E. 1968. Haemolymph esterases in the female larval honey bee, *Apis mellifera* L., during caste development. *Can. J. Zool.* 46:1013-1017.

Tripathi, R. K., Dixon, S. E. 1969. Changes in some hemolymph dehydrogenase isozymes of the female honey bee, *Apis mellifera* L. during caste development. *Can. J. Zool.* 47:763-770.

Tulloch, A. P. 1980. Beeswax - composition and analysis. *Bee World* 61:47-62.

Voss, E. G., Burdet, H. M., Chaloner, W. G., Demoulin, V., Hiepko, P., NcNeill, J., Meikle, R. D., Nicolson, D. H., Rollins, R. C., Silva, P. C., Greuter, W. 1983. *International Code of Botanical Nomenclature.* vol. 111. Regnum Vegetabile, Antwerpen.

Wagner, W. H. 1969. The role and taxonomic treatment of hybrids. *Bioscience* 19:785-789, 795.

Wilson, E. O., Brown, W. L. 1953. The subspecies concept and its taxonomic application. *Syst. Zool.* 2:97-111.

Woyke, J. 1977. The density of bristles covering the wings as discrimination value between African and other races of honey bee. In *African Bees: Taxonomy, Biology, and Economic Use*, ed. D. J. C. Fletcher, pp. 15-24. Pretoria: Apimondia.

3

GENETIC CHARACTERIZATION OF HONEY BEES THROUGH DNA ANALYSIS

H. Glenn Hall[1]

A paucity of genetic markers has limited understanding of honey bee genetics. Reviewed in this chapter are initial honey bee studies that have employed DNA as a source of markers. The increased genetic resolution provided by DNA has begun to resolve intriguing enigmas surrounding the African bee.

CHAPTER SUMMARY

The first part of the chapter is a general overview of DNA types and methods. The protocols described are those most applicable for finding and comparing genetic differences among organisms. Other steps required to ascertain the structure of functional genes are not covered. In addition to procedures already applied to honey bee research, approaches and considerations likely to be utilized in the near future are included. This information is intended to clarify rationale, terminology and experimental detail involved in the bee DNA work but is not essential to appreciate the results.

The second part of the chapter reviews the results from honey bee (*Apis mellifera* L.) DNA studies. Polymorphisms found to distinguish African and European bees are described. The DNA evidence points to significant maintenance of the African genotype in neotropical bees. These findings and mechanisms possibly responsible are discussed. The final section of the chapter addresses the use of DNA markers for regulatory identification. In this chapter, "African" and "European" are used to classify both Old and New World honey

---

[1]Professor Hall is in the Department of Entomology and Nematology, University of Florida, Gainesville, FL, 32611, USA.

bees according to their continents of origin. The term "Africanized bees" is limited to European maternal lines hybridized to African males.

THE VALUE OF HONEY BEE DNA MARKERS

Genetic relatedness of organisms, including honey bees, has been classically established through morphological taxonomy. Morphology, as well as physiology and behavior, are the end result of temporal and spatial expression of many genes and the interaction of multiple gene products. Regulation at many levels, including modulation by the environment, results in variation of physical parameters. Furthermore, survivability of genetic variation, giving rise to morphological distinctions, depends upon natural selection. Thus, morphology derives from complex circular processing of genetic information. As discussed in the previous chapter by Daly, morphological distinctions among honey bees are subtle. Variation leads to considerable overlap in distributions and may result in inaccurate identifications. Far preferable would be to establish genetic identity and relatedness by directly analyzing the genetic material.

Genetic changes, or mutations, altering the phenotype can sometimes allow identification of the genes responsible and determination of their function, pattern of inheritance, and linkage with other genes. Although mutations causing pronounced phenotypic changes can be recovered and maintained under controlled conditions, they are often adaptively disadvantageous. Generally such phenotypic markers that distinguish conspecific natural populations are limited in number. More useful for population studies are biochemical differences, such as in the amino acid composition of proteins (e.g. allozymes), which may or may not be expressed in the phenotype.

In Hymenopteran insects, however, protein variation is limited (Crozier, 1977; Lester and Selander, 1979; Graur, 1985). This may be due, in part, to their haploid-diploid sex-determination system, where even slightly detrimental changes may be strongly selected against in haploid males. In the social Hymenoptera, variation is further reduced because of the small population size of reproductive members. Characteristic of their order, honey bees have few protein electrophoretic variants (for more detail see Daly, Chapter 2). By studying a protein, rather than phenotypic characters, the analysis comes closer to the gene itself. Nevertheless, as exemplified by the Hymenoptera, any genetic analysis dependent upon the protein product of the gene is subject to the complicated influences affecting its existence.

Molecular genetic technology defines genes by their primary structure, the sequence of nucleotides in their DNA. This leads to identification of gene regulatory sequences and to detection of specific transcriptional products, so that temporal and spatial genetic activity can be followed. Finding variation in DNA is independent of the phenotype or proteins and, hence, independent of the multiple interactions responsible for expression and selection. Molecular genetic approaches powerfully interface with all aspects of genetics. By providing far more markers than have been available previously, honey bee DNA analyses can greatly enhance studies of systematics, transmission genetics, and population genetics.

A goal of honey bee DNA research is to find differences that characterize the subspecies, especially markers that distinguish African from European races. Alternate genetic forms arising from a polymorphism, known as alleles, differ in their effectiveness as subspecies-specific markers. After divergence of lineages, different allele frequencies result from mutation, selection, drift or migration. The frequency defines the probability that an allele comes from one group or the other. Specific alleles arise if they are retained in one group but lost in the other or if they derive from a new mutation in only one of the groups. Whatever their frequency, specific alleles are certain markers of their lineage, but at high frequencies they are more useful by representing more members of the group. Also, correspondingly low frequencies of the alternate non-specific alleles increase their probability as markers of the other group. If after divergence the events generating specific differences occur sufficiently early to allow fixation, the alleles become diagnostic for their respective lineages (Ayala and Powell, 1972). For honey bees, specific DNA alleles at high frequencies are being sought, as are more valuable, but more rare, diagnostic markers.

Once the specificities of the honey bee markers are determined, they can be utilized to follow interactions between the subspecies at the colony and population levels. Important questions surrounding the African honey bee pertain to its relationship to the European honey bee: how genetically divergent are the two, how do they interact, and how much genetic exchange occurs between them? Thus, not only will DNA analyses have a direct application in reliably distinguishing African and European honey bees and identifying their hybrids, but DNA studies will also provide basic genetic knowledge essential in designing control strategies.

PART I. DNA TYPES AND METHODS

CONSIDERATION OF DNA TYPES

Mitochondrial DNA

Mitochondrial DNA (mtDNA) has been most commonly used to determine genetic relatedness among organisms (reviewed by Brown, 1985; Wilson *et al.*, 1985; Avise, 1986; Avise *et al.*, 1987; Moritz *et al.*, 1987). Mitochondria of multicellular eukaryotes are present in the extranuclear cytoplasm and contain circular DNA molecules of about 16 to 17 kilobases in length (1 kilobase = 1000 nucleotide base pairs = 1kb; 17kb in the honey bee; Smith and Brown, 1988) that code for mitochondrial transfer RNA, ribosomal RNA, and enzymes for oxidative phosphorylation. In individual animals, with a few exceptions, all mtDNA molecules are the same. Many mitochondria are formed within cells. Thus, in effect, a small region of DNA is cloned, providing abundant amounts of a few genes. During egg maturation, the large number of mitochondria that accumulate give rise to those of the embryo. Sperm have relatively few mitochondria, and in the animals tested, including vertebrates and insects, no paternal contribution of these to the progeny has been detected.

Compared to the biparental transmission and recombination of nuclear DNA, maternal inheritance of mtDNA allows genetic divergence to be more easily followed. Nucleotide changes in mtDNA accumulate at a constant rate, so that the times when ancestral lineages diverged can be estimated. MtDNA is valuable in population studies, particularly when asymmetries in maternal gene flow are involved. In a number of cases, mtDNA is adaptively neutral and can even introgress across species boundaries (Ferris *et al.*, 1983; Takahata and Slatkin, 1984; Lamb and Avise, 1986; Marchant, 1988). However, as will be discussed later, recent results suggest that honey bee mtDNA may not be entirely neutral.

Single Copy Nuclear DNA

The major proportion of eukaryotic genomes is comprised of DNA sequences present as one or a few copies (Britten and Kohne, 1968; reviewed by Spradling and Rubin, 1981). Within this class of single or low copy DNA resides the sequences coding for messenger RNA and subsequently for proteins. However, only a fraction of low copy DNA has a coding function (Bishop, 1974; Bishop *et al.*, 1974; Izquierdo and Bishop, 1979). Interspersed among the coding regions (exons) of individual genes are non-coding introns which are excised to form the functional mRNAs (Green, 1986; Leff *et al.*, 1986; Padgett *et al.*, 1986). Within introns and non-coding regions that flank genes reside specific sequences that serve as promoters and enhancers of RNA transcription (Schibler and Sierra, 1987; Atchison, 1988; Wasylyk, 1988). These sequences hold the key to correct temporal and spatial expression of genes, and their elusive mechanisms of action are slowly being deciphered. The amount of DNA that has a regulatory role is not known but may account for only a minor proportion of the non-coding sequences.

Although the role of most low copy DNA remains a mystery, it is nevertheless valuable for the genetic characterization of organisms. Differences in non-coding DNA may not be as subject to evolutionary pressures as those in coding regions that must preserve amino acid sequences required for protein function (reviewed by Li *et al.*, 1985). Since DNA differences do not necessarily result in (or their detection depend upon) functional changes, they can potentially provide allele distinction at many loci within natural populations. Even without knowledge of their function, DNA alleles can be used to establish genetic relatedness and to follow population dynamics. It is this application with honey bees that will be emphasized later in the chapter.

DNA alleles have been used in classical approaches to determine the linear order of their loci along the length of chromosomes and to establish linkage with genes causing phenotypic effects. DNA markers are being mapped throughout the human genome to serve as linkage reference points (Donis-Keller *et al.*, 1987). Polymorphisms at nearby loci have helped identify genes causing cancer and other diseases (Gusella *et al.*, 1984; reviewed by Watkins, 1988).

Repetitive Nuclear DNA

In different organisms, different proportions of their nuclear genomes consist of repetitive sequences (Britten and Kohne, 1968; reviewed by Spradling and Rubin, 1981; Bouchard, 1982; Singer, 1982). Some functional genes are usually found as multiple copies, such as the genes coding for ribosomal RNA, but functions for most repetitive DNA, if they exist, have not been determined. Much of repetitive DNA is considered to be "selfish DNA", excess generated as an inevitable consequence of replication and evolution of the eukaryotic genome (Doolittle and Sapienza, 1980; Orgel and Crick, 1980; Loomis and Gilpin, 1986).

Repetitive DNA families vary in complexity, numbers of copies and distribution. Centromeric heterochromatin, for example, is largely comprised of highly reiterated, short, simple sequences, in tandem arrangement (reviewed by Brutlag, 1980; Singer 1982). Families of this DNA often have buoyant densities different than the bulk of genomic DNA and have been called "satellite DNA." The location of satellite DNA near centromeres has suggested a chromosomal structural role, but this distribution may just reflect preferential sequestering of inactive DNA due to a combination of factors (Stephan, 1987). Such sequences apparently do have a function as chromosome telomeres (reviewed by Zakian, 1989).

DNA classes not as highly repetitive, known as the "middle" repeats, are interspersed with single copy sequences in either a short or long arrangement (reviewed by Bouchard, 1982). The short pattern is found in humans, frogs and house flies, and the long pattern in fruit flies and honey bees (Davidson *et al.*, 1973; Manning *et al.*, 1975; Schmid and Deininger, 1975; Crain *et al.*, 1976; Cockburn and Mitchell, 1989). Other repetitive elements are comprised of variable numbers of tandem repeats (VNTR), at one or many loci (Nakamura *et al.*, 1987). The latter have been called "minisatellites" (Jeffreys *et al.*, 1985).

A number of repetitive sequences, present in low copy numbers (10 to 50), are transposons. These DNA elements excise from their positions in the genome and reinsert at different locations (reviewed by Shapiro, 1983, and Finnegan, 1985). One of the most notable transposable elements in *Drosophila*, the "P" element, facilitates the integration of foreign DNA into the genome, and its discovery has opened an avenue for genetic engineering (Spradling and Rubin, 1982; Rubin and Spradling, 1982; Pirrotta, 1988). Most interspersed repetitive sequences appear to be retrotransposons, nonfunctional DNA that has been reverse-transcribed from RNA and reinserted into the genome (reviewed by Rogers, 1985; Baltimore, 1985).

Sequences of the repetitive ribosomal RNA genes are strongly conserved among organisms, reflecting their basic vital function. The non-coding spacer regions in between, however, are highly polymorphic (reviewed by Gerbi, 1985). Because most other repetitive DNA has no coding function, considerable inter- and intra-specific variation can exist (Bouchard, 1982; Singer 1982; Dowsett, 1983; Miklos, 1985; Waye and Willard, 1987). Variation has been found in related sequences on different chromosomes (Arnold, 1986; Beauchamp *et al.*, 1979; Willard *et al.*, 1986). Perhaps within repetitive DNA resides the greatest potential for finding genetic differences among closely related organisms. In a number of animals, minisatellite DNA can distinguish individuals (Georges *et al.*, 1988; Kirchhoff, 1988) and is proving particularly valuable for human identification and forensic medicine (Gill *et al.*, 1985; Jeffreys *et al.*, 1985). This application has been popularly called "genetic fingerprinting."

ISOLATION OR DETECTION OF SPECIFIC DNA REGIONS

Bulk Isolation

Relatively simple procedures enable purification of DNA from other biological molecules. However, the similar physical properties of the four nucleotides limit bulk isolation of DNA types. In some cases, different A:T or G:C composition allows separation of highly repetitive nuclear DNA families as "satellites" by isopycnic centrifugation (reviewed by Brutlag, 1980). Other DNA types are isolated by virtue of their encapsulation. For example, mitochondrial and nuclear DNA are isolated by first separating the organelles. Additional purification capitalizes upon different densities of the circular mtDNA and linear nuclear DNA in the presence of intercalating dyes. Amenability to isolation is a major advantage of mtDNA. Once isolated, it is of a size more manageable than the lengths of nuclear DNA.

Restriction Enzymes, Clones and Probes

Bulk isolation procedures cannot separate or distinguish among the unique regions of the enormously long DNA molecule. Even the small genomes of viruses, bacteria, and mitochondria must be further fractionated to be analyzable. Fractionation became possible with the discovery of bacterial "restriction endonucleases," which have been largely responsible for the explosive new DNA technology (Meselson and Yuan, 1968; Roberts, 1982; Sambrook *et al.*, 1989).

These enzymes cut DNA at specific short sequences commonly consisting of four to six nucleotides. They reduce the long DNA molecules into specific fragments of discrete lengths, amenable to separation by electrophoresis through a gel matrix.

Beyond specific fragmentation of DNA, restriction enzymes facilitate isolation of the fragments through cloning. Many restriction enzymes cut DNA to create complementary, single-stranded, cohesive ends that, under the right conditions, will reanneal. This property is used to form recombinant DNA molecules between different organisms. When a particular DNA segment is inserted into the DNA of a bacterial plasmid or a bacteriophage (known as a vector), it will be replicated as the bacteria grow (Cohen *et al.*, 1973; Thomas *et al.*, 1974). Thus, a sufficient amount of the DNA insert can be obtained to analyze or utilize. A large proportion of a eukaryotic genome can be represented as fragments in a collection, or "library," of many separate clones (Maniatis *et al.*, 1978).

In lieu of physical isolation, specific DNA segments are recognized and distinguished from others by using a separate piece of cloned or isolated DNA as a probe. When an entire nuclear genome is digested with restriction enzymes, thousands of different-size fragments are generated. These can be separated by gel electrophoresis but cannot be resolved. A probe enables visualization of only a small portion of the genome at a time and allows comparison of the homologous region (locus) from one individual to the next. After electrophoretic separation, the DNA fragments are denatured into single strands and then transferred by capillary action, through the face of the gel, to become immobilized in the same relative positions on a membrane. The membrane is soaked for a period of time with a solution containing the probe, which is usually radioactively labelled. The probe will attach only to fragments on the membrane to which it is complementary, binding through base pairing, i.e. through molecular hybridization. The fragments that bind the probe are detected by exposure to X-ray film. The procedure is known as "Southern blotting" (Southern, 1975; methods in Sambrook *et al.*, 1989).

Amplification

A powerful method to isolate specific sequences of DNA is amplification through the polymerase chain reaction (PCR) (Saiki *et al.*, 1988). First, short single-stranded oligonucleotides (about 20 bases long) are artificially synthesized which represent sequences on opposite DNA strands, bracketing a region up to about 2kb to be amplified. These serve as primers for the heat resistant *Thermus aquaticus* (*Taq*) DNA polymerase. DNA containing the region of interest is thermally denatured to provide the single-stranded template for synthesis of the complementary strands. The primers are allowed to anneal to the template. Synthesis from each of the primers is oriented towards and continues past the position of the other primer. Through successive cycles of denaturation and polymerization, synthesis becomes exponential, because new strands serve as templates in subsequent cycles. The result is an abundant amplification of the segment between the primers. To utilize PCR, enough knowledge is needed about the sequence of the region to obtain the primers. Some DNA regions are sufficiently conserved so that sequences from different organisms can function as primers which do not necessarily have to match the template exactly to support the reaction (Kocher *et al.*, 1989).

IDENTIFYING AND EVALUATING DNA DIFFERENCES

DNA Sequencing

A central goal of DNA analyses is to determine the linear sequences of nucleotides that distinguish species and individuals. Sequencing DNA involves time-consuming procedures (Sanger *et al.*, 1977; Maxam and Gilbert, 1980), but advanced instruments facilitate the process (Smith *et al.*, 1986). Sequencing usually entails cloning the region of interest. To obtain and compare corresponding sequences among individuals, libraries must be made from each, from which the first clone is used to select the homologous clones. Alternatively, using PCR, homologous regions can be obtained which can be directly sequenced (Wong *et al.*, 1987; Innis *et al.*, 1988; Gyllensten, 1989). Despite such advances, it is still a major task to sequence homologous DNA segments of many individuals. However, PCR products are amenable to, and greatly facilitate, other types of comparative analyses (discussed in the following section) which justifies the sequencing necessary to obtain the primers.

Restriction Fragment Length Polymorphisms (RFLPs)

In fragmenting DNA, restriction enzymes can be used as a reflection of the nucleotide sequences that they recognize. This allows genetic distinction based on a small percentage of the genome without actually having to sequence the DNA (Kan and Dozy, 1978). If a sequence recognized by a specific restriction endonuclease is altered by a substitution of any of the nucleotides, it would not be cleaved. Likewise, nucleotide changes may create new sites. Differences in DNA that fall within the restriction site sequences, due to genetic divergence between organisms, would be reflected in the length of fragments generated by the enzymes and separated by electrophoresis.

FIGURE 1 illustrates how changes in restriction sites of one enzyme would alter the fragment patterns. Fragments generated by other enzymes would reflect positions of different restriction sites along the same region of DNA and would reveal changes in the additional sets of nucleotides. Changes in restriction fragment patterns result not only from point mutations, as depicted, but also from insertions, deletions or duplications that alter the length of fragments (Goodbourn *et al.*, 1983; Kreitman, 1983). In the nuclear genome, polymorphisms also reflect different locations of the same sequence (transposable or repetitive DNA), due to different flanking DNA. RFLP analyses can be performed on DNA that is isolated, cloned, amplified, or visualized by probes.

Isolated mtDNA is sufficiently small that the restriction fragments from the entire molecule can usually be distinguished. If a large amount of mtDNA is used, the fragments can be visualized directly by staining, but, when greater sensitivity is required, the ends of the fragments are radioactively labelled (Smith and Brown, 1988, 1990; labelling method in Sambrook *et al.*, 1989). Another approach is to use cloned mtDNA or an isolated preparation as a probe (Hall and Muralidharan, 1989). A probe eliminates the need to obtain pure mtDNA from each sample but requires the additional blotting and hybridization steps. Because of the relative abundance of mitochondrial sequences in total cellular DNA, probes carrying non-radioactive labels would be sufficiently sensitive. Such labels would include biotin-derivatized nucleotides, detected by a strepavidin-conjugated enzyme that generates a colored reaction product (Leary *et al.*, 1983). These labels are less sensitive than radioactivity but have a longer shelf life and would be more practical for routine testing.

Use of radioactively labelled, cloned probes is the primary means to analyze restriction fragment polymorphisms of nuclear DNA (Southern, 1975; methods in Sambrook *et al.*, 1989). To determine genetic relatedness among organisms, clones to specific genes are not necessary. Allele distinction at many loci scattered throughout and generally representative of the genome is needed. With a library of different probes, many homologous regions of the nuclear genome can be compared. The longer the probe DNA, the larger the number of possible restriction sites it can overlap and therefore detect. In searching for DNA differences, ideal probes would detect as many electrophoretically separated fragments as possible with none superimposed. The number of fragments depends upon the restriction enzyme used to digest the sample DNA. Restriction sites with the fewest nucleotides (four) are found at the highest frequency. The nucleotide sequence in DNA is not random, and polymorphisms at different restriction site sequences occur at different frequencies (Barker *et al.*, 1984).

Polymorphisms in either repetitive or single copy DNA are visualized depending upon the sequences comprising the probe. The probability of a random fragment containing repetitive DNA along with unique sequences increases with the probe size. Probes containing DNA represented only as single or low copy number sequences will reveal the discrete bands needed for restriction fragment analysis. Repetitive segments scattered throughout the genome would have different flanking DNA such that, after enzyme digestion, they would be present in many different sized fragments. Probes representing abundant repetitive sequences often reveal too many bands to be distinguished. However, when two or more restriction enzyme sites exist entirely within the repetitive

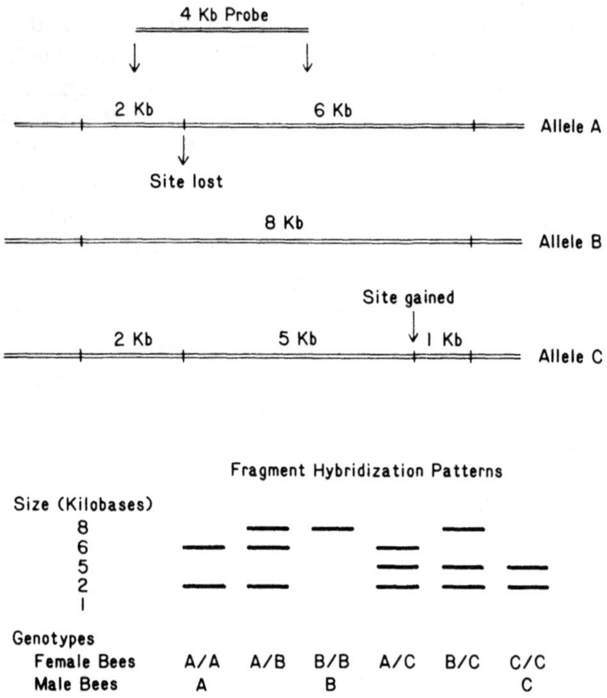

FIGURE 1. Scheme depicting the use of a probe to detect alleles with restriction fragment length polymorphisms and to allow genotype identification. The 4kb (kilobase - 1000 base pairs long) probe is complementary to part of the hypothetical locus depicted and was prepared by a restriction enzyme other than the one used to digest the complementary DNA. Allele A, for example, is defined by three restriction sites of a particular enzyme. The middle restriction site falls in a region overlapped by the probe. Thus, after separation of the fragments on a electrophoretic gel, a 2kb and a 6kb fragment will be detected by the hybridizing probe. To be visualized, only a portion of the fragment needs to be complementary to the probe. Allele B represents a nucleotide change that results in a loss of the center restriction site. Only one 8kb fragment will be detected by the probe. Allele C represents a nucleotide change that results in a gain of another restriction site within the outermost site converting the 6kb fragment to a 5kb fragment and a 1kb fragment. The probe will now detect the 2kb fragment and 5kb fragment, but not the 1kb fragment. The restriction fragments are shown for the genotypes. Heterozygous female bees display a combination of fragments from both parents (codominant expression). The hemizygous male bee carries only one allele or the other, yielding a restriction fragment pattern like that of a homozygous female.

sequence, the multiple copies will be reduced to the same size fragments. In tandemly arranged repetitive DNA, one or more restriction sites within the sequence can also generate the same size fragments. Upon electrophoresis, discrete bands will be formed, containing an amount of DNA proportional to the number of copies, as much as several thousand times greater than in bands from unique DNA. These may be detectable by probes carrying non-radioactive labels as discussed above.

The genomic location of a contiguous section of DNA, detected by a probe, is viewed as a locus. Substitutions at a single nucleotide position are allelic in the strictest sense. However, polymorphisms at different positions, within the region defined by the probe, can also be considered alleles, analogous to changes that cause amino acid substitutions at different points along a protein molecule.

Nuclear restriction fragments show codominant expression, enabling the identification of heterozygotes (Kan and Dozy, 1978; Gusella *et al.*, 1984). A polymorphism at one locus has only a 50% chance of being inherited. Thus, DNA alleles at a number of scattered loci are needed to identify hybrids after several generations. A polymorphism seen in discrete bands of a dispersed repetitive sequence could substitute for an entire collection of unique DNA polymorphisms. Even after many segregational and recombinational events, the proportion of a hybrid genome belonging to a parental type could be ascertained by the densities of the polymorphic fragment bands. Sequences tandemly duplicated at one genomic location would not be as useful, since all copies carrying a polymorphism could be lost after a single generation.

Evaluation of Restriction Site Differences

For a fragment to be characteristic of an organism, both restriction enzyme sites that form the fragment must be present. Loss of either site or the formation of a new site in between would alter the fragment and obscure the presence of sites that remain constant. Thus comparison of individual sites, rather than fragments, is preferable, although more difficult. In obvious cases where a polymorphism generates two short fragments whose lengths add up to that of a lost longer fragment, the relationship of the sites to the fragments can be easily deduced. With more complicated patterns, or in establishing the linear sequence of sites recognized by different enzymes, single and double enzyme digests are compared followed by puzzle-solving to piece the fragments together in the correct linear order (Danna, 1980; Sambrook *et al.*, 1989). This is too complex a procedure for each sample, but from the linear map of a few individuals, site alterations in other samples giving rise to different patterns can be determined. This analysis also clarifies whether insertions, deletions or duplications are responsible for fragment length polymorphisms.

In addition to the qualitative analysis of individual fragments, relatedness and divergence can be estimated by finding the percentage of either fragments or sites that the samples have in common (Nei and Li, 1979). The analysis applies only to polymorphisms resulting from nucleotide substitutions. This quantitative approach is not dependent upon specific or diagnostic differences and may have advantages where high individual variation results in complicated fragment patterns difficult to analyze qualitatively.

Sequence Specific Probes

Restriction fragment polymorphisms reflect relatively minor DNA changes--single nucleotide substitutions or short sequence rearrangements. Because most of the sequence is similar, but not necessarily identical, probes from the DNA of one individual can hybridize to and detect differences in another. With increasing divergence of organisms, greater nucleotide diversity results in extended lengths of uncommon sequences. Among closely related species, there is sufficient divergence so that probes of some repetitive sequences will only hybridize to the DNA of the homologous species (Dowsett, 1983; Cockburn, 1990). Among subspecies, such sequences would likely be rare (Dover, 1982) but, if found, would be valuable, particularly if they represented dispersed repetitive DNA. As described above for restriction fragment polymorphisms, the parental contribution to a hybrid genotype could be determined with a single probe. In this case, samples could be tested as dot blots, where total cellular DNA is simply spotted on a membrane, without the need to generate and separate restriction fragments. The number of sequence copies would be reflected in the density of DNA hybridization.

Oligonucleotide Probes

Under stringent reaction conditions, probes comprised of short lengths of nucleotides can exhibit differential hybridization to DNA sequences that differ by a single nucleotide (Caskey, 1987; Amselem *et al.*, 1988; Weisgraber *et al.*, 1988). To enhance such discrimination, the region to be tested is, at first, amplified. Thus, sequences must be determined for the probe and for the flanking PCR primers. This approach has important potential, since it enables allele distinction based on nucleotide differences not limited to restriction enzyme sites, and it allows dot blot analyses of a number of loci at a time.

PART II. HONEY BEE DNA ANALYSES

HONEY BEE MITOCHONDRIAL DNA

With mtDNA, the origins and divergence of honey bee maternal lineages can be followed despite convergence resulting from natural migrations and beekeeping practices.

In one study, European honey bee mtDNA was digested with several restriction enzymes and a single polymorphism was found (Moritz *et al.*, 1986). Since then, the honey bee mitochondrial genome has been extensively analyzed to establish the linear order of many restriction sites (D. R. Smith, 1988; Smith and Brown, 1988, 1990). Polymorphisms were found that resulted from insertions or duplications as well as nucleotide substitutions in restriction sites. Several polymorphisms distinguished groups of honey bee subspecies (D. R. Smith, 1988; Smith and Brown, 1988, 1990; Hall and Muralidharan, 1989, Smith *et al.*, 1989): an east European group (*A. m. ligustica* and *A. m. carnica)*, west European bees (*A. m. mellifera*) and African bees (*A. m. scutellata, A. m. capensis, A. m. intermissa*). The east European group showed greater divergence from African bees than did the west European bees (Smith and Brown, 1990), which coincided with phylogenetic relationships established from morphology (Ruttner, 1988).

MtDNA analyses will help clarify relationships previously based solely on morphology. For example, Spanish bees, classified as *A. m. iberica*, have phenotypic features resembling both *A. m. mellifera* and north African *A. m. intermissa* (Ruttner, 1988). Recently, bees from Spain were found to indeed represent these races as two distinct mtDNA types (Smith *et al.*, 1990). This finding brings into question whether the *A. m. iberica* designation is appropriate for a hybrid population or for either of the two matrilines. Among African races, the mtDNA was found to be very similar. However, with additional restriction enzymes, the *A. m. intermissa* present in Spain were distinguishable from bees of South Africa. The change in designation of the African bee introduced to South America from *A. m. adansonii* to *A. m.scutellata* (Ruttner, 1976) reflects their south and east, rather than west, African origin. With more detailed mtDNA mapping, separate matriline identities of these two subspecies and many of the others will likely be obtained.

Parts of the mitochondrial genome of European bees have been sequenced. This includes more than 1kb containing the large ribosomal subunit (probably from either *A. m. ligustica* or *A. m. carnica*; Vlasak *et al.*, 1987) and a 3kb region containing genes for two subunits of cytochrome c oxidase, and for the tRNAs of tryptophan, leucine, aspartate, and lysine (from *A. m. ligustica*; Crozier *et al.*, 1989). The sequences include polymorphic sites distinguishing European and African mtDNA and have proven to be a valuable resource for matriline identification. Drawing from the sequences, oligonucleotide primers were synthesized which enabled PCR amplification of mtDNA segments from several subspecies. Subsequent cleavage by restriction enzymes distinguished among the three groups of honey bee subspecies listed above (Hall and Smith, 1990). This procedure greatly facilitates matriline identification. Sequencing the rest of the honey bee mtDNA, making it analyzable by PCR, will enhance additional high resolution restriction site mapping.

As will be discussed below, a most exciting and revealing application of mtDNA analyses has been in following the population dynamics of African honey migration in both the Old World and the New World (Hall and Muralidharan, 1989; Smith *et al.*, 1989, 1990; Hall and Smith, 1990).

HONEY BEE NUCLEAR DNA

With nuclear DNA markers, diverged lineages can be distinguished but not as easily as with mtDNA. Unlike mtDNA, nuclear DNA can be used to follow convergence and hybridization among honey bee subspecies.

DNA Isolation, Probe Formation, and Polymorphisms

Honey bee nuclear DNA can be isolated from any stage but is conveniently obtained from larvae or pupae, because the soft tissue facilitates homogenization. Furthermore, brood originates from the colonies from which it is collected, whereas adults drift among hives. Uncapped larvae are easily squirted out of the comb with a stream of water or saline. Pupae are more difficult to collect but provide consistent amounts of DNA from each individual. DNA can be obtained from brood equilibrated for several days with several changes of 95% alcohol and shipped or transported at ambient temperatures. Equilibration is done at $4^{o}C$ to reduce DNA degradation.

As a source of probes, random fragments of honey bee nuclear DNA were cloned in *E. coli* bacteria, by insertion into plasmids pBR322 (Hall, 1986) and pUC19 (Muralidharan and Hall, 1990). To identify clones with repetitive DNA, bacterial colonies were hybridized to radioactively labelled, total nuclear, honey bee DNA (Hall, 1986). Because of the high concentration of repetitive sequences compared to single copy sequences in total DNA, only colonies carrying inserts represented as repetitive classes exhibited hybridization.

The honey bee genome contains approximately 10% repetitive DNA arranged in a long interspersion pattern (Jordan and Brosemer, 1974; Crain *et al.*, 1976). Thus, relatively large segments of bee DNA can be cloned, most containing little or no repetitive sequences. Although clones with 13kb DNA inserts have been obtained (Hall, 1990), the size range of inserts in plasmids is usually about 4 to 9kb. This is a good probe length for finding restriction fragment differences between African and European bee DNA, digested with four and five-base enzymes. Two to ten fragments are typically generated which are usually sufficiently separated on standard gels.

The first probes used with the intention of finding unique sequence polymorphisms showed only faint hybridization to total bee DNA (Hall, 1986, 1988, 1990; Muralidharan and Hall, 1990). To find characteristic differences between African and European honey bees, the strategy has been first to compare a few samples, each digested with nine separate restriction enzymes, using one probe at a time. Four and five base enzymes have been used to increase the probability of finding differences. Many of the DNA probes revealed polymorphisms. Sometimes up to seven of the nine restriction enzymes generated polymorphisms seen with a single probe. All multiple differences detected with single probes may not have represented independent nucleotide changes. A single DNA rearrangement, for example an insertion, would be seen as polymorphisms generated by different enzymes.

Specificity of the Nuclear DNA Polymorphisms

The enzyme-probe combinations that revealed polymorphisms in the initial screening were then tested against many more samples to determine population or subspecies specificity. Because a honey bee colony is comprised of a number of patrilines, samples of mixed worker siblings increase representation of the local population, although the queen genotype is over-represented. Use of mixed samples accelerates the screening of populations. The density of fragment bands would reflect the proportion of alleles among colony members. In mixed samples, the absence of fragments in some individuals can be obscured but will be revealed by the allelic fragments. Therefore, before using mixed samples to determine specificities, it is important to establish the allelic fragment relationships using individuals (discussed below).

Among a majority of polymorphisms found to have no apparent specificity, a number have been discovered that are either African (Muralidharan and Hall, 1990) or European specific (Hall, 1990). The markers found to date are not diagnostic (i.e. present in all members of one group and absent in the other; Ayala and Powell, 1972). Three European alleles, detected by three separate probes, have been the most thoroughly characterized. Each were found at about a 75% frequency in US bees but absent in South African bees (Hall, 1990). Possibly, the markers are diagnostic for certain European races. A limited number of *A. m. ligustica, A. m. caucasia,* and *A. m. mellifera* colonies from Europe were tested. Only the "European" alleles seen with two probes were found in only the Italian bees. Only the "European" allele seen with the other probe was found in the Italian and Caucasian bees, and largely absent in the *A. m. mellifera.*

Segregation of the Nuclear DNA Alleles

Polymorphisms of unique DNA sequences represent alleles of a locus and are inherited in a Mendelian fashion. It was through analysis of individual haploid drones that the relationships of the polymorphic fragments were established (Hall, 1990). Finding only one allelic fragment or the other in each drone indicated that a single locus was detected by the probe. Also by analysis of the parthenogenetically-derived drones from individual colonies, the genotypes of queens were determined and segregation of her alleles demonstrated. A maximum of two haplotypes from one queen further verified that one locus was represented. Allele frequencies among queens were obtained from the genotypes of their drone progeny. Because of the codominant expression of restriction fragments, homozygous and heterozygous worker genotypes were readily distinguished from which allele frequencies were also obtained. By comparing drones of a colony with sibling workers, the genotypes of drones with which the queen had mated could, in some cases, be determined.

Nature of the Nuclear DNA Polymorphisms

Most of the polymorphisms can be explained as single base changes, altering restriction enzyme recognition and cleavage. However, some represent additional fragments without a corresponding loss of an allelic fragment (Hall, 1990). A number of such polymorphisms, characterized in other organisms, are due to reiteration of short sequences (Bell *et al.*, 1982; Goodbourn *et al.*, 1983; Nakamura *et al.*, 1987). The gain of new fragments would be a function of restriction site location relative to the duplicated segment. Some fragment differences may be a result of deletions or insertions, but these should be detectable in digests of other enzymes. Among extra fragment polymorphisms, specific sequences may be present that have potential as probes for dot blot testing. However, in the honey bee, subspecies-specific insertions have yet to be found that have this promise.

One honey bee DNA probe has been discovered that detects a large number of alleles at a single locus (Hall, work in progress). The molecular basis of the high level of polymorphism has not been determined, but the electrophoretic fragment patterns suggest that variable numbers of tandem repeats (VNTR) (Nakamura *et al.*, 1987) may be responsible. Individual apiaries are characterized by high levels of heterozygosity, sufficient to identify several patrilines within colonies. This and similar probes, such as those specific to minisatellites, should be useful in following mating dynamics, introgression patterns, and genetic social structure within a colony.

The segregation behavior of the DNA alleles detected by this probe indicated that the apparent repetitive sequences were in a tandem arrangement at one locus. If scattered repetitive sequences were involved, varying amounts of the polymorphic fragments would be seen in individual drones. From another study, a polymorphism was described as African-specific and present in repetitive DNA (Severson *et al.*, 1988). Further characterization to ascertain genomic distribution, level of reiteration and polymorphism frequencies has not been reported.

REPRESENTATION OF EUROPEAN SUBSPECIES DNA IN US POPULATIONS

Although designated as different European races, most US bees probably have hybrid genotypes as a result of considerable cross-breeding. As discussed earlier, despite such mixing, maternal lineages of different subspecies can be identified using mtDNA. Among a majority of managed US colonies carrying east European mtDNA, a minority carried mtDNA of *A. m. mellifera* (Hall and Muralidharan, 1989; Smith *et al.*, 1989; Hall and Smith, 1990). Higher proportions of this mtDNA type were present among feral Arizona colonies (Hall and Smith, 1990; Smith, D.R. and Taylor, O.R. unpubl. data). Perhaps, these represent matrilines extending back to some of the earliest introductions.

Frequencies of nuclear DNA alleles that appear to be subspecies-specific may reflect the proportion of European races in US populations. The widespread distribution of the markers may reflect race mixing due to beekeeping. Perhaps the absence of these markers in a minority of US bees corresponds, in part, to the proportion of *A. m. mellifera* mtDNA that has been identified. American bees, derived from a limited number of founding populations, may have less genetic variability than populations in Europe and Africa. DNA analyses may eventually establish whether this possibility is true.

POPULATION DYNAMICS OF THE AFRICAN HONEY BEE FOLLOWED WITH DNA MARKERS

As mentioned above, a study of Spanish bees has revealed the presence of *A. m. mellifera*-like mtDNA and north African *A. m. intermissa* mtDNA, predominantly found in the north and south respectively of the Iberian peninsula (Smith *et al.*, 1990). This finding suggests that, after the original migration of the *A. m. mellifera* lineage from northwest African into northern Europe (Ruttner *et al.*, 1978), a secondary colonization took place long after divergence of the two subspecies. The secondary introgression may have been of *A. m. mellifera* from the north, retracing the direction of the earlier migration, or may have been due to another infiltration from Africa. The latter scenario is particularly intriguing, because it suggests that an early African-European bee encounter occurred in Europe similar to the event currently in progress in North America. Furthermore, there appears to be some limit to the northern spread of the African mtDNA type.

The mechanism by which African bees have spread in the Americas, either by paternal or maternal gene flow, has been controversial. The notorious swarming behavior of African bees, the long distance travel of the swarms, and the speed by which the migration has taken place has implicated maternal migration (Michener, 1975; Taylor, 1977, 1985). Conclusive genetic evidence was provided by two independent studies. Virtually all feral swarms near the expanding front of African bees in Mexico and colonies further behind the front, established from swarms, carried African mtDNA (Hall and Muralidharan, 1989; Smith *et al.*, 1989). Because mtDNA is maternally inherited, these results showed that the migrating force of African bees in the American tropics consists of African maternal lineages. The matrilines have remained unbroken despite many years and thousands of miles of travel through regions with abundant populations of managed European bees. These studies disproved the assumption that the primary mechanism of African bee spread has been through paternal introgression into extant European colonies (Rinderer *et al.*, 1985; Erickson *et al.*, 1986; Rinderer, 1986).

Additional PCR testing of neotropical bees has added support to the same conclusion (Hall and Smith, 1990). The few swarms found with European mtDNA were captured in regions just behind the migrating front, occupied by African bees for little more than a year. Additional analyses showed that the neotropical African bees did not carry the *A. m. intermissa* mtDNA type that may have been imported previously from Spain (Hall and Smith, 1990).

Many of the same samples tested for mtDNA were also analyzed for the European nuclear DNA markers described above. Levels of the European alleles were diminished in samples from managed European colonies near Tapachula, Mexico, where African bees had invaded fifteen months earlier (Hall, 1990). This finding coincides with the many observations of behavioral changes in European apiaries due to the queens mating with African drones (Michener, 1975; Collins, 1988). Despite this Africanization, the European maternal lines do not become part of the migrating African feral population and, unless actively maintained, probably disappear through attrition. Although paternal introgression has an impressive effect on apiaries, it has had little to do with the long range spread of African bees.

The European markers were present at low frequencies in the feral Mexican swarms that carried African mtDNA. Thus, there had been some minor European paternal gene flow into the colonies that had generated the swarms. Interestingly, however, the markers were almost completely absent in Venezuelan samples (Hall, 1990).

RETENTION OF THE AFRICAN HONEY BEE GENOTYPE

The term "Africanized bees" embraces the view that neotropical African bees are primarily hybrids of the extant European queens and African drones. The mtDNA results indicate that "Africanized bees" are largely confined to apiaries and, rather than serving a primary role, actually represent a dead end process. Neotropical African populations have also been called a "hybrid swarm," meaning that they represent the entire range of hybrid genotypes between the two types of bees (Rinderer, 1986). The asymmetries in both mtDNA and nuclear DNA in favor of African bees do not fit this definition. African colonies started as a minority and through their migration encountered an initial majority of managed European colonies. Yet their behavioral traits have not been attenuated (Michener, 1975; Taylor 1985, 1988; Fletcher, 1988). This has suggested, prior to DNA evidence, that strong selective mechanisms have maintained the African genotype in the tropics.

Such selective mechanisms would work by either limiting hybrid formation or selecting against hybrids once formed. If an African queen mates with a majority of African drones, one would be the likely father of the next generation queen. If so, the hybridization manifested in worker progeny of European drones would not be transmitted to the next generation. If faster developmental times (Taylor, 1988) or selective processes (Page and Erickson, 1984; Page *et al.*, 1989) favored queens of African paternity, this would be a strong force maintaining the African genotype. If such processes operate in European colonies, they could account for rapid and effective Africanization even where African drones are present as a minority.

Apparently, greater reproductive fitness in the tropics has enabled African bees, but not European bees, to establish massive feral populations. The resulting overwhelming numbers may be the primary reason for an apparent lack of hybridization. However, to adequately explain the paucity of European markers, African bees must also have greater fitness than African-European hybrids. European alleles were found at low frequencies in African swarms near the migrating front. Considerably greater accumulation of European alleles would be expected if hybrids, formed at the interface of African and European populations, simply moved ahead with the front and were not replaced. The migration process itself may select against introgressed European genes that reduce the propensity to swarm and the ability for long distance dispersal. Thus hybrids might fall behind the front and, in effect, would be replaced. Further behind the front, European DNA alleles are almost undetectable (Hall, 1990). This finding coincides with differences seen in allozyme frequencies (Smith *et al.*, 1989) and with temporal changes seen towards a more African morphology (Boreham and Roubik, 1987). These observations suggest that purging of European alleles continues as the African population becomes established. Effective selection against early generation hybrids would even tend to eliminate neutral European markers.

Like European bees, hybrids that might otherwise succumb in the wilds may survive in managed colonies over generations of backcrossing with feral African drones. Swarms generated from these Africanized colonies would be expected to be more adaptive to feral conditions. Thus, it is puzzling why such European maternal lines have not added significantly to the feral population over time. This brings into question whether higher reproductive capability of African bees could alone account for the paucity of European mtDNA. In other systems, introgression of mtDNA, even across species boundaries, is prevented only by complete reproductive isolation, inviable hybrids, or non-neutral mtDNA (Takahata and Slatkin, 1984). It has been speculated that European honey bee mitochondria may limit a metabolic ability needed for long distance dispersal, and thus be a factor against which there is continued selection (Hall and Muralidharan, 1989).

Other possible genetic incompatibilities have not been ruled out. First generation hybrids between African and European bees appear vigorous. However, the fitness of late generation hybrid colonies has not been adequately studied, when genetic incompatibilities are more likely to be expressed. Mitochondrial activities are influenced by nuclear factors and are impaired in interspecies hybrids (Liepins and Hennen, 1977; Miller *et al.*, 1986; Schmitz and Michaelis, 1988; discussed in Avise *et al.*, 1987, Moritz *et al.*, 1987). It is possible that, in hybrids of distantly related subspecies, mitochondrial activities could be impaired to some extent. This would be a disadvantage that also would tend to be manifested in late generation hybrids, especially if repeated backcrossing generated an increasingly heterologous nuclear-mitochondrial combination (Avise, 1986). With the complexity of honey bee social organization (Frumhoff and Baker, 1988; Robinson and Page, 1988), greater genetic heterogeneity among colony members might conceivably cause some disruption in the activities requiring communication and coordinated efforts.

As African bees approach temperate climatic regions of the US, more favorable to European bees, the two are expected to form a hybrid zone (Taylor and Spivak, 1984) as they apparently have in temperate regions of South America (Lobo *et al.*, 1989). Because of the large amount and different classes of nuclear DNA, many markers under different selective influences (neutral, non-neutral, or neutral linked to non-neutral) can potentially be found and will help to follow hybrid zone dynamics. Genotype frequencies and distributions, as well as introgression and linkage disequilibrium patterns may reveal the nature of selective processes--environmental, genetic, physiological or behavioral. Coordinated introgression of nuclear markers and mtDNA might point to the presence of a functional relationship.

DNA IDENTIFICATION FOR REGULATORY CONTROL OF THE AFRICAN HONEY BEE

Proposed methods for control of the African honey bee--quarantine, extermination, and certification of breeding stock--are dependent upon a reliable means of identification. Currently, the method most commonly used is morphometric analysis (Daly and Balling, 1978; Daly *et al.*, 1982; discussed in Daly, Chapter 2). Another promising approach is through analysis of cuticular hydrocarbons (Carlson, 1988; R. K. Smith, 1988). Methods based on the phenotype can be subject to environmental influences and are limited in their ability to distinguish hybrids, particularly after unknown numbers of generations. To date, morphometrics have been successful in following the migration of the African bee. The limited admixture of African and European bees, as indicated by nuclear and mtDNA findings, may have helped maintain the accuracy and success of morphometric identification. This accuracy may suffer with increased hybridization in temperate climates.

The mtDNA results indicate that African maternal lineages are the primary concern in the establishment of feral African populations. From a practical standpoint, emphasis on control of African maternal lineages would be the most effective initial step. Identification of mtDNA type is highly accurate and now can be done rapidly with the PCR method. However, mtDNA would not distinguish "Africanized bees," i.e. European maternal lines mated to African drones. These hybrids exhibit unpredictable defensive behavior and, in some cases, will need to be accurately identified. If there are no genetic incompatibilities, Africanized European matrilines may survive better in temperate climates and paternal gene flow into northern European populations could occur. The extent of introgression will dictate the need for widespread identification of hybrids.

To accurately identify hybrids, methods that can distinguish nuclear alleles are necessary. The principles involved in identification by protein variants and by DNA polymorphisms are basically the same. A number of polymorphisms need to be available and specific (preferably diagnostic) to serve as a reliable identification method (Page and Erickson, 1985). The more loci with distinguishing alleles, the more certain the identification of hybrids. With nuclear DNA, specific and perhaps diagnostic alleles can be found at widespread loci.

Phenotypic methods, such as morphometric and hydrocarbon analyses, have some advantages, and their levels of accuracy may be adequate depending on the action needed. DNA analyses would complement such methods and would be recommended when critical decisions are to be made costly to the beekeeper and the government (Stibick, 1984). As African bees become established in the US, their presence will have to be largely accepted. Only levels of control that are realistic and appropriate should be imposed on beekeepers, depending on their type of operation.

A central control measure would be regular requeening with European bees to limit the influx of African traits into apiaries. Maintenance of either African or European colonies in isolated, non-migratory apiaries would appropriately be at the discretion of the beekeeper. Migratory, package bee and queen rearing operations would have lower tolerances for levels of African introgression and would require certain levels of certification of their bees as European. Because queen production involves genetic propagation, breeders would require certification based on genetics and at the highest level of accuracy. This could only be provided by nuclear DNA analyses. The complexity of current nuclear DNA testing limits its use as a routine method. However, new methods, such as PCR applied to nuclear DNA as with mtDNA, will make certification on the basis of DNA feasible.

CONCLUSION

DNA analyses have valuable potential to advance understanding of honey bee genetics, and, hence, to address a number of important concerns regarding the African honey bee. Results from DNA research provide a detailed view of African-European bee genetic relationship not possible with other methods. Morphological and protein similarities have masked an underlying genetic variability only now being revealed by the bees' DNA. Some DNA differences are characteristic of subspecies, which allow accurate identification. Restriction fragment polymorphisms are providing insights into the dynamics of African-European bee interaction. By interfacing with many aspects of genetics, the DNA analyses will contribute to a deeper understanding of the African honey bee essential for future management and control.

Acknowledgements

The author thanks M. McMichael, A. Cockburn and D. Campton for thoughtful suggestions during manuscript preparation. The research by the author cited in this chapter was supported by the USDA Competitive Research Grants Office, the University of Florida Interdisciplinary Center for Biotechnology Research, and the Florida State Beekeepers Association. Florida Experiment Station Journal Series No. R-00961.

LITERATURE CITED

Amselem, S. Nunes, V., Vidaud, M., Estivill, X., Wong, C., d'Auriol, L., Vidaud, D., Galibert, F., Baiget, M., Goossens, M. 1988. Determination of the spectrum of ß-thalassemia genes in Spain by use of dot-blot analysis of amplified ß-globin DNA. *Am. J. Hum. Genet.* 43:95-100.

Arnold, M. L. 1986. The heterochromatin of grasshoppers from the *Caledia captiva* species complex. III. Cytological organisation and sequence evolution in a dispersed highly repeated DNA family. *Chromosoma* 94:183-188.

Atchison, M. L. 1988. Enhancers: mechanisms of action and cell specificity. *Ann. Rev. Cell Biol.* 4:127-153.

Avise, J. C. 1986. Mitochondrial DNA and the evolutionary genetics of higher animals. *Phil. Trans. R. Soc. London B* 312:325-342.

Avise, J. C., Arnold, J., Ball, R. M., Bermingham, E., Lamb, T., Neigel, J. E., Reeb, C. A., Saunders, N. C. 1987. Intraspecific phylogeography: the mitochondrial DNA bridge between population genetics and systematics. *Ann. Rev. Ecol. Syst.* 18:489-522.

Ayala, F., Powell, J. R. 1972. Allozymes as diagnostic characters of sibling species of *Drosophila. Proc. Natl. Acad. Sci. (USA)* 69:1094-1096.

Baltimore, D. 1985. Retroviruses and retrotransposons: the role of reverse transcription in shaping the eukaryotic genome. *Cell* 40:481-482.

Barker, D., Schafer, M., White, R. 1984. Restriction sites containing CpG show a higher frequency of polymorphism in human DNA. *Cell* 36:131-138.

Beauchamp, R., Mitchell, A., Buckland, R., Bostock, C. 1979. Specific arrangements of human satellite III DNA sequences in human chromosomes. *Chromosoma* 71:153-166.

Bell, G. I., Selby, M. J., Rutter, W. J. 1982. The highly polymorphic region near the human insulin gene is composed of simple tandemly repeating sequences. *Nature* 295: 31-35.

Bishop, J. O. 1974. The gene numbers game. *Cell* 2:81-86.

Bishop, J. O., Morton, J. G., Rosbash, M., Richardson, M. 1974. Three abundance classes in HeLa cell messenger RNA. *Nature* 250:199-204.

Boreham, M. M., Roubik, D. W. 1987. Population change and control of Africanized honey bees (Hymenoptera: Apidae) in the Panama canal area. *Bull. Entomol. Soc. Am.* 33:34-38.

Bouchard, R. A. 1982. Moderately repetitive DNA in evolution. *Int. Rev. Cytology* 76:113-193.

Britten, R. J., Kohne, D. E. 1968. Repeated Sequences in DNA. *Science* 161:529-40.

Brown, W. M. 1985. The mitochondrial genome of animals. In *Molecular Evolutionary Genetics*, ed. R. J. MacIntyre, pp. 95-130. New York: Plenum Press.

Brutlag, D. L. 1980. Molecular arrangement and evolution of heterochromatic DNA. *Ann. Rev. Genetics* 14:121-144.

Carlson, D. A. 1988. Africanized and European honey-bee drones and comb waxes: analysis of hydrocarbon components for identification. In *Africanized Honey Bees and Bee Mites*, ed. G. R. Needham, R. E. Page, M. Delfinado-Baker, C. E. Bowman, pp. 264-274. Chichester, England: Ellis Horwood Limited.

Caskey, C. T. 1987. Disease diagnosis by recombinant DNA methods. *Science* 236:1223-1229.

Cockburn, A. F. 1990. A simple and rapid technique for identification of large numbers of individual mosquitoes using DNA hybridization. *Arch. Insect Biochem. Physiol.* 14:191-199.

Cockburn, A. F., Mitchell, S. E. 1989. Repetitive DNA interspersion patterns in Diptera. *Arch. Insect Biochem. Physiol.* 10:105-113.

Cohen, S. N., Chang, A. C. Y., Boyer, H. W., Helling, R. B. 1973. Construction of biologically functional bacterial plasmids *in vitro*. *Proc. Natl. Acad. Sci. (USA)* 70:3240-3244.

Collins, A. M. 1988. Genetics of honey-bee colony defense. In *Africanized Honey Bees and Bee Mites*, ed. G. R. Needham, R. E. Page, M. Delfinado-Baker, C. E. Bowman, pp. 110-117. Chichester, England: Ellis Horwood Limited.

Crain, W. R., Davidson, E. H., Britten, R. J. 1976. Contrasting patterns of DNA sequence arrangement in *Apis mellifera* (honeybee) and *Musca domestica* (housefly). *Chromosoma* 59:1-12.

Crozier, R. H. 1977. Evolutionary genetics of the Hymenoptera. *Ann. Rev. Entomol.* 22:263-268.

Crozier, R. H., Crozier, Y. C., Mackinlay, A. G. 1989. The Co-I and Co-II region of honey-bee mitochondrial DNA: evidence for variation in insect mitochondrial evolutionary rates. *Molec. Biol. Evol.* 6:399-411.

Daly, H. V., Balling, S. S. 1978. Identification of Africanized honeybees in the Western Hemisphere by discriminant analysis. *J. Kansas. Entomol. Soc.* 51:857-869.

Daly, H. V., Hoelmer, K., Norman, P., Allen, T. 1982. Computer-assisted measurement and identification of honeybees (Hymenoptera, Apidae). *Ann. Entomol. Soc. Am.* 75:591-594.

Danna, K. J. 1980. Determination of fragment order through partial digests and multiple enzyme digests. *Methods Enzymol.* 65:449-467.

Davidson, E. H., Hough, B. R., Amenson, C. S., Britten, R. J. 1973. General interspersion of repetitive with nonrepetitive sequence elements in the DNA of *Xenopus*. *J. Mol. Biol.* 77:1-23.

Donis-Keller, H. *et al.* 1987. A genetic linkage map of the human genome. *Cell* 51:319-337.

Doolittle, W. F., Sapienza, C. 1980. Selfish genes, the phenotype paradigm and genome evolution. *Nature* 284:601-603.

Dover, G. 1982. Molecular drive: a cohesive mode of species evolution. *Nature* 299:111-117.

Dowsett, A. 1983. Closely related species of *Drosophila* can contain different libraries of middle repetitive DNA sequences. *Chromosoma* 88:104-108.

Erickson, E. H., Erickson, B. J., Young, A. M. 1986. Management strategies for "Africanized" honey bees: Concepts strengthened by our experiences in Costa Rica. Part I. *Gleanings Bee Cult.* 114: 456-459.

Ferris, S., Sage, R., Huang, C., Nielsen, J., Ritte, U., Wilson, A. 1983. Flow of mitochondrial DNA across a species boundary. *Proc. Natl. Acad. Sci. (USA)* 80:2290-2294.

Finnegan, D. J. 1985. Transposable elements in eukaryotes. *Inter. Rev. Cytol.* 93:281-326.

Fletcher, D. J. C. 1988 Relevance of the behavioral ecology of African bees to a solution to the Africanized-bee problem. In *Africanized Honey Bees and Bee Mites*, ed. G. R. Needham, R. E. Page, M. Delfinado-Baker, C. E. Bowman, pp. 55-61. Chichester, England: Ellis Horwood Limited.

Frumhoff, P. C., Baker, J. 1988. A genetic component to division of labour within honey-bee colonies. *Nature* 333:358-361.

Georges, M., Lequarre, A. -S., Castelli, M., Hanset, R., Vassart, G. 1988. DNA fingerprinting in domestic animals using four different minisatellite probes. *Cytogenet. Cell. Genet.* 47:127-131.

Gerbi, S. A. 1985. Evolution of ribosomal DNA. In *Molecular Evolutionary Genetics*, ed. R.J. MacIntyre, pp. 419-518. New York: Plenum Press

Gill, P., Jeffreys, A. J., Werrett, D. J. 1985. Forensic application of DNA "fingerprints." *Nature* 318:577-579.

Goodbourn, S. E. Y., Higgs, D. R., Clegg, J. B., Weatherall, D. J. 1983. Molecular basis of length polymorphism in the human ζ-globin gene complex. *Proc. Natl. Acad. Sci. (USA)* 80:5022-5026.

Graur, D. 1985. Gene diversity in Hymenoptera. *Evolution* 39:190-199.

Green, M. R. 1986. Pre-mRNA splicing. *Ann. Rev. Genetics* 20:671-708.

Gusella, J. F., Tanzi, R. E., Anderson, M. A., Hobbs, W., Gibbons, K., Raschtchian, R., Gilliam, T. C., Wallace, M. R., Wexler, N. S., Conneally, P. M. 1984. DNA markers for nervous system diseases. *Nature* 225:1320-1326.

Gyllensten, U. B. 1989. PCR and DNA sequencing. *Biotechniques* 7:700-708.

Hall, H. G. 1986. DNA differences found between Africanized and European honeybees. *Proc. Natl. Acad. Sci. (USA)* 83:4874-4877.

Hall, H. G. 1988. Characterization of the African honey-bee genotype by DNA restriction fragments. In *Africanized Honey Bees and Bee Mites*, ed. G. R.

Needham, R. E. Page, M. Delfinado-Baker, C. E. Bowman, pp. 287-293. Chichester, England: Ellis Horwood Limited.

Hall, H. G. 1990. Parental analysis of introgressive hybridization between African and European honey bees using nuclear DNA RFLPs. *Genetics* 125:611-621.

Hall, H. G., Muralidharan, K. 1989. Evidence from mitochondrial DNA that African honey bees spread as continuous maternal lineages. *Nature* 339:211-213.

Hall, H. G., Smith, D. R. 1990. Distinguishing African and European honey bee matrilines using amplified mitochondrial DNA. Submitted for publication.

Innis, M. A., Myambo, K. B., Gelfand, D. H., Brow, M. A. D. 1988. DNA sequencing with *Thermus aquaticus* DNA polymerase and direct sequencing of polymerase cain reaction-amplified DNA. *Proc. Natl. Acad. Sci. (USA)* 85:9436-9440.

Izquierdo, M., Bishop, J. O. 1979. An analysis of cytoplasmic RNA populations in *Drosophila melanogaster,* Oregon R. *Biochem. Genet.* 17:473-497.

Jeffreys, A. J., Wilson, V., Thein, S. L. 1985. Hypervariable "minisatellite" regions in human DNA. *Nature* 314:67-73.

Jordan, R. A., Brosemer, R. W. 1974. Characterization of DNA from three bee species. *J. Insect Physiol.*20:2513-2520.

Kan, Y. W., Dozy, A. M. 1978. Polymorphism of DNA sequence adjacent to human beta-globin structural gene: relationship to sickle mutation. *Proc. Natl. Acad. Sci. (USA)* 75:5631-5635.

Kirchhoff, C. 1988. GATA tandem repeats detect minisatellite regions in blowfly DNA (Diptera: Calliphoridae). *Chromosoma* 96:107-111.

Kocher, T. D., Thomas, W. K., Meyer, A., Edwards, S. V., Pääbo, S., Villablanca, F. X., Wilson, A. C. 1989. Dynamics of mitochondrial DNA evolution in animals: Amplification and sequencing with conserved primers. *Proc. Natl. Acad. Sci. (USA)* 86:6196-6200.

Kreitman, M. 1983. Nucleotide polymorphism at the alcohol dehydrogenase locus of *Drosophila melanogaster*. *Nature* 304:412-17.

Lamb, T., Avise, J. C. 1986. Directional introgression of mitochondrial DNA in a hybrid population of tree frogs: The influence of mating behavior. *Proc. Natl. Acad. Sci. (USA)* 83:2526-30.

Leary, J. L., Brigati, D. J., Ward, D. C. 1983. Rapid and sensitive colorimetric method for visualizing biotin-labeled DNA probes hybridized to DNA or RNA immobilized on nitrocellulose: bio-blots. *Proc. Natl. Acad. Sci. (USA)* 80:4045-4049.

Leff, S. E., Rosenfeld, M. G., Evans, R. M. 1986. Complex transcriptional units: diversity in gene expression by alternative RNA processing. *Ann. Rev. Biochem.* 55:1091-1117.

Lester, L. J., Selander, R. K. 1979. Population genetics of haplodiploid insects. *Genetics* 92:1329-1345.

Li, W. -H., Luo, C. -C., Wu, C. -I. 1985. Evolution of DNA sequences. In Molecular *Evolutionary Genetics,* ed. R.J. MacIntyre, pp. 1-94. New York: Plenum Press.

Liepins, A., Hennen, S. 1977. Cytochrome oxidase deficiency during development of amphibian nucleocytoplasmic hybrids. *Devel. Biol.* 57:284-292.

Lobo, J. A., Del Lama, M. A., Mestriner, M. A. 1989. Population differentiation and racial admixture in the Africanized honeybee (*Apis mellifera* L.). *Evolution* 43:794-802.

Loomis, W. F., Gilpin, M. E. 1986. Multigene families and vestigial sequences. *Proc. Natl. Acad. Sci. (USA)* 83:2143-2147.

Maniatis, T., Hardison, R. C., Lacy, E., Lauer, J., O'Connell, C., Quon, D., Sim, G. K., Efstratiadis, A. 1978. The isolation of structural genes from libraries of eucaryotic DNA. *Cell* 15:687-701.

Manning, J. E., Schmid, C. W., Davidson, N. 1975. Interspersion of repetitive and non-repetitive DNA sequences in the Drosophila melanogaster genome. *Cell* 4:141-155.

Marchant, A. D. 1988. Apparent introgression of mitochondrial DNA across a narrow hybrid zone in the *Caledia captiva* species-complex. *Heredity* 60:39-46.

Maxam, A. M., Gilbert, W. 1980. Sequencing end-labeled DNA with base-specific chemical cleavages. *Methods Enzymol.* 65:499-560.

Meselson, M., Yuan, R. 1968. DNA restriction enzyme from *E. coli. Nature* 217:1110-1114.

Michener, C. D. 1975. The Brazilian bee problem. *Ann. Rev. Entomol.* 20:399-416.

Miklos, G. L. G. 1985. Localized highly repetitive DNA sequences in vertebrate and invertebrate genomes. In *Molecular Evolutionary Genetics,* ed. R. J. MacIntyre, pp. 241-322. New York: Plenum Press.

Miller, S. G., Huettel, M. D., Davis, M., Weber, E. H., Weber, L. A. 1986. Male sterility in *Heliothis virescens* X *H. subflexa* backcross hybrids: evidence for abnormal mitochondrial transcripts in testes. *Mol. Gen. Genetics* 203:451-461.

Moritz, C., Dowling, T. E., Brown, W. M. 1987. Evolution of animal mitochondrial DNA: relevance for population biology and systematics. *Ann. Rev. Ecol. Syst.* 18:269-292.

Moritz, R. F. A., Hawkins, C. F., Crozier, R. H., Mackinlay, A. G. 1986. A mitochondrial DNA polymorphism in honeybees (*Apis mellifera* L.). *Experentia* 42:322-324.

Muralidharan, K., Hall, H. G. 1990. Prevalence of African DNA RFLP alleles in neotropical African honeybees. *Arch. Insect Biochem. Physiol.* in press.

Nakamura, Y., Leppert, M., O'Connell, P., Wolff, R., Holm, T., Culver, M., Martin, C., Fujimoto, E., Hoff, M., Kumlin, E., White, R. 1987. Variable number of tandem repeat (VNTR) markers for human gene mapping. *Science* 235:1616-1622.

Nei, M., Li, W.-H. 1979. Mathematical model for studying genetic variation in terms of restriction endonucleases. *Proc. Natl. Acad. Sci. (USA)* 76:5269-5273.

Orgel, L. E., Crick, H. C. 1980. Selfish DNA: the ultimate parasite. *Nature* 284:604-607.

Padgett, R. A., Grabowski, P. J., Konarska, M. M., Seiler, S., Sharp, P. A. 1986. Splicing of messenger RNA precursors. *Ann. Rev. Biochem.* 55:1119-1150.

Page, R. E. Jr., Erickson, E. H. Jr. 1984. Selective rearing of queens by worker honey bees: Kin or nestmate recognition. *Ann. Entomol. Soc. Am.* 77:578-580.

Page, R. E. Jr., Erickson, E. H. Jr. 1985. Identification and certification of Africanized honey bees. *Ann. Entomol. Soc. Am.* 78:149-158.

Page, R. E. Jr., Robinson, G. E., Fondrk, M. K. 1989. Genetic specialists, kin recognition and nepotism in honey-bee colonies. *Nature* 338:576-579.

Pirrotta, V. 1988. Vectors for P-mediated transformation in Drosophila. In *Vectors - A Survey of Molecular Cloning Vectors and Their Use,* ed. R. L. Rodriguez, D. T. Denhardt, pp. 437-456. Boston: Butterworth.

Rinderer, T. E. 1986. Africanized bees: the Africanization process and potential range in the United States. *Bull. Entomol. Soc. Am.* 32:222-227.

Rinderer, T. E., Hellmich, R. L., Danka, R. G., Collins, A. M. 1985. Male reproductive parasitism: a factor in the Africanization of European honey-bee populations. *Science* 228:1119-1121.

Roberts, R. J. 1982. Restriction and modification enzymes and their recognition sequences. *Nucleic Acids Res.* 10:R117.

Robinson, G. E., Page, R.E. 1988. Genetic determination of guarding and undertaking in honey-bee colonies. *Nature* 333:356-358.

Rogers, J. H. 1985. The origin and evolution of retroposons. *Inter. Rev. Cytology* 93:187-279.

Rubin, G. M., Spradling, A. C. 1982. Genetic transformation of *Drosophila* with transposable element vectors. *Science* 218:348-353.

Ruttner, F. 1976. The races of bees in Africa. *Proc. 25th Int. Cong. Apiculture.* Apimondia.

Ruttner, F. 1988. *Biogeography and Taxonomy of Honey Bees.* Berlin: Springer-Verlag.

Ruttner, F., Tassencourt, L., Louveaux, J. 1978. Biometrical-statistical analysis of the geographic variability of *Apis mellifera* L. *Apidologie* 9:363-381.

Saiki, R. K., Gelfand, D. H., Stoffel, S., Scharf, S. J., Higuchi, R., Horn, G. T. 1988. Primer-directed enzymatic amplification of DNA with a thermostable DNA polymerase. *Science* 239:487-491.

Sambrook, J., Fritsch, E. F., Maniatis, T. 1989. *Molecular cloning. A laboratory manual.* Cold Spring Harbor: Cold Spring Harbor Laboratory Press. 2nd ed.

Sanger, F., Nicklen, S., Coulson, A. R. 1977. DNA sequencing with chain-terminating inhibitors. *Proc. Natl. Acad. Sci. (USA)* 74:5463-5467.

Schibler, U., Sierra, F. 1987. Alternative promoters in developmental gene expression. *Ann. Rev. Genetics* 21:237-257.

Schmid, C. W., Deininger, P. L., 1975. Sequence organization of the human genome. *Cell* 6:345-358.

Schmitz, U. K, Michaelis, G. 1988. Dwarfism and male sterility in interspecific hybrids of *Epilobium*. 2. Expression of mitochondrial genes and structure of the mitochondrial DNA. *Theor. Appl. Genet.* 76:565-569.

Severson, D. W., Aiken, J. M., Marsh, R. F. 1988. Molecular analyses of North American and Africanized honey bees. In *Africanized Honey Bees and Bee Mites,* ed. G. R. Needham, R. E. Page, M. Delfinado-Baker, C. E. Bowman, pp. 294-302. Chichester, England: Ellis Horwood Limited.

Shapiro, J. A. 1983. *Mobile Genetic Elements.* New York: Academic Press.

Singer, M. F. 1982. Highly repeated sequences in mammalian genomes. *Inter. Rev. Cytology* 76:67-112.

Smith, D. R. 1988. Mitochondrial DNA polymorphisms in five Old World subspecies of honey bees and in New World hybrids. In *Africanized Honey Bees and Bee Mites,* ed. G. R. Needham, R. E. Page, M. Delfinado-Baker, C. E. Bowman, pp. 303-312. Chichester: Ellis Horwood Limited.

Smith, D. R., Brown, W. M. 1988. Polymorphisms in mitochondrial DNA of European and Africanized honeybees (*Apis mellifera*). *Experientia* 44:257-260.

Smith, D. R., Brown, W. M. 1990. Restriction endonuclease cleavage site and length polymorphisms in mitochondrial DNA of *Apis mellifera mellifera* and *A. m. carnica* (Hymenoptera: Apidae). *Ann. Entomol. Soc. Am.* 83:81-88.

Smith, D. R., Taylor, O. R., Brown, W. W. 1989. Neotropical Africanized honey bees have African mitochondrial DNA. *Nature* 339: 213-215.

Smith, D. R., Palopoli, M. F., Garney, L., Cornuet, J.-M., Solignac, M., Brown, W. M. 1990. Geographical overlap of two mitochondrial genomes in Spanish honey bees (*Apis mellifera iberica*). *J. Hered.* in press.

Smith, L. M., Sanders, J. Z., Kaiser, R. J., Hughes, P., Dodd, C., Connell, C. R., Heiner, C., Kent, S. B. H., Hood, L. E. 1986. Fluorescence detection in automated DNA sequence analysis. *Nature* 321:674-679.

Smith, R.-K. 1988. Identification of Africanization in honey bees based on extracted hydrocarbons assay. In *Africanized Honey Bees and Bee Mites,* ed. G. R. Needham, R. E. Page, M. Delfinado-Baker, C. E. Bowman, pp. 275-280. Chichester, England: Ellis Horwood Limited.

Southern, E. M. 1975. Detection of specific sequences among DNA fragments separated by gel electrophoresis. *J. Molec. Biol.* 98:503-517.

Spradling, A. C., Rubin, G. M. 1981. Drosophila genome organization: Conserved and dynamic aspects. *Ann. Rev. Genetics* 15:219-64.

Spradling, A. C., Rubin, G. M. 1982. Transposition of cloned P elements into *Drosophila* germ line chromosomes. *Science* 218:341-347.

Stephan, W. 1987. Quantitative variation and chromosomal location of satellite DNAs. *Genet. Res.* 50:41-52.

Stibick, J. N. L. 1984. Animal and plant inspection service strategy and the African honey bee. *Bull. Entomol. Soc. Am.* 30:22-26.

Takahata, N., Slatkin, M. 1984. Mitochondrial gene flow. *Proc. Natl. Acad. Sci. (USA)* 81:1764-1767.

Taylor, O. R. 1977. The past and possible future spread of Africanized honeybees in the Americas. *Bee World* 58:19-30.

Taylor, O. R. 1985. African Bees: potential impact in the United States. *Bull. Entomol. Soc. Am.* 31:15-24.

Taylor, O. R. 1988. Ecology and economic impact of African and Africanized honey bees. In *Africanized Honey Bees and Bee Mites,* ed. G. R. Needham, R. E. Page, M. Delfinado-Baker, C. E. Bowman, pp. 29-41. Chichester, England: Ellis Horwood Limited.

Taylor, O. R., Spivak, M. 1984. Climatic limits of tropical African honeybees in the Americas. *Bee World* 65:38-47.

Thomas, M., Cameron, J. R., Davis, R. W. 1974. Viable molecular hybrids of bacteriophage Lambda and eukaryotic DNA. *Proc. Natl. Acad. Sci. (USA)* 71:4579-4583.

Vlasak, I., Burgschwaiger, S., Kreil, G. 1987. Nucleotide sequence of the large ribosomal RNA of honeybee mitochondria. *Nucleic Acids Res.* 15:2388.

Wasylyk, B. 1988. Enhancers and transcription factors in the control of gene expression. *Biochim. Biophys. Acta* 951:17-35.

Watkins, P. 1988. Restriction fragment length polymorphism (RFLP): applications in human chromosome mapping and genetic disease research. *Biotechnology* 6:310-319.

Waye, J. S., Willard, H. F. 1987. Nucleotide sequence heterogeneity of alpha satellite repetitive DNA: A survey of alphoid sequences from different human chromosomes. *Nucleic Acids Res.* 15:7549-7569.

Weisgraber, K. H., Newhouse, Y. M., Mahley, R. W. 1988. Apolipoprotein E genotyping using the polymerase chain reaction and allele-specific oligonucleotide probes. *Biochem. Biophys. Res. Comm.* 157:1212-1217.

Willard, H., Waye, J., Skolnick, M., Schwartz, C., Powers, V., England, S. 1986. Detection of restriction fragment length polymorphisms at the centromeres of human chromosomes by using chromosome-specific α-satellite DNA probes: Implications for development of centromere-based genetic linkage maps. *Proc. Natl. Acad. Sci. (USA)* 83:5611-5615.

Wilson, A. C., Cann, R. L., Carr, S. M., George, M., Gyllensten, U. B., Helm-Bychowski, K. M., Higuchi, R. G., Palumbi, S. R., Prager, E. M., Sage, R. D., Stoneking, M. 1985. Mitochondrial DNA and two perspectives on evolutionary genetics. *Biol. J. Linnean Soc.* 26:375-400.

Wong, C., Dowling, C. E., Saiki, R. K., Higuchi, R. G., Erlich, H. A., Kazazlan, H. H. 1987. Characterization of β-thalassaemia mutations using direct genomic sequencing of amplified single copy DNA. *Nature* 330:384-386.

Zakian, V. A. 1989. Structure and function of telomeres. *Ann. Rev. Genet.* 23:579-604.

PART TWO

The Spread of Africanized Bees and the Africanization Process

4

INTERDEPENDENCE OF GENETICS AND ECOLOGY IN A SOLUTION TO THE AFRICAN BEE PROBLEM

David J.C. Fletcher[1]

Recently, I proposed a solution to the problems that are expected to arise when African bees reach the United States (see Fletcher, 1988). This solution, radically different to others that have been suggested, was based on general biological principles and on information about the behavioral ecology of African bees in Africa and the Americas. It also included a certain amount of speculation. In this chapter I shall review the arguments I employed, but I shall reinforce them with some new information and new insights.

THE NEED FOR A NEW APPROACH

It has been known for about three decades that African bees are headed for the United States and during that time a variety of problems to which they will likely give rise in this country has been identified (see, for example, McDowell, 1984; Taylor, 1985). We are nevertheless ill prepared to deal with the potential problems and one does not have to look far to find the reasons. While the bees were still in the southern hemisphere, they were too far away to bother about. Then, as reports of "killer bees" persisted and movies were made about them, the whole thing took on an air of science fiction and a dismissive attitude developed; there was no real problem, only media hype. On the other hand, if there really was a problem, the bees could be stopped at the isthmus, or if not, they would never make it through Mexico, because they would be genetically diluted out of existence by the high density of European bees already there. Or again, if they did get through, they could be exterminated locally in the United States while their undesirable characteristics were ameliorated by interbreeding them with

[1]Professor Fletcher is in the Department of Biology and Program in Animal Behavior, Bucknell University, Lewisburg, PA, 17837, USA.

European bees. Many beekeepers have also taken refuge in the belief that they will be able to virtually ignore the presence of African bees and continue to stock their apiaries with European bees regardless. Currently, it seems to be generally accepted that African bees will soon cross into the United States, so that only the last two hopes in the long line of false hopes remain alive.

Of course, the prevalence of these attitudes does not mean that nothing has been done. The expanding distribution of the bees has been very competently tracked (see, for example, Taylor, 1977, 1985, and Kerr *et al.*, 1982), increasingly sophisticated means of identifying them have been developed (Chapter 3) and a great deal has been learned about their behavior by both South and North American researchers, as summarized by various contributors to this volume. Most importantly, recent genetic studies in the Americas by Smith *et al.* (1989), using mitochondrial (mt) DNA, by Hall and Muralidharan (1989) using both mtDNA and nuclear DNA restriction fragment polymorphisms, and by Taylor *et al.* (1988) using enzyme polymorphisms, have provided dramatic new insights into the true identity of Africanized bees and into the process of Africanization of formerly European populations of honey bees. These authors have shown that the African bees of South and Central America are direct descendants of the *Apis mellifera scutellata* originally introduced into Brazil, i.e., the initial population has expanded by producing reproductive swarms headed by African queens. As expressed by Hall and Muralidharan, "Since the African bee introduction in 1956, there have been, as a very rough estimate, 150 generations that together have migrated a total of more than five thousand miles to reach Mexico. For the swarms to carry African mtDNA, not a single intervening generation could have been the result of an African drone mating to a European queen." The alternative mechanism for the spread of African genes, the interbreeding of established European bees with the introduced Africans to produce a hybrid swarm containing both parental types and a continuum of intermediate forms, seems not to have occurred. African drones do mate with European queens, but the resulting European mother lines, according to Hall and Muralidharan, "do not contribute significantly, if at all, to the African bee migrating force."

Information on what goes on behind the moving front is less securely based, but indications are that in the absence of repeated introductions of European genetic stock, the feral population of honey bees would quickly become indistinguishable from the *scutellata* bees of South Africa. According to Smith *et al.* (1989), gene flow from the feral African bees to managed European apiaries is well documented. This gene flow occurs mainly via the mating of European queens with the far more numerous African drones, and to a small extent by the direct takeover of European colonies by African swarms. The numerically superior African drones also ensure that a substantial portion of the African population remains pure (Taylor, 1986). To determine whether queens carrying European mtDNA survive in the feral population, Smith *et al.* (1989) sampled colonies of bees from various locations behind the front. Nine of ten colonies from Ribeirão Preto, Brazil (where African *scutellata* were introduced), 12 of 12 colonies from Acarigua, Venezuela, and 38 of 39 colonies from Tapachula, Chiapas, Mexico not far behind the front were of the African type. Additionally, the nuclear DNA studies of Hall and Muralidharan (1989) have shown that some Europeanization of the feral African population does occur as African bees migrate into areas with managed apiaries of European bees, but that the European marker genes disappear over time. Putting all the information from the different sources together, Taylor *et al.* (1988) concluded that as a population the Africanized bees, "retain little if any mtDNA, and nuclear DNA, acquired through matings with European drones, is rapidly lost through selection and dilution. Therefore, emphasis on a hybrid origin and current hybrid nature of these bees is misleading since it implies that the population contains substantial components of the European genome"; and they add that since the African population tends to maintain its genetic integrity, it should be recognized as a distinct biological unit and should be referred to as "African derived" or "neotropical African." Page (1989) has sounded a cautionary note by pointing out that although the mtDNA results are convincing, studies of the co-occurring distributional patterns of both mtDNA and nuclear DNA throughout South and Central America are still needed for confirmation of the above conclusions.

The new genetic data show that it will almost certainly not be possible to modify the neotropical African bees genetically by crossing them with European bees, except, as pointed out by Taylor *et al.* (1988), where their populations are of extremely low density and vagility and are numerically inferior to European bees, as may be expected to occur close to the geographical limits of their distribution. This, I believe, removes what has hitherto been regarded as the number one contender from the list of possible ways of meeting the challenge posed by the African bee. It does so, because no matter what genetic modifications are made to the bees via the techniques of instrumental insemination and natural matings in isolated places, the resulting hybrids will have to compete in many regions with high density feral populations of African bees by which they will be swamped and eliminated. In other words, the all-important ecological context in which the "battle" is to be waged between African bees and Afro-European hybrids is being ignored. This is likely to prove a costly mistake. What other options are there?

It was always clear that the establishment of a bee regulated zone in Central America (see Rinderer *et al.*, 1987) would not halt the northward spread of the bees, so there is no known way of preventing them from reaching the United States. It is certain also that to treat them simply as a pest and to attempt to control them like any other insect pest would be doomed from the start and for the same reasons that a bee regulated zone could not be made to work. The bees are too mobile and have too high a reproductive rate for their populations to be effectively controlled, but of equal importance, there will always be incoming migrants from the south, so that any attempt to suppress the population in the United States would be like deciding to manage a leaky boat by perpetually bailing out the inflowing water.

There appears to me to be only one viable solution: to accept that the neotropical African bee will be here to stay and to institute a selective breeding program to improve its apicultural qualities in the same way as was done with European bees before it.

I was first led to propose this solution by two observations involving African *scutellata* bees that seemed to me to be particularly significant. The first was that wherever populations of African *scutellata* and bees of European origin have been brought together in places where either one could survive on its own, the African bees have always dominated. In Africa in particular, introduced European genes seem always to have been relatively rapidly purged from local populations (Fletcher, 1978a). The second observation was that in a number of their behavioral characteristics, African bees in the New World seemed to differ markedly from the parental *scutellata* population in Africa, and I ascribed this to impoverishment of the neotropical African gene pool (Fletcher, 1988). While the new genetic evidence referred to above substantiates the first observation, there are no genetic data to either support or deny the second; in this case the evidence remains behavioral and deductive.

DIFFERENCES BETWEEN SUBSPECIES IN AFRICA

In Africa, *Apis mellifera* is a very diversified species. Ruttner (1986) recognized ten geographical subspecies based on multivariate morphometric analyses, but he emphasized that this number reflects only the present level of knowledge and that the honey bees of large parts of the African continent have still to be investigated. Of the ten subspecies, six (*litorea, scutellata, monticola, adansonii, yemenitica,* and *capensis*) occur on the mainland south of the Sahara, three (*sahariensis, intermissa,* and *major*) in North Africa, one (*lamarckii*) both south of the Sahara and in North Africa, and one (*unicolor*) on the island of Madagascar. Some of these are geographically isolated, others are not. Between those on the mainland south of the Sahara there are no geographical barriers, but then sub-Saharan Africa encompasses a very wide range of ecological conditions from true desert through semi-desert, grassland and savannah, to equatorial rain forest, as well as from tropical coastal vegetation to alpine meadow, so it is not surprising to find that among honey bees occupying these habitats, distinct ecotypes that we call subspecies have evolved (Fletcher, 1978a).

Differences between some of the subspecies are very marked indeed. For example, in E. Africa, *A. m. scutellata* is essentially replaced by *A. m. monticola* at higher elevations. *Monticola* is a large, black bee that inhabits the cool forests between about 2400 and 3000 m. where frequent mists occur, but from 2400 m down to about 1500 m, i.e., through the lower forest and upper cultivated zone, intermediate forms occur, which, according to Smith (1961), offers ample evidence of crossing between the two subspecies. However, typical *scutellata* colonies may also be found above 2400 m. Smith believed these to be migrants that return to lower altitudes in periods of dearth and very cold weather in the upper forest. Thus, his account is consistent with the existence of two markedly different ecotypes of *Apis mellifera*.

A second example concerns the Cape bee, *A. m. capensis*, from the southwestern corner of the African continent. The workers of this extraordinary bee lay eggs that develop into females, either workers or queens (Onions, 1912; Anderson, 1963; Ruttner, 1977), yet the subspecies is not geographically isolated from *scutellata*, the workers of which can lay only male eggs; neither is it reproductively isolated, because when I took *capensis* bees to Pretoria, 1600 km north of their natural home, virgin queens mated freely with *scutellata* drones, the only drones available to them, and produced hybrid workers. The most likely selection pressure favoring the maintenance of the ability to produce queens from worker eggs in the *capensis* population is the extensive loss of queens on their mating flights due either to the notoriously windy conditions of the area or to predation by the swift, *Apus melba*, or both. The unique ability of the *capensis* workers to lay female eggs gives colonies that are left queenless another chance to rear a new queen; it is evident, therefore, that *capensis* is better adapted than *scutellata* to the ecological conditions of the south-western Cape.

So, without mountains or deserts or any other physical barrier to separate them, bees with very different characteristics occur in Africa, characteristics that reflect the local conditions.

CHARACTERISTICS OF THE *SCUTELLATA* SUBSPECIES

Most North American researchers seem to be agreed that when direct comparisons are made between African bees in South and Central America and European honey bees, the African bee may be characterized as follows: (1) they defend their nests more vigorously, i.e., they sting more readily; (2) they swarm more prolifically; (3) they frequently abscond; (4) their feral colonies are smaller; (5) they store less honey. There is less agreement concerning their overwintering ability, i.e. their cold tolerance, as we shall see.

Two important questions arise: how valid are these generalizations and, assuming that there is some substance to them, what are the prospects for changing such characteristics to make the African bees less of a threat to the public, and to North American apiculture and agriculture? Much that is relevant to answering these questions may be learned by considering the behavioral ecology of the parental *A. m. scutellata* population in Africa.

Stinging

Since stinging behavior defines the African bee problem for most people, I shall discuss this aspect first.

Stinging is, of course, a defensive act, and because the sting itself is barbed, the bee cannot remove it once it has penetrated the skin of a human, so that the bee dies soon after it tears itself loose leaving the sting and other tissues behind. Massed attacks can therefore be very costly to a colony, which may lose a substantial part of its worker force. Evidently, the selection pressures favoring the evolution of a barbed sting and mass attacks must have been intense. What were these selection pressures?

Honey bees are indigenous to Africa and since honey seems to have been the only source of sweetness known to humans until recent times, one may be fairly certain that there has been a very long history of humans and their ancestors interacting with bees. The barbed sting probably evolved as a defensive weapon against large, mainly mammalian, predators, since the barbs rarely hold when honey bees sting each other while defending their food stores against robber bees.

I do not know how many such large predators there may have been in the evolutionary history of honey bees in Africa, but in modern times there seem to have been only two of importance, ratels (*Mellivora capensis*) and humans. Ratels, or honey badgers, are fairly common and have powerful claws with which they tear open nests. However, the stings of the bees are virtually ineffective against them, which points to humans as the main selective factor in the maintenance of determined defensive behavior. The depredations of humans are certainly severe enough. The available data show that the numbers of nests and colonies destroyed by honey hunters and harvesters of honey from bark and log hives runs to many millions each year (see Fletcher, 1978a).

Since the costs to a colony of mass attacks are high, one would expect the threshold for release of the behavior to be high also and, contrary to popular belief, this is indeed the case; it appears that physical disturbance of the nest itself is necessary to provoke such attack.

If one considers the benefit to cost ratio of mass attack, one might also expect that the threshold of response to even strong physical disturbance will vary according to circumstance. As we have seen, the cost to a colony can be very high, but so too can the benefits. A well established nest is very valuable to the bees. The wax combs represent a substantial investment of energy and time, as do the young bees, which may number thousands, being reared in them, as well as the honey and pollen stored in them. The greater this investment, the more vigorously one might expect the bees to defend it. By contrast, at the other extreme, a swarm cluster hanging in a tree prior to moving to a new nest site behaves very differently. The costs of stinging now outweigh the benefits, since there is no nest with combs to defend. Every bee that dies represents a loss not only of its valuable honey load needed to build new combs, but also of a labor unit now worth much more than in an established colony, because there will be no replacement bees until the first young ones have been reared and grown old enough to become foragers (Fletcher, 1986). Consequently, swarming bees are often very hard to provoke and one can usually hive them without wearing even a bee veil for protection.

During my 20 years of experience with the *scutellata* bee in Africa, I found that a number of environmental factors may lower the threshold of response of colonies to physical disturbance:

•1. *High temperatures.* The temperature of a bee's body varies with the surrounding temperature, and bees are more active at higher temperatures, so that if disturbed they tend to react more quickly and persistently.

•2. *Robbing by other bees.* African bees are persistent robbers and in apiaries their attempts to enter other hives keep the guard bees of all colonies on the alert. These then tend to attack any humans that happen to be in the vicinity.

•3. *Certain crop plants.* The *scutellata* bee is often outright vicious on crops that yield a large quantity of pollen as well as nectar, e.g., *Eucalyptus* trees and aloes. Perhaps one reason is that colonies can grow to a very large size on such crops, but other factors also seem to be involved.

•4. *Strong odors.* Both the odors of animals and certain plants are often associated with mass bee attacks, e.g., the odors of dogs, farm animals and sweaty beekeepers, as well as the odors given off by a variety of herbaceous plants occurring as weeds in apiaries when these are cut or crushed underfoot. It seems possible that when strong enough, such odors may act as primary stimuli, but more usually they serve in a secondary capacity, providing defenders already aroused by physical disturbance with a means of orienting to the source of the disturbance (Fletcher, 1978a).

Other stimuli are generally well known to beekeepers and include rapid movement, dark coloration of objects, and alarm pheromones released by stinging bees. None of the stimuli I have listed are unique to *scutellata*; it is more a question of these bees having a lower threshold of response to them than do bees of European origin.

Swarming

Swarming is characterized by the rearing of new queens and by colony fission. It is the method whereby colonies, as opposed to individuals, are reproduced. Therefore, maximizing reproductive success means maximizing the number of swarms produced per unit time. Moreover, the extraordinarily high death rate suffered by colonies in Africa as a result of predation constitutes an intense selection pressure favoring high swarming frequency. Two main strategies are displayed by *scutellata* in maximizing swarming frequency, and which of these strategies is adopted depends largely on ecological circumstances:

•*1*. Where honey flows occur more or less simultaneously over wide areas as dictated by marked seasonal effects on plant growth, colonies occupy large, well-protected cavities, grow to a large size, and swarm several times in rapid succession. The old queen leaves with the first (prime) swarm and one or more virgin queens accompanies each of the secondary or after-swarms (Fletcher, 1978b). The number of after-swarms produced during a swarming cycle depends on the size and condition of the parent colony and on local weather conditions.

•2. Where honey flows occur successively, as they do for example in areas where seasonal changes are less well marked, an alternative reproductive strategy is for colonies to occupy temporary nest sites in smaller cavities, migrate with the honey flows, and swarm at shorter intervals. However, because they do not have enough time to develop very large colonies, the number of after-swarms they are able to produce is lower.

It will not have escaped the reader's notice that the first strategy of staying put in a large and secure nest cavity and concentrating reproductive effort on the production of after-swarms is well adapted to living in temperate climates with long winters, while the second strategy is suited to tropical and subtropical conditions. This explains, in part, the differences between African and European bees. However, one must beware of generalizing too far. There is an important asymmetry between the two groups of bees; the first strategy appears to be fixed for bees of European origin (they do not follow honey flows, even if they are taken to the right ecological setting), whereas African bees can and do adopt either strategy, although, there appear to be differences between ecotypes (strains?) in the frequency with which they do so. It seems distinctly possible that genetic differences are involved, and this has important implications in a North American context.

Absconding

Absconding is a non-reproductive process that, unlike swarming, does not involve queen rearing and colony division. Instead, the whole colony abandons its nest site (usually leaving little more than empty combs and a few stray bees) and re-establishes itself in a new location. It is a common behavior among honey bees of the tropics and subtropics, but it is rare among the bees of North America and their relatives in Europe. It has a variety of causes (Fletcher, 1975-76, 1986; Winston *et al.*, 1979), but here I shall discuss two only.

The first is a need to escape from adverse circumstances, such as destruction of the nest site by large predators, attack by means of wasps or ants, locally inadequate food or water resources, etc. The second is attraction to resource-rich areas, i.e., the migratory reproductive option I have just discussed. This strategy involves the trade-off of present relative security, if the nest site is of high quality, against the hazards of reaching the resource-rich area and securing a nest site upon arrival that is adequate for the completion of a swarming cycle. Examples of this behavior follow:

•*1*. In the Natal Province of South Africa, where I lived for a number of years, an annual two-way migration by the bees is highly visible. They leave the Muden river valley for the *Eucalyptus grandis* plantations (which are at a higher altitude) in late summer, and fly back into the valley the following spring.

•2. In Natal and Zululand, the shrub, *Isoglossa eckloniana,* flowers every 7-10 years and may provide an extraordinary abundance of nectar. The numbers of colonies that migrate into the area is reported to exceed the available nest sites to the extent that they build their combs among rocks and trees that offer little protection. It is not known how the bees locate these resource-rich areas, but I have speculated (Fletcher, 1978a) that they do so by flying upwind in response to strong floral odors, which are probably carried many miles, until they reach the source of the odor.

Colony size and honey storage

Clearly, both colony size and the quantity of honey stored are closely associated with the reproductive strategy adopted. Colonies that migrate with the honey flows and occupy small, temporary nest sites will not usually grow to more than a moderate size and, since most of the nectar and pollen they forage is turned into bees rather than stores, neither do they accumulate large honey stores. The converse, of course, applies to non-migratory colonies. Detailed measurements have not been made on *scutellata* colonies in Africa, but Winston *et al.* (1983) measured more than 70 feral colonies in French Guiana, Peru and Venezuela. At the time of first swarming, those in the first two countries had an average comb area of 8,000 cm^2, whereas colonies of European bees had almost 16,000 cm^2. Those in Venezuela had an intermediate amount (11,000 cm^2). In French Guiana and Peru feral colonies had an average honey store of only about 5 kg and in Venezuela 5-7 kg.

Biological conditions do not favor a migratory reproductive strategy everywhere within the distributional range of *scutellata* in Africa. For example, in areas such as the foothills of the Drakensberg mountains in South Africa, food resources appear to be adequate to maintain a limited number of resident colonies, but not to attract migrating colonies (Fletcher, 1988). More importantly, however, temporal rather than spatial differences in resource availability may favor sometimes one strategy and sometimes the other in the same area. Thus, a protracted dry season may normally favor the development of large colonies with substantial stores, but at irregular intervals exceptionally abundant resources (e.g., the flowering of *Isoglossa eckloniana*) may become available in certain localities, and those colonies able to exploit them opportunistically by migrating to them enjoy a reproductive advantage. A mixed strategy is therefore evident in such areas.

Overwintering

Not unreasonably, it is commonly assumed that subspecies living in tropical and subtropical climates are less cold tolerant than those living in temperate climates, and from this premise it is further assumed that the northward spread of African bees in the United States will be limited by an inferior ability to overwinter. This important issue has been debated for at least a decade, but the data needed to settle it are still lacking. This is because there is a law prohibiting the importation of honey bees, so all we have to go on is an astonishingly wide divergence of opinion based on indirect evidence. One prediction (Roubik, 1986) is that the bees will not even reach the United States, another (Taylor and Spivak, 1984) is that they will be limited to areas far south, and a third (Dietz, 1986) is that they will overwinter as far north as Canada. Obviously, the only way to get reliable data is to bring African bees into the country to find out by direct observation where they can and where they cannot survive. Nevertheless, there is something to be learned from the distribution of the *scutellata* in southern Africa.

The queens taken by Dr. Kerr from the Transvaal to Brazil (25 of those that escaped were from the Transvaal and one from Tabora in Tanzania) were supplied by two beekeepers, Mr. E. Schnetler and Mr. W. Crisp, both of whom practiced migratory beekeeping between the highveld and the lowveld. The bees were therefore adapted to surviving at least short periods of sub-freezing temperatures as well as temperatures of over 38°C. I attempted to find out more about the cold tolerance of the South African *scutellata* by studying them in the Drakensberg mountains. There are no honey bees at the top of these mountains (highest point: 3480 m), where snow sometimes stays for three months of the year, but they are fairly common in the foothills. The highest altitude at which I found perennial colonies was 1950 m, where snow cover may last for one week to ten days at a time. In summer I found foragers as high as 2500 m, but according to the local people, colonies that attempt to establish themselves between 2100 and 2500 m do not survive the winters. However, I am not satisfied that low temperatures are responsible for this, as no honey flows of sufficient volume occurred to provide food reserves adequate for overwintering (Fletcher, 1988). It seems to me that honey bees that are able to withstand sub-freezing temperatures at all may well be able to survive fairly long periods of such temperatures given adequate food supplies, but this is mere speculation. Data are urgently needed on the overwintering ability of African bees within the United States and Canada.

DIFFERENCES BETWEEN NEOTROPICAL AND AFRICAN *SCUTELLATA* BEES

It is of interest to note first that genetically "pure" African *A. m. scutellata* may not have been imported into Brazil in the first place, because mated queen bees of European origin were imported many times from North America into the Transvaal province of South Africa, the geographical source of the neotropical African population (see Fletcher, 1978a). Drones of the *scutellata* bee are dark in color, whereas occasional yellow drones still appear in that population, indicating that at least a minor part of the European genome may have persisted in the parental *scutellata* gene pool. However, they were obviously overwhelmingly African, otherwise there would have been no "Africanized bee problem" in the New World.

One may be tempted to conclude from the above that there are only very minor genetic differences between the African derived bees that will enter the United States from Mexico and the parent African population, but such a conclusion would certainly be mistaken. There is, in fact, a high probability that African bees in Mexico differ in important ways from the African populations from which they were derived. For example, as previously discussed, colonies in French Guiana, Peru, and Venezuela are reported to attain only a small size and to be poor honey producers (but see Spivak *et al.* 1989). By contrast, in Africa, both feral and managed colonies in frame hives can become extremely large. For example, E. Schnetler (pers. comm.) maintained many of his 2,000 colonies in the Transvaal in Langstroth hives averaging two brood chambers and three supers, with regular replacement of supers during honey flows. He obtained 256 kg (565 lbs) of honey from one such colony in one year on a fixed site (Schnetler, 1946). In Tanzania, colonies in frame hives yield 54-90 kg (120-200 lbs) of honey per year (Smith, 1966).

How could differences between *scutellata* bees in the Americas and *scutellata* bees in Africa have arisen? An obvious candidate cause is hybridization, but it now appears likely that this affects the characteristics of only part of a population and only in the short term (Hall and Muralidharan, 1989). Other causes must be sought for the major, long-term changes. Among the possibilities are the following:

•*A founder effect.* I have suggested (Fletcher, 1988) that some of the differences may be due to sampling of the parental *scutellata* gene pool, i.e., to a founder effect. At first blush this may not seem very plausible, since a founding population of 26 queens (Kerr, 1967), each mated with up to 18 drones (Adams *et al.*, 1977) would probably contain considerable genetic variation, but we should remember that the geographical areas sampled were limited and that within the total distributional area of *scutellata*, a much wider range of habitats occurs. It would therefore be consistent with my main theme to suggest that much additional variation occurs within these habitats. One trait certainly seems to have been left behind in Africa, and that is a high defensive threshold, i.e., docility. In my experience and that of others in southern Africa, a small proportion of colonies (probably not more than 1-2%) tends to remain very docile even when they are extremely populous and have large honey stores to defend. Moreover, in Zimbabwe, M. Schmolke (pers. comm.) has been able to propagate bees of a gentle strain, and it is significant that he did some of his queen rearing in a suburban area.

•*Selection on the advancing front.* In addition to a loss of genetic diversity through sampling, I have also suggested (Fletcher, 1988) that African bees are subject to strong selection pressures in South and Central America. As the bee spreads, the colonies that abscond and/or swarm the most are the ones constantly in the van, i.e., they form the moving front, so that absconding and swarming have been selected for continuously during the 30 plus years since the bees escaped. Such selection may be expected to have produced bees that swarm while their colonies are still relatively small, have shorter swarming cycles, store little honey, and abscond as soon as local conditions become adverse. In other words, the second of the two reproductive strategies discussed previously has been selected on the advancing front at the expense of the first. It appears, therefore, that the African bees that will soon enter the United States from Mexico are very inferior from the apicultural point of view as a result of continuous colonizing selection that accentuates the least desirable characteristics of the subspecies.

•*Selection in the tropics.* There is a strong tendency in the literature on honey bees, especially that relating to Africanized bees, to draw a sharp distinction between temperate and tropical subspecies. While such categorization is useful and enlightening in some contexts, it obscures the fact that the change from temperate to tropical conditions occurs along a continuum and with it the adaptations of the bees. This becomes important when one considers that the greater part of the Transvaal lies outside the tropics and has a climate moderated by high altitude. For example, the latitude of Pretoria and Johannesburg is about 26°S and these cities are situated on the highveld at altitudes of 1400 and 1753 m (4593 and 5751 ft). The more tropical part of the province lies in the eastern lowveld below an escarpment, where altitudes in the Kruger National Park typically vary between 250 and 500 m (820 and 1640 ft). Most commercial beekeeping occurs on the highveld and along the escarpment and both of the beekeepers who supplied the African *scutellata* queens to Kerr (see later this section) were based in Pretoria on the highveld. It is thus evident that while it is correct in many respects to refer to the *scutellata* bees of the Transvaal as tropical bees, they are not nearly as tropical as the honey bees of, say, Gabon or the Congo, from which they are likely to differ adaptively in numerous ways.

By the time they reach the Texas border through Mexico from their point of introduction in southern Brazil, the Transvaal *scutellata* bees will have crossed the entire tropics from south to north, a journey that will have taken them upward of 33 years. If natural selection means anything at all, it is hard not to conclude that the African bees now at the tropic of cancer are more tropical genotypically and phenotypically than those that left Pretoria.

•*Attrition of patrilines.* As already noted, honey bee queens are polyandrous; they mate with up to 18 drones and with an average of about 10 (Adams *et al.*, 1977). It is very probable that during the lifetime of a queen all potential patrilines are represented among her worker offspring, but it is very unlikely that more than a small proportion of them are represented among her daughter queens that become the queens of reproductively successful colonies. Since the egg from which each daughter queen develops is fertilized by a single sperm, only one patriline is propagated in each parent colony after the departure of the prime swarm and in each after-swarm. Moreover, at each fertilization the probability of success among the competing patrilines is proportional to their relative contributions to the sperm complement of the mother queen, and these contributions may be very unequal. Thus, even if more than one swarm were produced during a swarming cycle and the original mother queen survived several such cycles as the queen of successive prime swarms, there would certainly have been a very high level of attrition of original patrilines during the initial stages of establishment of the feral African population. This loss of patrilines would then have been further accentuated by failure of some of the swarms themselves to survive long enough to reproduce.

•*Attrition of matrilines.* While there is little doubt concerning a severe reduction in the number of imported patrilines, the same level of confidence does not apply to the loss of matrilines. It is not impossible that all 26 of the colonies that escaped with imported African queens established themselves successfully in new nest sites and produced swarms that were also successful, yet this is most unlikely. Among the factors that would have militated against a 100% success rate are the following. 1) The 26 colonies that left the experimental apiary were probably all abscondees rather than prime swarms, unless the departures took place over a long period, which means that at the time local food resources may have been poor. Such conditions would have impaired the colonies' chances of survival. 2) Feral colonies are subject to greater risks from predation and other hazards than are colonies in apiaries. It is also worth noting here that African bees are particularly vulnerable, since they often occupy inferior nest sites that afford them little protection.

In addition to the probable loss of a number of matrilines at the time the feral population was founded, it is likely that fewer than 26 matrilines were represented among the escaping colonies anyway. The queens from South Africa (25 of the escapees) were supplied by two beekeepers, Mr W. Crisp and Mr E. Schnetler, both of whom reared queens in their apiaries, Crisp for requeening purposes and Schnetler (pers. comm.) for royal jelly production. Most of the 100+ queens that were obtained by Dr. Kerr were supplied by Crisp (P. Mountain, pers. comm.) and only 14 by Schnetler (Schnetler, pers comm). No information is available on how many breeder queens may have been used to produce these South African queens.

To summarize this section, there is strong empirical and deductive evidence to suggest that the African bees that will enter the United States differ in many important characteristics from those of their parental population in southern Africa. These differences, of course, are of degree rather than of kind and may be ascribed partly to a founder effect and partly to a post-introduction impoverishment of the neotropical gene pool via the mechanisms of selection on the advancing front, selection under tropical conditions, the attrition of introduced patrilines, and a likely attrition of introduced matrilines. Whatever genetic changes have occurred, they are detrimental from the practical standpoint of the North American beekeeper. The question is, what can be done about this?

PROPOSED SOLUTION TO THE AFRICANIZED BEE PROBLEM

I do not believe it is possible to prevent further spread of the bees; they are far too mobile and prolific to be held back for long, if at all, by means of a bee regulated zone. I also do not believe that attempts to maintain apiaries of European bees in those areas of the United States where African bees can survive will last for long, if only for economic reasons. This approach would not in any case solve the public health problem. The logic I have attempted to establish suggests a very different approach.

Ideally one would like to begin by making direct comparisons in the Americas between the *scutellata* bees in Africa and the Americas in order to assess to what extent the original African gene pool has been altered by selection, but time and other constraints would probably preclude this and it is not essential. The short-cut approach would be to accept the existing deductive evidence that the neotropical African gene pool is impoverished and to make a decision to restore and enrich the gene pool through new importations of *scutellata* from Africa. This would entail making careful selections of material in Africa and its re-evaluation in this country before release. Releases would need to be made repeatedly both at and behind the front. No other strategy is likely to modify the African gene pool more rapidly or more favorably. But this would represent only the first stage of a rational program.

At no time during and after the release would attempts be made to prevent the natural spread of the African bees or to modify them by hybridization with other subspecies, as such attempts would be futile in the longer term and would thus have the effect of prolonging the problem. We know already from the distribution of subspecies in sub-Saharan Africa that where subspecies are not separated by geographical barriers, they grade into one another in more-or-less stable zones. We need to know as soon as possible where this zone will be across North America and, realistically, we shall have to obtain that information from the bees themselves.

The final phase of the program would consist of the re-establishment of selective breeding programs that would take advantage of the genetic variation in the restored *scutellata* gene pool. By the end of the natural stabilization phase, the best locations for carrying out such programs will be known and, as with all artificial selection programs, rapid initial progress could be expected until much of the existing genetic variation has been used up.

In summary, what this program calls for is recognition of the fact that the neotropical African bees are inferior in comparison to the parent *scutellata* bees in Africa, and that the fastest way to modify them would be to release selected *scutellata* genes into the population. This would, of course, require either exemption from, or adjustments to, existing statutes that currently make it illegal to import live honey bees. The program further calls for restraint while the bees find their own ecological limits, and finally, for what amounts essentially to a re-enactment of the success story of American apiculture, i.e., the improvement of the apicultural qualities of the bees through selective breeding.

The discovery by Hall and Muralidharan (1989; see also Hall, 1990) that hybrids between *scutellata* and European bees do not persist, but are replaced over time by essentially pure *scutellata*, increases the probability that there will be fairly well defined boundaries between the areas occupied by African bees and those occupied by European bees. It is most unfortunate that, as discussed earlier, we do not know where the northernmost limits of the African bees will be (although the information could be obtained in a relatively short time), because as long as this information is lacking, it will not be possible for beekeepers involved in the package bee and queen breeding businesses to plan for the future, since the majority of them will not know whether they will be in a *scutellata* area or not. Their uncertainty also means uncertainty for every beekeeper who depends on them for bees and queens and hence for those branches of agriculture that depend upon honey bees for crop pollination, so the problem is a national rather than a local one.

To be obliged to speculate when one could make important decisions based on scientific data that would be easy to obtain does not make sense. The inaction is the result of a total lack of support for bringing colonies of African bees into the United States for experimental purposes, even though they would be kept under strict controls to prevent their escape. This unfavorable political climate could be moderated by the dissemination of information about why a change of policy is needed. Great benefit could be derived from doing this now, even at this very late stage.

The solution to the African bee problem that I have advocated must be considered in relation to this question of where the putative interzone between African and European bees will be. The following points are relevant:

•*1.* It is not known whether the single Tanzanian queen that escaped along with the 25 queens from the Transvaal survived long enough to contribute to the neotropical *scutellata* gene pool via drones or swarms, but whether or not she contributed, for all practical purposes the present population is derived from the Transvaal. If, therefore, coadapted gene complexes are important, as we may conservatively assume they are, it would make sense to get the new genetic material from the same region. On the basis of such thinking there could, of course, be reason to exclude even the Drakensberg area as a potential source of breeding material.

•2. If the new genetic material were obtained from countries further north in southern Africa, e.g., from Zimbabwe, where we know some desirable traits occur, or from countries in Central and East Africa, i.e., north of the Zambezi river, one would be taking a chance on disrupting coadapted gene complexes. On the other hand, there are two reasons for not dismissing such potential sources out of hand. Firstly, introduction of a *scutellata* ecotype that is more tropically adapted might conceivably have the effect of keeping the interzone between African and European bees in North America further south than it might otherwise be and would also be of greater value to our neighbors across the border. Secondly, as already discussed, during their spread through the tropics and subtropics of Central and South America, the neotropical African bees have already been subject to selection by those conditions, i.e., gene frequencies are likely to have been altered in favor of the "tropical end of the spectrum," thereby making them more compatible with bees from the African tropics anyway.

The main objective of this chapter has been, firstly, to point out that the currently dominant theoretical approach to managing African bees in North American, namely, amelioration of their undesirable traits by interbreeding them with bees of European origin, is based on a set of assumptions that do not appear to be supported by the available evidence and, secondly, to offer an alternative approach to the problem. The alternative is based on a different set of assumptions that take into account newly published genetic information by several investigators and various aspects of the behavioral ecology of African subspecies of honey bees, especially *Apis mellifera scutellata*. My main premise is a very simple one: that to be successful, attempts to modify honey bee populations genetically must remain cognizant of the ecological milieu in which the modified bees are expected to function.

LITERATURE CITED

Adams, J., Rothman, E. D., Kerr W. E., Paulino, Z. L. 1977. Estimation of the number of sex alleles and queen matings from diploid male frequencies in a population of *Apis mellifera*. *Genetics* 86:583-596.

Anderson, R. H. 1963. The laying worker of the Cape honey bee, *Apis mellifera capensis*. *J. Apic. Res.* 2:85-92.

Dietz, A. 1986. The potential limit of survival for Africanized bees in the United States. In *Proc. Africanized Honey Bee Symp. (Atlanta)*, pp. 87-100. Chicago: American Farm Bureau Federation.

Fletcher, D. J. C. 1975-76. New perspectives in the causes of absconding in the African bee, (*Apis mellifera adansonii* L.) I & II. *S. Afr. Bee J.* 47:11-14; 48:6-9.

Fletcher, D. J. C. 1978a. The African bee, *Apis mellifera adansonii*, in Africa. *Ann. Rev. Entomol.* 23:151-171.

Fletcher, D. J. C. 1978b. Vibration of queen cells by honey bees and its relation to the issue of swarms with virgin queens. *J. Apic. Res.* 17:14-26.

Fletcher, D. J. C. 1986. Management of absconding and swarming. In *Proc. Africanized Honey Bee Symp. (Atlanta)*, pp. 73-76, Chicago: American Farm Bureau Federation.

Fletcher, D. J. C. 1988. Relevance of the behavioral ecology of African bees to a solution to the Africanized bee problem. In *Africanized Honey Bees and Bee Mites*, ed. G. R. Needham, R. E. Page Jr., M. Delfinado-Baker, C.E. Bowman, pp. 55-61. Chichester, England: Ellis Horwood Limited.

Hall, H. G. 1990. Parental analysis of introgressive hybridization between African and European honey bees using nuclear DNA RFLPs. *Genetics* 125:611-621.

Hall, H. G., Muralidharan K. 1989. Evidence from mitochondrial DNA that African honey bees spread as continuous maternal lineages. *Nature* 339:211-213.

Kerr, W. E. 1967. The history of the introduction of African bees to Brazil. *S. Afr. Bee J.* 39:3-5.

Kerr, W. E., Del Rio, S. de L., Barrionuevo, M.D. 1982. Distribução da abelha africanizada em seus limites ão sul. *Ciência e Cultura* 34:1439-1442.

McDowell, R. 1984. The Africanized Honey bee in the United States: What Will Happen to the US Beekeeping Industry? *USDA Agric. Econ. Rep. No. 519.* Washington, D. C.: Government Printing Office.

Onions, G. W. 1912. South African "fertile worker bees." *Agric. J. Union. S. Afr.* 3:720-728.

Page, R. E. 1989. Neotropical African bees. *Nature* 339:181-182.

Rinderer, T., Wright, J. E., Shimanuki, H., Parker, F., Erickson, E., Wilson, W. T. 1987. The proposed honey-bee regulated zone in Mexico. *Am. Bee J.* 127:160-164.

Roubik, D. 1986. Long-term consequences of the Africanized honey bee invasion: implications for the United States. In *Proc. Africanized Honey Bee Symp. (Atlanta)*, pp. 46-52. Chicago: American Farm Bureau Federation.

Ruttner, F. 1977. The problem of the Cape bee (*Apis mellifera capensis* Escholtz): parthenogenesis - size of population - evolution. *Apidologie* 8:281-294.

Ruttner, F. 1986. Geographical variability and classification. In *Bee Genetics and Breeding*, ed. T. E. Rinderer, pp.23-56. Orlando, Fla: Academic Press.

Schnetler, E. A. 1946. Honey production record. *S. Afr. Bee J.* 21:10.

Smith, D. R., Taylor, O. R., Brown, W. M. 1989. Neotropical Africanized honey bees have African mitochondrial DNA. *Nature* 339:213-215.

Smith, F. G. 1961. The races of honey bees in Africa. *Bee World* 42:255-260.

Smith, F .G. 1966. Beekeeping as a forest industry. *E. Afr. Agric. J.* 31:350-355.

Spivak, M., Batra, S., Segreda, F., Castro, A. L., Ramirez, W. 1989. Honey production by Africanized and European honey bees in Costa Rica. *Apidologie* 20:207-220.

Taylor, O. R. 1977. The past and possible future spread of Africanized honey bees in the Americas. *Bee World* 58:19-30.

Taylor, O. R. 1985. African bees: potential impact in the United States. *Bull. Entomol. Soc. Am.* 31:15-24.

Taylor, O. R. 1986. A possible solution to the African bee problem: a proposal to maximize genetic dilution of African bees through the development of a European stock with late-flying drones. In *Proc. Africanized Honey Bee Symp. (Atlanta)*, pp. 55-72. Chicago: American Farm Bureau Federation.

Taylor, O. R., Long, M. Rowell, G. A. 1988. Genetic differences between European and African bees in Mexico. *Am. Bee J.* 128:809.

Taylor, O. R., Spivak, M. 1984. Climatic limits of tropical African honey bees in the Americas. *Bee World* 65:38-47.

Winston, M. L., Otis, G. W., Taylor, O. R. 1979. Absconding behaviour of the Africanized honey bee in South America. *J. Apic. Res.* 18:85-94.

Winston, M. L., Taylor, O. R., Otis, G. W. 1983. Some differences between temperate European and tropical African and South American honey bees. *Bee World* 64:12-21.

5

THE PROCESSES OF AFRICANIZATION

Thomas E. Rinderer and Richard L. Hellmich II[1]

The rapid and widespread colonization by Africanized bees of much of South America and all of Central America (Rinderer, 1988) is perhaps the most remarkable biological event of this century. From a reported accidental release of 26 absconding swarms of African bees (*Apis mellifera scutellata*, Ruttner, 1986) in 1956 (Kerr, 1967), the population has grown to many millions of colonies which currently occupy about 20 million square km.

What has enabled the Africanized bee populations to grow so large and to occupy such a large area so quickly? What enables them to occupy such a wide biomic range? What enables them to seemingly displace existing populations of European bees? What are the evolutionary implications for populations of bees in the Americas? The intention of this discussion is to provide an ecological and evolutionary perspective on honey bees which will contribute to our understanding of these questions. Perhaps this perspective will not prove to be correct in all its aspects. Nonetheless, it is consistent with currently known data and provides a framework for a variety of testable hypotheses.

OLD WORLD ORIGINS

In order to appreciate the interactions of Africanized and European bees in the Americas, it is important to understand something of the ecology of the Old World parental stocks. The parental subspecies in Europe and Africa are members of very different ecosystems. Their adaptations to the principal constraining features of these ecosystems are the behavioral and physical antecedents of the genetic revolution now occurring in American honey bee populations.

---

[1] Drs. Rinderer and Hellmich are with the USDA--ARS, Honey Bee Breeding, Genetics and Physiology Research Laboratory, 1157 Ben Hur Road, Baton Rouge, LA, 70820, USA.

The diversity of ecosystems in Africa is tremendous. Predictably, the honey bees of Africa are equally diverse. The darkest, lightest, smallest, largest, most defensive and least defensive Western honey bees are all found in Africa. Ruttner (1986) lists 11 subspecies that collectively display over 70% of the total morphological variation of the species. These subspecies also display large behavioral variation of the species. For example, some of these subspecies abscond so regularly that some authors consider the behavior to be migratory, while other subspecies are not known to abscond at all (Ruttner, 1986).

The parental African subspecies of Africanized bees (*A. m. scutellata*) occupies an arid to semi-arid zone ranging from tropical East Africa to subtropical South Africa. The generally tropical nature of this area leads the casual observer to speculate that since the area is tropical, the bees are only adapted to areas with high temperatures. However, a closer inspection shows that the area is generally composed of highlands which may have cool to cold temperatures, especially at night. One part of the range, the Drakensburg mountains in South Africa, has extended periods of snow cover (Fletcher, 1978). Because of the highland nature of *A. m. scutellata's* range, and because this subspecies does not have a common name, we use the term "Highland bee" when referring to it.

Rather than temperature, the chief climatic feature which appears to limit the Highland bee is rainfall. Rainfall, or more precisely, variability in rainfall, has probably been the primary selective factor in the evolution of the subspecies. The northern extension of the area is considered to have two annual rainy seasons: the "short rains" of October and November, and the "long rains" of March, April and May (Norton-Griffiths *et al.*, 1975). However, the ecological essence of these rains is their unpredictability. They may or may not occur each year; if they occur, they may not occur with strong seasonality or everywhere in the area; and where they occur, their duration and amount is highly varied (Kendrew, 1961; Griffiths, 1972, 1976; Norton-Griffiths *et al.*, 1975).

Similar conditions extend southward. One climatologist (Kendrew, 1961) considers that "most of South Africa except the highest altitudes is arid or semi-arid, the rain being scanty and uncertain in amount and duration." Another (Schulze, 1972) points out that in 11 of the first 60 years of this century, South Africa sustained droughts over large tracts, and writes of a strong unpredictable regional variability in rainfall. This unpredictability is reflected in large fluctuations in South African honey production prior to widespread use of irrigation (Fletcher and Johannsmeier, 1978).

We interpret the distinctive characteristics of the Highland subspecies to be the evolutionary products of the area's unpredictable rains. The vehicle delivering this selection pressure is plant growth and flowering which, in this zone, are completely dependent upon rainfall. Nectar and pollen availabilities are thereby equally dependent upon the unpredictable rains.

Europe

Compared to Africa, Europe is a small continent. Nevertheless, it contains a diversity of ecosystems. There are sufficient differences in ecology and geography to produce five subspecies of honey bees (Ruttner, 1986). Italian bees (*A. m. ligustica*) probably predominate in the ancestry of North American honey bee stocks, and Iberian bees (*A. m. iberica*) probably predominated in the ancestry of South American honey bee stocks prior to their Africanization (Morse *et al.*, 1973; Oertel, 1980). However, all European subspecies have been sources for germplasm introductions to the Americas (Oertel, 1980). Consequently, we are interested in a wide view of European honey bee biology and evolution. Despite diversity, certain generalizations can be made which describe important selection factors for almost all European honey bee populations.

As with Africa, climate is the pivotal selection factor in Europe. Moisture patterns are important, but in almost all of Europe, moisture is coupled with winter periods of varied lengths. Winter clearly promotes the evolution of adaptive honey bee behavior which extends to activities throughout the year. Annual snow melt in spring contributes a reliable annual increase in soil moisture. This predictable moisture, combined with a generally reliable rainfall leads to a dependable, predictable, annual period of plant growth and flowering (Cantu, 1977; Furlan, 1977; Schuepp and Schirmer, 1977). The continent's "rainfall is usually adequate for agriculture, and fluctuations from one growing season to another are rarely excessive" (Bourke, 1984). Thus, in Europe, bees are adapted to climatic regimes which are highlighted by the principle of predictability; predictable winters are followed by predictable periods of plant growth and flowering which are supported by predictable snowmelt and predictable rains.

We interpret the distinctive characteristics of European bees to be evolutionary products of Europe's predictable seasonal cycle. A predictable and sharply varying photoperiod cycle is correlated with European seasonality and has been shown to be an important cue in the regulation of colony population cycles (Morse, 1975; Avitabile, 1978; Kefuss, 1978).

The climatic patterns observed by climatologists in Europe and the highlands of Africa are reflected in apicultural reports of resource availability. Crane (1975) cites several examples of comparatively predictable patterns of nectar and pollen availability in Europe, and Fletcher (1978) and Smith (1951, 1953, 1958, 1960) describe examples of unpredictable patterns for South and East Africa.

Foraging

A variety of studies have led to a model of the regulation of the annual cycle of European honey bee nectar foraging (see Figure 7, Rinderer and Collins, Chapter 12) (Rinderer and Baxter, 1978, 1979, 1980; Rinderer, 1981, 1982a, b, c, 1983; Rinderer and Hagstad, 1984, Rinderer *et al.*, 1984). This model focuses on volatiles from empty comb as the chief regulator of the intensity and efficiency of European honey bee foraging. This seasonably varying regulation is well suited to optimizing the foraging of European honey bees throughout the predictably varying seasonal cycle. A full explanation of the model can be found in Chapter 12.

Several experiments (Rinderer, 1982a, b, c, 1985; Rinderer *et al.*, 1984, 1985a, 1986; Pesante, 1985; Pesante *et al.*, 1987) have led to a model (see Figure 11, Rinderer and Collins, Chapter 12) of Africanized honey bee foraging which stresses a far greater unreliability in nectar and pollen availability. When collecting nectar, Africanized bees are essentially more opportunistic than European bees and behave more like the European bees of autumn. Regardless of season, given the opportunity of a subsistence level of nectar resource availability, Africanized bees will forage, store the resources they collect, and increase their chances of survival. When better resources are available, they will forage, but with less intensity than European bees.

The pollen collection of Africanized bees strongly contrasts with that of European bees. Africanized bees collect more pollen over time, partially because they devote a greater proportion of their foraging cohort to pollen collection (Pesante, 1985; Danka *et al.*, 1987).

Defensive Behavior

The same contrast between resource predictability and resource unpredictability which has led to the evolution of different foraging patterns in European and Highland bees may also underlie some of the differences in defensive behavior between the ecotypes. Interestingly, the foraging regulator, empty comb and its volatiles, also causes changes in the intensity of colony defense (Collins and Rinderer, 1985). More empty comb, which in the natural history of feral bees signals reduced amounts of honey stores, increases the intensity of nest defense by both Africanized and European bees.

The massive defensive responses of Highland bees contrast sharply with the far less intensive responses of European bees (Collins *et al.*, 1982). Certainly, selective pressure from pests and predators is key to the existence of defensive behavior. However, the intensity of the Highland bee's nest defense (Chandler, 1976; Guy, 1976; Nightingale, 1976) and that of its Africanized progeny in the

Americas (Collins *et al.*, 1982) must be primarily rooted elsewhere. A comparison of the honey bee pest and predator complexes of the highlands of Africa and Europe since it was occupied by bees after the last ice-age does not reveal widespread differences (Caron, 1978; De Jong, 1978; Clauss, 1985). Humans have probably been the most efficient and the most abundant honey bee predators in both areas. Beyond humans, similar numbers and kinds of both vertebrate and invertebrate predators of honey bees have existed in Europe and the Highlands of Africa: European brown bears (*Ursus arctos*) are probably equal in destructive ability to ratels (*Mellivora capensis*), and bee wolves (*Phalothus triangalum*) are as notorious as banded bee pirates (*Palarus latifrons*). Perhaps only safari ants (*Anomma spp.*) lack a European equivalent. However, honey bee stinging behavior is not considered an effective defense against ants (De Jong, 1978).

There has likely been a difference in the selective pressures applied by humans in the two areas. In Europe, humans have developed skills in managing perennial colonies of bees for an annual harvest. Where multi-seasonal bee management exists, humans tend to destroy those colonies that sting frequently and to preferentially manage those colonies that sting less. Such artificial selection has probably reduced the overall defensiveness of European bees. In contrast, traditional beekeeping with Highland bees generally involves the periodic trapping of swarms followed by the general destruction of all colonies without regard to their defensiveness during harvest (Smith, 1960; Nightingale, 1976; Kigatiira, 1985). This practice, employed because of the Highland bee's tendency to abscond rather than to establish perennial colonies, is non-selective and cannot be expected to change the general intensity of defensive responses in the honey bee population. Differential human management and selection of bee populations have probably increased fundamental differences between the intensities of the defensive responses of European and Highland bees.

In addition, intensity of nest defense can be viewed as being derived through evolutionary processes which balance the costs and benefits of defense (Seeley, 1985). These costs are not fixed, even for European bees within their seasonal cycle. A notable condition which reduces nest defense is a good immediate availability of harvestable resources (Seeley, 1985), and a notable condition which increases nest defense is the ending of a harvest (Collins *et al.*, 1980). These relationships suggest that part of the balance between the cost and the benefits of defense is measured in the potential for existing bees of the colony to become productive foragers (FIGURE 1). One cost of massive defense is the loss of potential foragers in future harvests. Where harvests in the near future are highly predictable, the costs of losing potential foragers is greater. This increased cost probably accounts for the general reduction of colony defense during nectar flows. It also probably accounts for some of the differences between European bees and Highland bees. The Highland bees, subject to a

Relationship of Resource Predictability and the Intensity of Nest Defense

Chance for a Current or Impending Harvest	Defense: Costs	Defense: Benefits	Intensity of Defense
↑	↑	↓	↓
↓	↓	↑	↑

FIGURE 1. Relationship of resource predictability and the intensity of nest defense. ↑= high, ↓= low.

regime of resource unpredictability, harvest resources that are made more valuable by the unpredictability of their replacement.

Through evolutionary time, selective forces should favor bees that conserve collected resources well enough to survive long dearths. Part of this conservation is likely to increase colony defense. The costs of increased defense would be lower, since the potential value of defenders as future foragers is less when the predictability of a future harvest is less (FIGURE 1).

Absconding

A third characteristic difference between European and Highland bees is their tendency to abscond (Ruttner, 1986). European bees rarely abscond. In contrast, Highland bees and their Africanized progeny display the absconding characteristic to a marked degree (Smith, 1960).

Absconding can be organized in two categories. First, swarms which recently have occupied a nest will readily abscond if disturbed. We consider disturbance-induced absconding to be an alternate form of defense; locations that require early active defense often will require frequent active defense. The second category of absconding is resource related. During periods of extended dearth, colonies will often abscond after converting existing food reserves into adult bees. The adaptive value of such absconding is clear in the context of African highland floral resources. Locally unreliable rainfall will cause some large areas to have dearths when nearby areas have abundant pollen and nectar. Absconding in such conditions provides a chance of securing a colony's survival and future reproduction. In Europe, resource related absconding has far less value since European floral resources generally do not have this coarse-grained variability.

Swarming

The patterns of swarming displayed by European and Highland bees also reflect the contrasting patterns of predictable and unpredictable resource availability. Established colonies of European bees can be expected to issue prime swarms one to three times a season (Seeley, 1977; Winston, 1980). An important constraint on this number is the need for both the parent colony and the swarm to harvest sufficient stores during the predictable but time-limited flowering periods to survive the long and also predictable dearth from late summer or autumn to spring. Only very rarely will newly cast swarms themselves produce a swarm during the same season. Thus, European colonies of honey bees generally survive an annual cycle before they reproduce.

In contrast, Highland bees and their Africanized progeny will produce prime swarms up to four times annually (Winston *et al.*, 1981; Otis, 1982) if resources are available. Colonies do not necessarily have ample reserve stores prior to swarming and issued swarms may themselves swarm during the same season (Winston *et al.*, 1981). Apparently, continued swarm production by Africanized bees is dependent upon the continuing availability of resources in the field.

COLONY POPULATION CYCLES

The colony population growth pattern of European honey bees is a remarkable feature of their natural history (FIGURE 2). Often in mid-winter, long before any flowering, colonies begin expanding their brood nests in

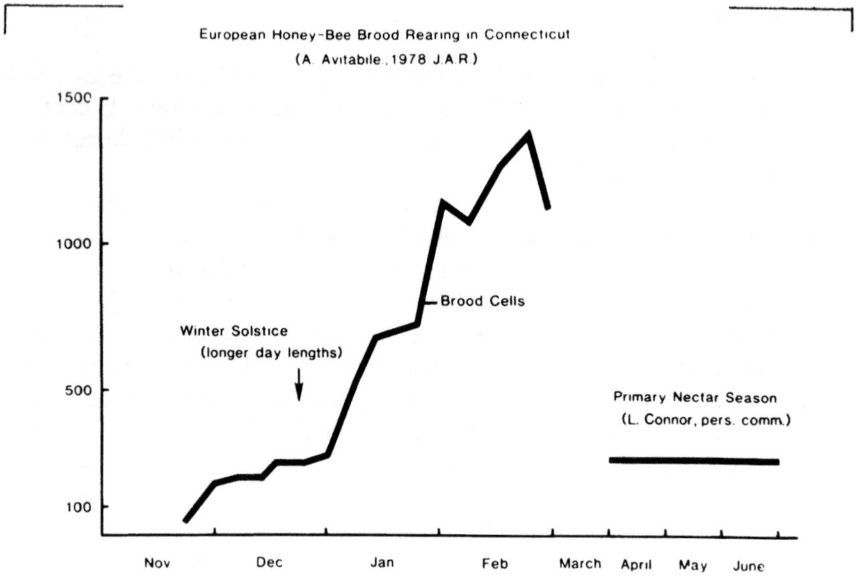

FIGURE 2. European honey bee brood rearing relative to winter solstice in Connecticut.

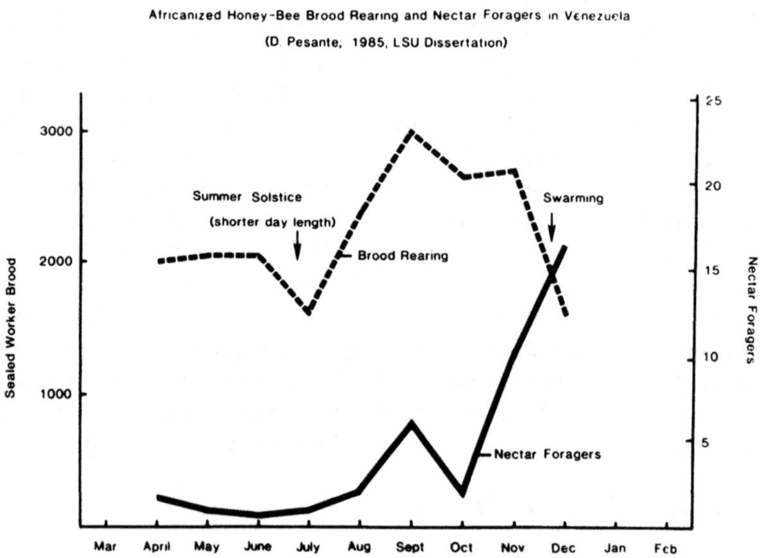

FIGURE 3. Africanized honey bee brood rearing and nectar foragers relative to summer solstice in Venezuela.

response to photoperiod cues (Morse, 1975; Avitabile, 1978; Kefuss, 1978). This early commitment of food resources to the production of large numbers of new colony members is an excellent indication that European bees have evolved in stable, predictable ecosystems. Later in the season, even during the modest nectar flows of autumn, European colonies reduce the intensity of their brood rearing and often discard their reproductive investment in drones. Again, these activities are evolutionarily developed responses to the stability and predictability of European ecosystems.

No data exist on the colony population cycles of Highland bees but good data are available from Africanized bees (Winston *et al.*, 1981; Otis, 1982; Pesante, 1985). Africanized bees develop large colony populations in remarkably short periods. Interestingly, they do so only in response to the immediate existence of harvestable resources (FIGURE 3). During harvests they show a marked preference for pollen in comparison to European bees (Pesante, 1985; Danka *et al.*, 1987). Development of large brood nests and the reduction of brood nests are triggered by the immediate presence or absence of harvestable resources (Pesante, 1985). Rapid expansion is founded on a comparatively high fecundity (Smith, 1958) and shorter individual development times (Smith, 1958; Harbo *et al.*, 1981). Such colony population development events are clearly indicative of bees which have evolved to thrive in unpredictable ecosystems.

Population Densities

There are no exhaustive studies of the sizes of feral honey bee populations from anywhere in the world. However, fragmentary information (Smith, 1953; Kerr, 1971; Seeley and Morse, 1976; Seeley, 1977; Fletcher, 1978; Taber 1979; Rinderer *et al.*, 1981, 1982) from South America, Africa, and North America suggest that in favorable periods, the population densities of European honey bees are far less than those of Highland or Africanized bees. However, personal experience (T.E.R.) in Africa at the end of a widespread drought suggests that, during periods of adversity, the populations of Highland bees can become quite low. A general inference from the scattered data is that European honey bee populations are reasonably stable and that their densities are generally intermediate between the high and low densities of Highland and Africanized populations.

Comparative Demography

Although the characterization of organisms as comparatively "K" or "r-selected" is often imperfect, it still has use in formulating hypotheses and summarizing general comparisons. Honey bees, and especially European honey bees, are considered to be "K-selected" (Seeley, 1978). However, among honey bee subspecies, the Highland bee is, for many characteristics, a comparative "r-strategist" (MacArthur and Wilson, 1967; Danka *et al.*, 1987). According to Wilson (1975), "r-strategists (1) discover habitat quickly, (2) reproduce rapidly and use up the resources before the habitat disappears, and (3) disperse in search of other new habitats as the existing one becomes inhospitable." The absconding and swarming rates of the Highland bee qualify it well for the label of comparative "r-strategist" within the subspecies of the western honey bee. Demographic theory holds that "r-selected" organisms are adapted to exploiting variable, unpredictable, or ephemeral resources. The food sources of the Highland bee show these characteristics to a marked degree and are a direct consequence of variable and unpredictable rainfall patterns.

The perspective of Africanized bees as "r-selected" is useful for comparisons with European bees and in understanding the broad range of Africanized honey bee adaptability. However, western honey bees generally fit demographic descriptions of "K-selected" species. As such, Africanized bees are well adapted to develop stable populations in conditions of stable resources.

The Adaptability of Honey Bees to Tropical America

The climates of tropical America are highly varied but they generally provide a difficult environment for European honey bees. However, this mismatch is poorly explained by the thought that temperate bees could not possibly be well-adapted to tropical areas. The climates of the American tropics show marked wet and dry seasonal cycles that have a predictability reminiscent of the predictability of the European climates. The natural histories of native social bees in the American tropics are similar in many respects to the natural histories of European honey bees. But the organization of resource predictability that has influenced their evolution is much different. The seasonal rhythm of nectar availability reported for Piracicaba, Brazil differs from the nectar-flow rhythms of Europe (Amaral, 1957) (FIGURE 4). Although Piracicaba is at 22°S latitude, it has its major nectar flow from July to September. Thus, weak photoperiod cues guide European bees to increase the sizes of their brood nests only immediately prior to the nectar flow rather than several months earlier as is the case in Europe. Seasonality in Acarigua, Venezuela has completely opposite photoperiod cues. In Venezuela, at 10°N latitude, resources become available in August or September and are at a peak from November to February. Earlier, in the rainy season of April to August, European bees do so poorly that we used exclusively Africanized bees in our colonies in Venezuela during this period. They conserve resources well in this period, while European bees do not. The extent to which European bees are successful in such areas reflects their limited capacity to adjust to unpredictable resource variation rather than their much stronger capacity to exploit the predictable floral resources of Europe. Other areas of tropical America, as in the Amazon Basin and the Brazilian Highlands, are ecologically different from each other, but are also so different from European

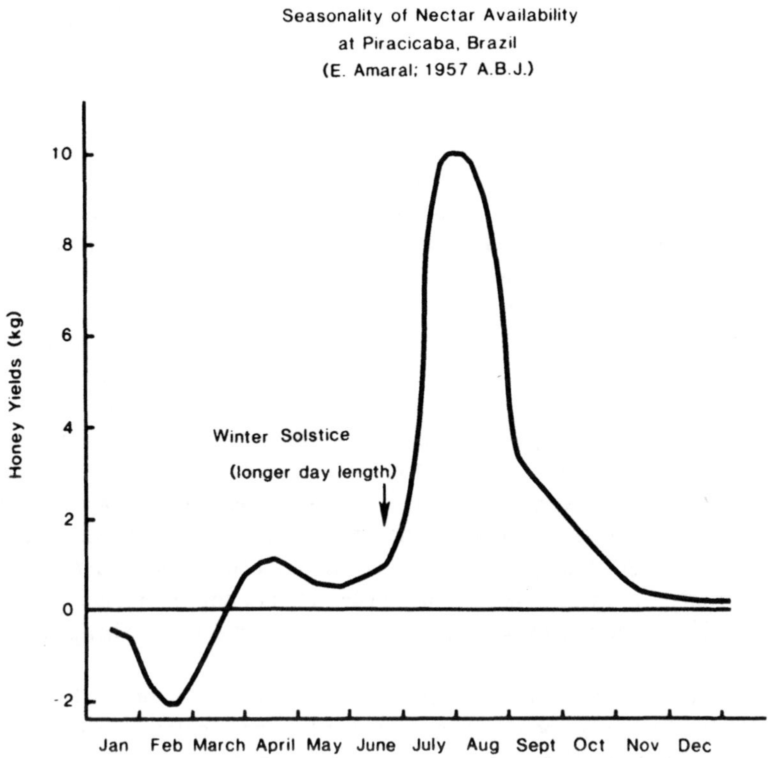

FIGURE 4. Seasonality of nectar availability at Piracicaba, Brazil.

seasonal patterns that they seem to be beyond the range of European honey bee adaptability.

In contrast, Africanized honey bees in the varied and unique ecosystems of tropical America have proved to be highly successful. This success rests, not on general tropical adaptations, but rather on their highly evolved capacity to successfully exploit unpredictable and varied resource conditions. The differing resource periodicities of Piracicaba and Acarigua are both well within the limits of Africanized honey bee adaptability. The unpredictability of the Brazilian highlands is reminiscent of highland Africa and the food resource patterns of Amazonia resemble a poor season in highland Africa. The oddities of the African highland rains have been a foundation for the evolution of adaptive characteristics (foraging, defense, swarming, absconding and colony population development patterns) which have enabled Africanized bees to occupy nearly the entire biomic diversity of tropical America.

THE GENETICAL NATURE OF HONEY BEES IN THE AMERICAS

European Honey Bees

All bees in the Americas are derived from imports. Early importations were probably relatively few in number due to the difficulties of transport by sailing ships. In North America these importations were from northern Europe and were of representatives of *A. m. mellifera.* This was the chief commercial honey bee in North America until the mid 19th century. It was then discovered that bees from Italy, *A. m. ligustica,* were comparatively resistant to the brood disease European foulbrood. A major campaign by the U.S. and Canadian Departments of Agriculture promoted the "Italianization" of commercial apiaries (Oertel, 1980). This effort met with resounding success and in a short time Italian bees were used in the vast majority of commercial enterprises. Most likely, these bees were an admixture of representatives of the two subspecies. The success of this program spurred increased importation of bees from Europe which included bees from a variety of populations, including other subspecies (Morse *et al.*, 1973; Oertel, 1980; Sheppard, 1989). The honey bees of North America are derived from all of these imported honey bees and are now a population of bees which itself has varied sub-populations (H. V. Daly and W. S. Sheppard, pers. comm.).

The European bees of South and Central America are probably derived mostly from the Iberian peninsula as were the European colonists who brought them. The Iberian peninsula contains populations of both *A. m. iberica* and *A. m. mellifera* (Ruttner, 1986; Sheppard, 1989). Representatives of other subspecies have been imported occasionally to South and Central America. However, these importations have been insufficiently large to strongly alter the character of the already existing admixtures of *A. m. iberica* and *A. m. mellifera.*

Africanized Bees

What happened genetically when representatives of the Highland subspecies began colonizing South America is central to understanding the process of Africanization. A morphological comparison by Buco *et al.* (1987) of South American, Africanized and South African (Highland) honey bees helped clarify several features of the process of Africanization.

First, although the samples of Africanized bees which were studied came from west-central Venezuela, their descriptive statistics were nearly identical to the descriptive statistics for samples of Africanized bees studied by Daly and Balling (1978). The Daly and Balling survey of Africanized bees included a larger number of samples from a much larger geographical area. This comparison suggests that the processes of Africanization consistently lead to similar Africanized bee populations which are maintained for at least several years.

Second, the Africanized bees from South America differed significantly from South African representatives of the Highland subspecies for 19 of the 24 morphometric characteristics analyzed. Africanized bees were consistently larger than Highland bees, although the variances of the two groups were generally similar. The differences found in comparisons of individual characteristics were reflected in a multi-variate discriminant function analysis which completely separated the two populations. The posterior probability of group membership for bees of the Mexican highlands and for Venezuelan Africanized bees for all cases indicated a correct and unambiguous classification when using the derived discriminant function, suggesting the presence of European genes in Africanized populations. A recently completed extensive morphological study involving populations of honey bees from South America, North America, Europe, Africa and Arabia confirms this conclusion (Rinderer *et al.*, unpubl. data) and indicates that about one-third of the morphological characteristics of Africanized populations are derived from European parentage.

Similar evidence comes from a study of honey bees (Lobo *et al.*, 1989) from a transect across the tropics (seven samples) and subtropics (four samples) in Brazil and Uruguay. Lobo *et al.* studied the morphology of forewings in these samples and the frequencies of allelomorphs for several enzyme loci. Both lines of investigation led Lobo *et al.* to conclude that the populations they studied were racial mixtures that differed in the percentages of admixture among these various sampling sites. An increase in the contribution of European parental populations seemed to occur on a North to South cline. Hall (1990) studied populations of feral bees from Mexico and Venezuela using three nuclear DNA probes that detected restriction fragment length polymorphisms of non-repetitive DNA. Although this system of analysis is not yet sufficiently precise to permit the inference of percentages of parentage in hybrid genomes, hybridization was detected in both the Mexican and the Venezuelan Africanized populations.

The conclusion that Africanized bees are a mixture of European and Highland populations contrasts with the conclusions of Hall and Muralidharan (1989) and Smith *et al.* (1989). These investigators reported the frequencies of African and European mitochondria in various samples of bees. Hall and Muralidharan reported that 10 of 10 feral Venezuelan colonies, 19 of 19 feral swarms from Mexico and 3 of 3 feral swarms from Costa Rica all contained African mitochondrial DNA. Smith *et al.* reported that 9 of 10 Brazilian samples, 12 of 12 Venezuelan samples and 38 of 39 Mexican samples contained bees having African mitochondrial DNA. Mitochondrial DNA is only maternally inherited and not subject to sexually derived hybridization as in nuclear DNA. Nonetheless, these authors concluded that European bees have not contributed significantly to the feral Africanized gene pool. However, Page (1989) suggested that given the small presence of European honey bees in

South America prior to Africanization, European mitochondria should be rare if the South American feral population resulted from the expansion of a large and rapidly growing population of African maternal descent. Page's explanation of the data showing low frequency of European mitochondrial DNA in Latin America seems correct. More recently Sheppard (pers. comm.) has found a much higher frequency of European mitochondria in Africanized bees from tropical northern Argentina. Moreover, the bees from that region show a wide range of morphological intergrades.

Thus, the feral bees of South America constitute a population which is genetically quite different from the parental population of Highland bees in South Africa. The most likely explanation for these differences is that Africanized bees in the Americas fit the classic taxonomic definition of a subspecific hybrid swarm (King, 1974) which has resulted from matings with European bees followed by subsequent crossing and back-crossing. The generally intermediate morphometric character of Africanized bees and their hybrids with European bees suggests that a large number of loci are involved in the determination of honey bee morphology (Rinderer *et al.*, 1990). Predictable Mendelian events associated with crossing and then back-crossing between Highland and European bees would produce the Africanized bees measured by Buco *et al.* (1987). Thus, Africanized bees clearly show the influence of both their European and their African parentage.

It is important to understand that Africanized bees are hybrids. The processes of Africanization resist changes in gene frequencies towards the European types and yet there is sufficient mating between Africanized and European bees that extensive measurable hybridization exists throughout the Africanized population. Since Africanized bees only resist cross breeding but are not insulated from it, the processes of Africanization are imperfect. Consequently, (1) hybrid populations form throughout the range of bees in the New World, (2) programs by agencies, groups or individual beekeepers that promote the mitigation of Africanization through hybridization are likely to be successful, and (3) variation derived from hybridization can provide a good basis for the development of quality strains using selection programs throughout the New World.

KNOWN PROCESSES OF AFRICANIZATION

Human Transport

Human transport of Highland bees to Brazil in 1956 was responsible for the start of Africanization. Indeed such transport of bees, either accidental or intentional, remains a major threat to further spread of these bees. There are several reports of Africanized bees being transported on ships into U. S. ports (M. Holmes, pers. comm.). A recent example of an accidental introduction involved bees presumably accompanying the commercial shipment of other goods into the Bakersfield, California area (Gary *et al.*, 1985). As many as 11 colonies of Africanized bees were found in the area after 24,000 colonies were inspected.

Despite the many undesirable traits of Africanized bees, there is always the threat that these bees could be intentionally introduced into new areas by misinformed people. Some beekeepers want more defensive bees in order to curtail colony theft and vandalism. Others believe that these bees will increase their honey production. Both of these arguments have little foundation. In Venezuela, for example, theft and vandalism of Africanized colonies, usually during the night, occur at a high rate; and honey yields per colony are at least 50% less than they were prior to Africanization (W. Vogel, pers. comm.).

Biological Processes

The biological processes of Africanization may be divided into two main categories. Those involved with the initial spread of Africanized bees into an area, and those involved with reducing the genetic contribution of existing European honey bees to the general bee population.

Africanized honey bees spread into new areas primarily by the movement of absconding swarms. These swarms are generally those that are induced to abscond by resource related conditions. A period of resource unavailability in the field, coupled with reduced stores in a nest induce Highland and Africanized colonies to abscond. Evidence from Kenya (Kigatiira, 1985) suggests that absconding swarms will move along a vector for up to 60-80 km. They make several stops along the way and forage, sometimes for a number of days. Presumably, successful foraging cues such swarms to establish colonies in nearby nesting sites. Once absconding swarms arrive in an area, other characteristics of Africanized bees account for the large scale Africanization of the area and its existing bee populations.

There is a suggestion that mating populations of European bees show certain subtle behavioral changes towards the Africanized type prior to the discovery of Africanized bees within an area (G. Vogel, pers. comm.). Perhaps the reproductive influence through crosses of a few colonies of Africanized bees in an area is more easily observed than the colonies themselves when no special effort is made to find them.

Once in an area, the adaptive characteristics of Africanized bees lead to a rather rapid Africanization of the overall honey bee population. Generally, these traits are r-selected reproductive characteristics. High swarming rates, described earlier, rapidly lead to large feral population of Africanized bees. In and of itself, this comparatively large fecundity is a strong force in the Africanization process and can account in large measure for the Africanization of mixed bee populations (Ruttner, 1986).

Also, social reproductive parasitism has contributed to the success of these bees, and particularly to the general displacement of previously existing populations of honey bees with European origins. Parasitism is widely known among the Apoidea. It ranges from cleptobiosis (food robbing) to cleptoparasitism (typically involving solitary bee species) to true social parasitism, wherein a reproductive female usurps the position of the queen of a parasitized host colony (Wilson, 1971; Michener, 1974). Social parasitism can be facultative or obligate, temporary or permanent, intraspecific or interspecific (Wilson, 1971). Both queens and drones of Africanized honey bees have been observed to parasitize conspecific colonies. Queens engage in temporary facultative parasitism, a phenomenon first described by Wheeler (1904) for ants. A queen which normally is free-living invades a conspecific nest; the parasitism is temporary in that the usurped nest eventually is inhabited only by offspring of the parasite. Such parasitism is regarded evolutionarily as the least specialized form of social parasitism (Taylor, 1939). Drone (male) parasitism has not been reported for other social insects, and does not fit into Taylor's (1939) evolutionary scheme. The behavior is best described as permanent facultative parasitism, since the host queen continues to produce offspring while the colony supports the drone parasites.

Female social reproductive parasitism has long been an activity associated with Africanized bees, but the phenomenon has not been studied on a rigorous experimental basis. Anecdotal reports (Anonymous, 1972; Michener, 1975) typically cite a small cluster of a few to several hundred worker bees, including a mated Africanized queen, landing near the entrance of a hive. Workers from the swarm enter the hive, kill the resident queen, and the parasitizing queen assumes the egg-laying role in the nest. Thus, the usurped colony becomes fully Africanized (i.e., all the bees are Africanized, and no further hybridization with European bees has occurred) after about nine weeks.

Our experience in Venezuela has been that the frequency of queen parasitism increases when colonies are queenless or have failing queens, are small, or have been stressed by manipulation. In one preliminary experiment (unpubl. obs.), queenless colonies of either Africanized or European bees usually accepted both Africanized and European queens when the queens, in small artificial swarms, were placed at the colony entrances. Queenright colonies never accepted foreign queens. European colonies whose queens were confined in cages accepted queens of either type, but similarly prepared Africanized colonies generally did not accept queens of either type.

The sources of parasitizing queens are unknown. It is possible that the small clusters of invading bees are absconding swarms or afterswarms. An observation of parasitism among European bees in Baton Rouge in autumn suggests an absconding swarm as the source. In Africa, migrating swarms of *A. m. scutellata* amalgamate and then cast off extra queens with small groups of workers (Kigatiira, 1988); if similar swarm mechanics occur in the neotropics, cast queens may also parasitize colonies.

Social reproductive parasitism by queens can contribute to the Africanization process in several ways. Complete Africanization soon occurs when an Africanized queen usurps a European colony. An invaded colony then may contribute to further Africanization through swarms, drones, and additional queens. Recently parasitized colonies may unknowingly be transported to previously uninfested areas; frequent colony inspections and marked European queens are required to observe such occurrences. Africanized queens may be able to overwinter in European colonies in regions with harsh winters, and subsequently contribute to wider Africanization during the spring and summer. In any event, years of reproductive parasitism by Africanized queens may explain why nuclear assays (morphology, isozymes and DNA probes) detect different levels of hybridization for Africanized populations than is detected by mtDNA analysis.

Parasitism by Africanized drones has been the subject of much greater experimental attention than has queen parasitism (Rinderer *et al.*, 1985b, Rinderer *et al.*, 1987). Originally, inspections of an experimental apiary of genetically marked European colonies in Venezuela revealed that the majority of drones in the colonies were Africanized. This observation led to a series of experiments to study the causes and magnitude of this parasitism and its effects on host-colony drone production.

These experiments showed that Africanized drone honey bees migrate into European honey bee colonies in large numbers, but Africanized colonies only rarely host drones from other colonies. This migration leads to a strong reproductive advantage for Africanized bees since it both inhibits European drone production and enhances Africanized drone production.

In addition to social parasitism by drones, Africanized colonies produce and maintain far more drones than do European colonies (Rinderer *et al.*, 1987). The combined effects of drone parasitism and differential production and maintenance resulted in 91% of the total drone population being Africanized in an experiment using two apiaries having equal numbers of Africanized and European colonies.

Such disproportionate numbers of drones would enhance the positive assortive mating tendencies (Kerr and Bueno, 1970) of Africanized bees and impair those of European bees. Over several generations this condition alone would result in substantial Africanization. Maximally, a few Africanized colonies could produce all or nearly all of the drones in a mating area. In this case, substantial Africanization would result in two or three generations.

All of these processes of Africanization are interactive rather than additive. For example, increased numbers of swarms and afterswarms will lead to increased frequencies of queen parasitism. Queen parasitism increases the levels of drone parasitism as does increased numbers of swarms.

An additional hypothesis has been advanced to further explain the Africanization of European apiaries (Taylor, 1985; Taylor and Rowell, 1988; Taylor *et al.*, in press). It is based on the correct observation that Africanized drones experience the peak of their mating flight somewhat later in the day than European drones and the incorrect interpretation of European literature that honey bee queens fly further than do drones from their nests during mating flights (Rinderer, 1986). Mating flight differences may enhance the Africanization process but not in the way suggested by the hypothesis. Mating flight range differences may be asymmetrical for European and Africanized reproductives with queens generally flying shorter distances than drones (Konopacka, 1968, 1970; Ruttner and Ruttner, 1972; Böttcher, 1975). If Africanized reproductives generally fly shorter distances than European reproductives, conditions may be established whereby all European queens mating from an apiary have a high likelihood of mating with Africanized drones flying from the same apiary (Rinderer, 1986). Although there may be numerous other differences between Africanized and European bees that enhance the Africanization process, they probably are all secondary to the important phenomenon of absconding swarms, differential reproductive rates, and social reproductive parasitism. These three categories of biological events are sufficient to account for the observed characteristics of the Africanization process.

CONCLUSION

Natural selection was invoked by Ruttner (1986) to explain the occurrence of Africanized bees in Argentina at 39° South latitude (Dietz *et al.* 1985; Krell *et al.*, 1985). The range and the intensity of the intrusion of Highland honey bee traits into populations of bees in the temperate zones of the Americas will probably depend, in part, upon natural selection. Several questions remain wholly or partially unanswered. In American temperate zones do the populations of European bees, which are themselves hybrid swarms (King, 1974) of varied parentage, already contain highly adapted feral ecotypes as do the populations of Europe? Do the varied American ecosystems provide, on a biome by biome basis, resource availability patterns more tuned to exploitation by Africanized or European honey bees or by some intermediate form? Are adaptive traits in honey bees based upon such a complicated non-additive genetic basis that increased fitness can only occur in rare segregates in hybrid populations? Or, are adaptive traits in honey bees based upon additive genetical underpinnings and simpler non-additive genetical events as is the case with hygienic behavior (Rothenbuhler, 1964)? How much will human activity enhance or retard the development of new ecotypes of bees?

The possibilities exist for natural selection to produce, rather rapidly, new ecotypes of honey bees in the Americas. The ecology of temperate, as well as of tropical America, is rich and varied. In addition, the introduction of Highland bees and the spread of their Africanized progeny represents an infusion of additional genetic variation. The operation of natural and artificial (i.e., breeding) selection on this expanded genetic variation in the varied American biomes has the potential to produce novel populations of bees very rapidly.

LITERATURE CITED

Amaral, E. 1957. Honey bee activities and honey plants in Brazil. *Am. Bee J.* 97:394-395.

Anonymous. 1972. *Final Report, Committee on the African honey bee.* Washington, D. C.: Natl. Res. Counc., Natl. Acad. Sci. 95pp.

Avitabile, A. 1978. Brood rearing in honey bee colonies from late autumn to early spring. *J. Apic. Res.* 17:69-73.

Böttcher, F. K. 1975. Beitrage zur Kenntis des Parrungsfluges der Honigbiene. *Apidologie* 6:233-281.

Bourke, A. 1984. Impact of climatic fluctuations on European agriculture. In *The Climate of Europe: Past, Present and Future,* ed. H. Flohn, R. Fantechi, pp. 269-314. Dordrecht, Holland: D. Reidel Publishing Co.

Buco, S. M., Rinderer, T. E., Sylvester, H. A., Collins, A. M., Lancaster, V. A., Crewe, R. M. 1987. Morphometric differences between South American, Africanized and South African (*Apis mellifera scutellata*) honey bees. *Apidologie* 18:217-222.

Cantu, V. 1977. The climate of Italy. In *The Climates of Europe,* ed. C. C. Wallen, pp. 127-174. Vol. 6 of *The World Survey of Climatology.* ed. in chief, H. E Lansberg, New York: Am. Elsevier Pub. Co.

Caron, D. M. 1978. Other insects. In *Honey Bee Pests, Predators, and Disease,* ed. R. A. Morse, pp.158-185. New York: Cornell University Press.

Chandler, M. T. 1976. The African Honey bee - *Apis mellifera adansonii:* the biological basis of its management. In *Apiculture in Tropical Climates,* ed. E. Crane, pp. 61-68. London: Int. Bee Res. Assoc.

Clauss, B. 1985. The status of the banded bee pirate, *Palarus latifrons,* as a honey bee predator in southern Africa. *Proc. of 3rd Int. Conf. Apic. in Trop. Climates, Nairobi.* pp. 57-159.

Collins, A. M., Rinderer, T. E. 1985. Effect of empty comb on defensive behavior of honey bees. *J. Chem. Ecol.* 11:333-338.

Collins, A. M., Rinderer, T. E., Harbo, J. R., Bolten, A. B. 1982. Colony defense by Africanized and European honey bees. *Science* 218:72-74.

Collins, A. M., Rinderer, T. E., Tucker, K. W., Sylvester, H. A., Lackett, J. J. 1980. A model of honeybee defensive behavior. *J. Apic Res.* 19:224-231.

Crane, E. 1975. The flowers honey comes from. In *Honey: A Comprehensive Survey,* ed. E. Crane, pp. 3-76. London: Heinemann Press.

Daly, H. V., Balling, S. S. 1978. Identification of Africanized honey bees in the Western Hemisphere by discriminant analysis. *J. Kans. Entomol. Soc.* 51:857-869.

Danka, R. G., Hellmich, R. L., Rinderer, T. E., Collins, A. M. 1987. Diet selection ecology of tropically and temperately adapted honey bees. *Anim. Behav.* 35:1858-1863.

De Jong, D. 1978. Insectes: Hymenoptera (Ants, Wasps, and Bees). In *Honey Bee Pests, Predators, and Diseases,* ed. R. A. Morse, pp. 138-157. New York: Cornell University Press.

Dietz, A., Krell, R., Eischen, F. A. 1985. Preliminary investigations on the distribution of Africanized honeybees in Argentina. *Apidologie* 16:99-108.
Fletcher, D. J. C. 1978. The African bee, *Apis mellifera adansonii*, in Africa. *Ann. Rev. Entomol.* 23:151-171.
Fletcher, D. J. C., Johannsmeier, M. F. 1978. The status of beekeeping in South Africa. *S. Am. Bee J.* 50:5-20.
Furlan, D. 1977. The climate of southeast Europe. In *The Climates of Europe*, ed. C. C. Wallen, pp. 185-224. Vol. 6 of *The World Survey of Climatology*. ed. in chief, H. E. Lansberg, New York: Am. Elsevier Pub. Co.
Gary, N. E., Daly, H. V., Lock, S., Race, M. 1985. The Africanized honey bee: ahead of schedule. *Calif. Agr.* 39:4-7.
Griffiths, J. F. 1972. Eastern Africa. In *The Climate of Africa*, ed. J. F. Griffiths, pp. 318-347. Vol. 10 of *The World Survey of Climatology*, ed in chief, H. E. Landsberg, New York: Am. Elsevier Pub. Co.
Griffiths, J. F. 1976. *Climate and the Environment*. Boulder, Colo: Westview Press.
Guy, R. D. 1976. Commercial beekeeping with *Apis mellifera adansonii* in intermediate and movable-frame hives. In *Apiculture in Tropical Climates*, ed. E. Crane, pp. 31-37. London: Int. Bee Res. Assoc.
Hall, H. G. 1990. Parental analysis of introgressive hybridization between African and European honey bees using nuclear DNA RFLP's. *Genetics* 125:611-621.
Hall, H. G., Muralidharan, 1989. Evidence from mitochondrial DNA that African honey bees spread as continuous material lineages. *Nature* 339:211-213.
Harbo, J. R., Bolten, A. B., Rinderer, T. E., Collins, A. M. 1981. Development periods for eggs of Africanized and European honey bees. *J. Apic. Res.* 20:156-159.
Kefuss, J. A. 1978. Influence of photoperiod on the behavior and brood-rearing activities of honey bees in a flight room. *J. Apic. Res.* 17:137-151.
Kendrew, W. G. 1961. *The Climates of the Continents*. London: Oxford Univ. Press. 5th ed.
Kerr, W. E. 1967. The history of the introduction of African bees to Brazil. *S. Afr. Bee J.* 39:3-5.
Kerr, W. E. 1971. Contribuição a ecogenetica de algumas especies de abelhas. *Ciênc. Cult. São Paulo* 23:89-90.
Kerr, W. E., Bueno, D. 1970. Natural crossing between *Apis mellifera adansonii* and *Apis mellifera ligustica*. *Evolution* 24:145-155.
Kigatiira, I. K. 1985. *Aspects of the ecology of the African honeybee*. PhD Dissertation. University of Cambridge, England.
Kigatiira, K. L. 1988. Amalgamation in tropical honey bees. In *Africanized Honey Bees and Bee Mites*, ed. G. R. Needham, R. E. Page, M. Delfinado-Baker, C. E. Bowman, pp. 61-71. Chichester, England: Ellis Horwood Limited.
King, R. C. 1974. *A Dictionary of Genetics*. London: Oxford Univ. Press. 2nd ed.
Konopacka, Z. 1968. Mating flights of queen and drone honeybees. *Pszczelnicze Zesz. Nauk.* 12:1-30.
Konopacka, Z. 1970. Studies on distance of mating flights of honey bee queens and drones, and the necessary isolation of mating stations for preventing mismatings. *Final report. Research Institute of Pomology, Pulway, Poland.*
Krell, R., Dietz, A., Eischen, F. A. 1985. A preliminary study on winter survival of Africanized and European honey bees in Cordoba, Argentina. *Apidologie* 16:109-118.
Lobo, J. A., Del Lama, M. A., Mestriner, M. A. 1989. Population differentiation and racial admixture in the Africanized honey bee (*Apis mellifera*). *Evolution* 43:794-802.

MacArthur, R. H., Wilson, E. O. 1967. *The Theory of Island Biogeography*. New Jersey: Princeton University Press.
Michener, C. D. 1974. *The Social Behavior of the Bees: A Comparative Study*. Cambridge, Mass: Harvard Univ. Press.
Michener, C. D. 1975. The Brazilian bee problem. *Ann. Rev. Entomol.* 20:399-416.
Morse, R. A. 1975. *Bees and Beekeeping*. Ithaca, NY: Cornell Univ. Press,
Morse, R. A., Burgett, M., Ambrose, J. T., Conner, W. E., Fell, R. D. 1973. Early introductions of African bees in Europe and the New World. *Bee World* 54:57-60.
Nightingale, J. M. 1976. Traditional beekeeping among Kenya tribes, and methods proposed for improvement and modernization. In *Apiculture in Tropical Climates*, ed. E. E. Crane, pp. 15-22. London: Int. Bee Res. Assoc.
Norton-Griffiths, M., Herlocker, D., Pennycuick, L. 1975. The patterns of rainfall in the Serengeti ecosystem, Tanzania. *E. Afr. Wildlife J.* 13:347-374.
Oertel, E. 1980. History of beekeeping in the United States. In *Beekeeping in the United States. USDA Agric. Handbook No. 335*. pp. 2-9.
Otis, G. W. 1982. Population biology of the Africanized bee. In *Social Insects in the Tropics*, ed. P. Jaisson, pp. 209-219. Paris: Université Paris-Nord.
Page, R. E., Jr. 1989. Neotropical African bees. *Nature* 339:181-182.
Pesante, D. 1985. *Africanized and European honey bee (Apis mellifera L.) pollen and nectar foraging behavior and honey production in the Neotropics*. PhD Dissertation. Louisiana State Univ., Baton Rouge.
Pesante, D., Rinderer, T. E., Collins, A. M. 1987. Differential nectar foraging by Africanized and European honeybees in the neotropics. *J. Apic. Res.* 26:210-216.
Rinderer, T. E. 1981. Volatiles from empty comb increase hoarding by the honey bee. *Anim. Behav.* 29:1275-1276.
Rinderer, T. E. 1982a. Maximal stimulation by comb of honeybee (*Apis mellifera*) hoarding behavior. *Ann. Entomol. Soc. Am.* 75:311-312.
Rinderer, T. E. 1982b. Regulated nectar harvesting by the honeybee. *J. Apic. Res.* 21:74-87.
Rinderer, T. E. 1982c. Sociochemical alteration of honey bee hoarding behavior. *J. Chem. Ecol.* 8:867-871.
Rinderer, T. E. 1983. Regulation of honey bee hoarding efficiency. *Apidologie* 14:87-92.
Rinderer, T. E. 1985. Africanized honey bees in Venezuela: honey production and foraging behavior. In *Apiculture in Tropical Climates. Vol. 3*, ed. M. Adee, pp. 112-116. Gerrards Cross, England: Int. Bee Res. Assoc.
Rinderer, T. E. 1986. Africanized bees: the Africanization process and potential range in the United States. *Bull. Entomol. Soc. Am.* 32:222-227.
Rinderer, T. E. 1988. Evolutionary aspects of the Africanization of honey bee populations in the Americas. In *Africanized Honey Bees and Bee Mites*, ed. G. R. Needham, R. E. Page, M. Delfinado-Baker, C. E. Bowman, pp. 13-29. Chichester, England: Ellis Horwood Limited.
Rinderer, T. E., Baxter, J. R. 1978. Effect of empty comb on hoarding behavior and honey production of the honey bee. *J. Econ. Entomol.* 71:757-759.
Rinderer, T. E., Baxter, J. R. 1979. Honey bee hoarding behaviour: effects of previous stimulation by empty comb. *Anim. Behav.* 27:426-428.
Rinderer, T. E., Baxter, J. R. 1980. Hoarding behavior of the honey bee: effects of empty comb, comb color, and genotype. *Environ. Entomol.* 9:104-105.
Rinderer, T. E., Bolten, A. B., Collins, A. M., Harbo, J. R. 1984. Nectar-foraging characteristics of Africanized and European honey bees in the neotropics. *J. Apic. Res.* 23:70-79.

Rinderer, T. E., Collins, A. M., Bolten, A. B., Harbo, J. R. 1981. Size of nest cavities selected by swarms of Africanized honeybees in Venezuela. *J. Apic. Res.* 20:160-164.

Rinderer, T. E., Collins, A. M., Hellmich, R. L. II, Danka, R. G. 1986. Regulation of the hoarding efficiency of Africanized and European honey bees. *Apidologie* 27:227-232.

Rinderer, T. E., Collins, A. M., Hellmich, R. L.,II, Danka, R. G. 1987. Differential drone production by Africanized and European honey bee colonies. *Apidologie.* 18: 61-68.

Rinderer, T. E., Collins, A. M., Tucker, K. W. 1985a. Honey production and underlying nectar harvesting activities of Africanized and European honey bees. *J. Apic. Res.* 23:161-167.

Rinderer, T. E., Daly, H. V., Sylvester, H. A., Collins, A. M., Buco, S. M., Hellmich, R. L., Danka, R. M. 1990. Morphometric differences among Africanized and European honey bees and their F_1 hybrids (Hymenoptera: Apidae). *Ann. Entomol. Soc. Am.* 83:346-351.

Rinderer, T. E., Hagstad, W. A. 1984. The effect of empty comb on the proportion of foraging honeybees collecting nectar. *J. Apic. Res.* 23:80-81.

Rinderer, T. E., Hellmich, R. L. II, Danka, R. G., Collins, A. M. 1985b. Male reproductive parasitism: a factor in the Africanization of European honey bee populations. *Science* 228:1119-1121.

Rinderer, T. E., Tucker, K. W., Collins, A. M. 1982. Nest cavity selection by swarms of European and Africanized honeybees. *J. Apic. Res.* 21:98-103.

Rothenbuhler, W. C. 1964. Behavior genetics of nest cleaning in honey bees IV. Responses of F_1, and backcross generations to disease-killed brood. *Am. Zool.* 4:111-123.

Ruttner, F. 1986. Geographical variability and classification. In *Bee Genetics and Breeding,* ed. T. E. Rinderer, pp. 23-56. Orlando, Fla: Academic Press.

Ruttner, H., Ruttner, F. 1972. Untersuchungen uber die Flugaktivitat und das Paarungsverhalten der Drohnen. V. Drohnensammelplatze und Paarungsditanz. *Apidologie* 3:203-232.

Schuepp, M. Shirmer, H. 1977. *Climates of Central Europe,* ed. C. C. Wallen, pp. 3-74. Vol. 6 of *World Survey of Climatology*, ed. in chief, H. E. Landsberg. New York: Am. Elsevier Pub. Co.

Schulze, B. R. 1972. South Africa. In *Climates of Africa,* ed. J.F. Griffiths, pp. 501-586. Vol. 10 of *World Survey of Climatology,* ed. in chief, H. E. Landsberg. New York: Am. Elsevier Pub. Co.

Seeley, T. D. 1977. Measurements of nest cavity volume by the honey bee, *Apis mellifera. Behav. Ecol. Sociobiol.* 2:201-227.

Seeley, T. D. 1978. Life history strategy of the honey bee, *Apis mellifera. Oecologia.* 32:109-118.

Seeley, T. D. 1985. *Honeybee Ecology: A Study of Adaptation in Social Life.* New Jersey: Princeton Univ. Press.

Seeley, T. D., Morse, R. A. 1976. The nest of the honeybee (*Apis mellifera* L.) *Insectes Soc.* 23:495-512.

Sheppard, W. S. 1989. A history of the introduction of honey bee races into the United States. *Am. Bee J.* 121:617-619, 664-667.

Smith, D. R., Taylor, O. R., Brown, W. M. 1989. Neotropical Africanized honey bees have African mitochondrial DNA. *Nature* 339:213-215.

Smith, F. G. 1951. Beekeeping observations in Tanganyika 1950/51. *E. Afr. Agric. J.* 17:84-87.

Smith, F. G. 1953. Beekeeping in the tropics. *Bee World* 34:233-245.
Smith, F. G. 1958. Beekeeping observations in Tanganyika 1949-1957. *Bee World* 39:29-36.
Smith, F. G. 1960. *Beekeeping in the Tropics*. London: Longmans.
Taber, S. 1979. A population of feral honey bee colonies. *Am. Bee J.* 119:842-847.
Taylor, L. H. 1939. Observations on social parasitism in the genus *Vespula* Thompson. *Ann. Entomol. Soc. Am.* 31:14-24.
Taylor, O. R. 1985. African bees: potential impact in the United States. *Bull. Entomol. Soc. Am.* 31:15-24.
Taylor, O. R., Jr., Rowell, G. A. 1988. Drone abundance, queen flight distance, and the neutral mating model for the honey bees *Apis mellifera*. In *Africanized Honey Bees and Bee Mites*, ed. G. R. Needham, R. E. Page, M. Delfinado-Baker, C. E. Bowman, pp. 173-183. Chichester, England: Ellis Horwood Limited.
Taylor, O. R., Kingsolver, R. W., Otis, G. W. A neutral mating model for honeybees (*Apis mellifera* L.) *J. Apic. Res.* In press.
Wheeler, W. M. 1904. A new type of social parasitism among ants. *Bull. Am. Mus. Nat. Hist.* 20:347-375.
Wilson, E. O. 1971. *Insect Societies*. Cambridge, Mass: Harvard Univ. Press.
Wilson, E. O. 1975. *Sociobiology: The New Synthesis*. Cambridge, Mass: Harvard Univ. Press.
Winston, M. L. 1980. Swarming, afterswarming, and reproductive rate of unmanaged honeybee colonies (*Apis mellifera*). *Insectes Soc.* 27:391-398.
Winston, M. L., Dropkin, J., Taylor, O. R. 1981. Demography and life history characteristics of two honey bee races (*Apis mellifera*). *Oecologia* 48:407-413.

6

AFRICANIZED BEES: NATURAL SELECTION FOR COLONIZING ABILITY

Francis L.W. Ratnieks[1]

Africanized bees are frequently viewed in terms of the relative contributions of African and European races of honey bees in their ancestry, as evidenced by morphological (Daly and Balling, 1978), chemical (Carlson, 1988; Smith 1988), and genetic analyses (Hall and Muralidharan, 1989; Smith *et al.*, 1989). In contrast, this chapter emphasizes the role that natural selection within the American continent may have played on the characteristics of Africanized bees at different phases of colonization. In particular, I develop the idea proposed by Fletcher (1988, Chapter 4), that natural selection at the colonizing front leads to selection for traits that increase colonizing ability (e.g., absconding of colonies, rapid generation time, and long distance swarm movement) and contrast this with natural selection behind the colonizing front. Two further ideas relating to selection for colonizing ability will also be explored: 1) the possibility that swarm movement is itself not random, but is oriented away from the colonizing front, thereby increasing the rate of colonization; 2) the possibility that selection for colonizing ability in Africanized bees reduces the introgression of genes from European honey bees into the gene pool of the colonizing front of Africanized bees.

TRAITS WHICH AID COLONIZATION BY AFRICANIZED BEES

The rapid colonization of tropical America by Africanized bees is thought to be the result of long distance swarm movement, aided by a short generation time (i.e., the time between episodes of reproductive swarming), and absconding (i.e., the abandonment of the nest site by the whole colony with the formation of a

---

[1] Dr. Ratnieks is in the Department of Entomology, University of California, Berkeley, CA,94720, USA.

non-reproductive swarm) (Winston, 1979; Winston *et al.*, 1979; Otis, 1982; Winston *et al.*, 1983; Taylor, 1985; Fletcher, 1988). Because swarms are almost impossible to mark and recapture or follow (although Kigatiira [1984] managed to follow swarms for several miles across the Kenyan savannah using vehicles) this conclusion is primarily based on the inability of other methods, such as short range swarm dispersal, dispersal by humans, or gene flow via drone mating flights, to account for the rapid spread of Africanized bees. Supporting evidence for long range swarm movement is provided by observations on individual swarms, which have been seen moving considerable distances in Costa Rica by observers on hilltops (D.H. Janzen, pers. comm.), and by the colonization of islands up to 32 km offshore (Roubik, 1989; Roubik and Boreham 1990). In addition, observations on Africanized swarms in Chiapas, Mexico (Ratnieks *et al.*, in prep.) indicate that workers perform wag tail dances of extreme length, and of variable tempo but similar orientation, prior to the departure of the swarm. These dances are similar to those noted by Koeniger and Koeniger (1980) on migrating swarms of *Apis dorsata.*

Absconding and long distance swarm movement, as occurs in some African races of *A. mellifera* (Fletcher, 1978; Kigatiira, 1984), is considered to be an adaptation to the pattern of food availability in Africa which, due to patchy and limited rainfall and variation in vegetation type and altitude, exhibits unpredictable variation on a scale over which swarms can travel (Fletcher, 1978; Seeley, 1985). Unlike European honey bees which must pass the winter--a period of predictable and uniform food unavailability--feeding on stored honey, reproductive and absconding swarms in Africa may have the possibility of moving to a more suitable location nearby.

Long distance swarm movement fits into the general view of migration as a life history strategy for escape in space (Southwood, 1978). Both theory (Dingle, 1986) and empirical evidence (Southwood, 1962; Vepsäläinen, 1978) suggest that greater vagility and migratory behavior occur in organisms that exploit unstable and patchy resources. Opportunistic migration is reported for other African insects (e.g., the African armyworm moth *Spodoptera exempta* [Gatehouse, 1986]), and insects inhabiting other areas with spatially highly variable food resources (e.g., the butterflies *Kricogonia lyside* and *Libytheana bachmanii* in Texas [Gilbert, 1985]).

NATURAL SELECTION FOR COLONIZING TRAITS AT THE COLONIZING FRONT

How will traits that aid colonization, such as long distance swarm movement, be affected by natural selection? Consider a trait, *c*, that affects colonization in a population of organisms colonizing a new geographic area. Organisms that move further into the new area will represent a selected subset of the original population which is phenotypically above the original population mean for trait *c*. Provided trait *c* is heritable then the progeny of colonizers will also show increased colonizing ability. This process is shown in FIGURE 1, and can be stated as the following evolutionary principle: *the process of colonization selects for colonizing ability in the organisms that form the colonizing front.*

Furthermore, within the colonizing front, colonizing ability will be consistently selected for over many generations, if the suitable geographic area into which the organisms are spreading is large so that colonization takes many generations. The colonizing front of Africanized bees has been consistently selected for above average colonizing ability over the approximately 100 generations (i.e., 30 years with several generations per year) that have occurred since their introduction to Brazil. As a result they should strongly possess those traits which enhance colonizing ability and for which heritable variation exists. Of course, colonizing traits cannot increase indefinitely, and increased mortality due to overlong swarm movement or reduced survival of low population swarms (Otis, 1982) etc., will exert stabilizing selection at some point.

The major assumption of the above argument is the existence of heritable genetic variation for colonizing traits. A number of considerations suggest that such variation would exist in Africanized bees. Variation within African bees apparently occurs between and within races (Fletcher 1988, Chapter 4). Furthermore, any introgression of European genes into the colonizing front would increase this variation, because European bees have lower colonizing ability (see below). In other migratory insects, traits that affect colonizing ability have high heritability (e.g., *Oncopeltus fasciatus* [Dingle, 1968; Dingle *et al.*, 1986]; *Spodoptera exempta* [Gatehouse 1986]). Theoretical studies indicate that variation in migratory tendency can be maintained by variation in environmental quality (reviewed in Dingle, 1986), and by inversely frequency dependent fitness of migratory and non-migratory types (Greenwood, 1990).

Conditions selecting for increased colonizing ability have been suggested for other insects. Leslie (1990) showed that *O. fasciatus* collected from Maryland and Iowa had different life history characteristics than their presumed source populations in Florida and Texas, possibly because selection for colonizing ability occurs during the annual process of colonizing northern states from southern winter refugia. In particular, the northern populations showed delayed reproduction (Leslie, 1990), presumably because of the intercalation of a period of migratory behavior prior to reproduction. In the aphid *Acyrthosiphon pisum*, fields located centrally on the Ontario peninsula contained clones producing higher proportions of alates than peripheral fields (Lamb and Mackay, 1983). In explanation, it was suggested that peripheral fields received fewer migrants, therefore selecting against alate production relative to central fields (Lamb and Mackay, 1983). Selection for increased migratory behavior may also occur in the grasshopper *Melanoplus sanguinipes* (McAnelley, 1985). Unlike range expansion into virgin territory, in which colonizing ability can potentially be

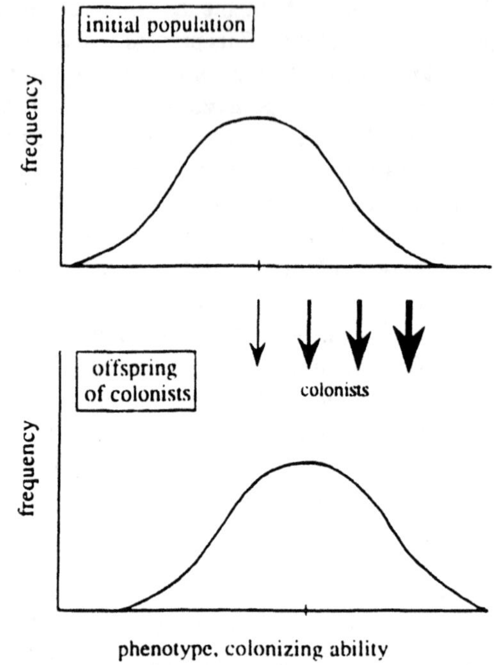

FIGURE 1. Selection for colonization ability. Mean colonization ability of the offspring of colonists will be higher than that in the initial population if colonizing ability is heritable.

selected for over many generations, microevolution for increased colonizing ability in the above examples occurs, at most, for a few generations in the mosaic of favorable and unfavorable habitat patches or the seasonally expanding and contracting range.

SPECIFIC CONSIDERATIONS, SWARM ORIENTATION

Swarm movement can be a seasonal return migration, as in Sri Lankan *A. dorsata* (Koeniger and Koeniger, 1980), or migration for opportunistic exploitation of food resources that become available in a non-predictable way. In African bees in Africa, Fletcher (1978, Chapter 4) suggests that both occur, depending on the location. In neither case would a single orientation for swarm movement be continually favored. However, during the colonization of a new geographic area of suitable habitat from a single point of introduction, such as tropical America, consistent orientations will be continually selected for during the colonization process. In particular, colonization will select for dispersal away from the point of introduction, and away from the colonizing front (FIGURE 2). If the orientation of long distance swarm movement has non-zero

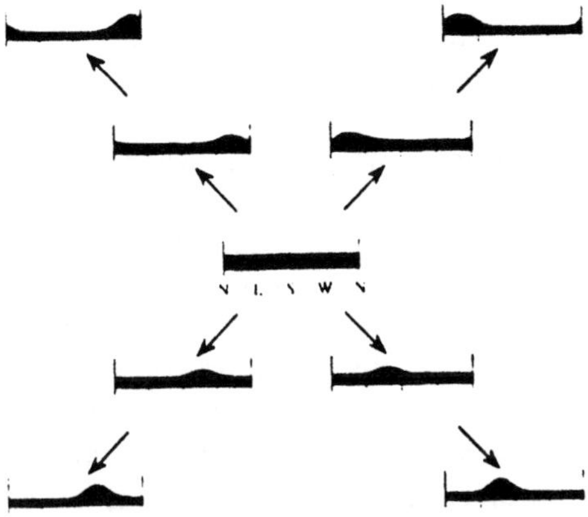

FIGURE 2. Selection for oriented swarm movement away from the colonizing front, if movement direction is heritable.

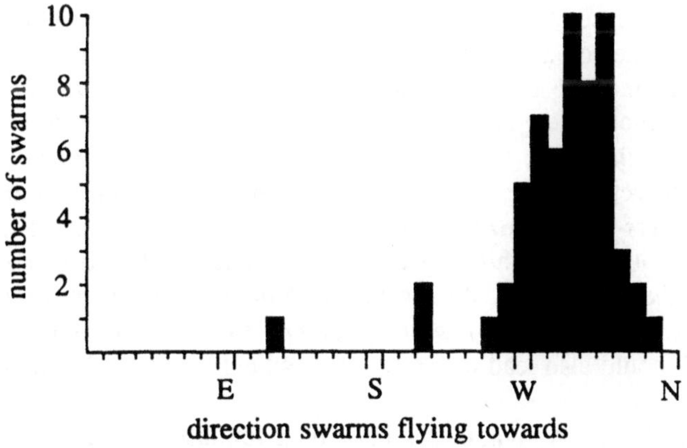

FIGURE 3. Movement orientations of Africanized swarms in the Tapachula region of Chiapas, Mexico. Data were collected from January 20 to March 23, 1988 by asking local people who had seen swarms passing overhead to point in the direction the swarms were headed. This direction was then measured with a magnetic compass. (The two swarms at 210° were both reported by a single person.)

heritability this will result in the evolution of different orientations in the different colonizing fronts. In a land mass such as Central America, which is an isthmus oriented in a SE-NW direction, the process of colonization from SE (Panama) to NW (Mexico) would lead to the colonizing front being selected for NW dispersal. Genetic variation not leading to NW dispersal would be "left behind" as colonization proceeded.

There have been no studies to determine whether the dispersal orientation of honey bee swarms is heritable, which is a necessary condition for its evolution by natural selection, but there is convincing evidence for heritable migratory orientation in other animals. For example, there is heritable migration orientation between winter and summer ranges by *Sylvia* warblers in Europe (reviewed in Berthold, 1988). Similarly, Perdeck (1958, 1967) showed that the winter migration of juvenile starlings is direction- not goal-oriented, suggesting a preferred orientation in naive birds. In salmon, there is good evidence for genetic control of orientation with respect to water current direction (rheotaxis), namely whether fry swim upstream or downstream to nursery areas in the inflow or outflow of lakes (reviewed in Smith, 1985). In the seasonally migrating North American Monarch butterfly a genetic component to migratory orientation also occurs (Brower, 1985). Naive butterflies that developed in the north-eastern United States fly south-west to overwintering sites in Mexico during the fall (Gibo, 1986; Schmidt-Koenig, 1985), and actively maintain their bearings and compensate for deviations caused by crosswinds (Schmidt-Koenig, 1985).

Data collected in Chiapas, Mexico, from January-March 1988 (FIGURE 3) indicate that the colonizing front of Africanized bees exhibited highly directional swarm movement to the NW. In addition to the genetic hypothesis outlined above, alternative explanations are also possible for the oriented swarm movement in this area: *geographic funneling*-- the movement of swarms down the corridor between the Pacific ocean and the Sierra Madre mountain range; *density effects*--as colonization of Central America has occurred from the SE there will initially be a higher density of colonies to the SE of an area being colonized than to the NW. As a result, random movement of swarms could lead to the pattern in FIGURE 3; *resource gradient*--a decrease in colony density from SE to NW could also lead to a resource gradient which could be perceived by swarms.

These alternative hypotheses have their limitations. The coastal plain in Chiapas and Guatemala, for example, is fairly broad, ca. 40 km between the coast and the 500 m contour, so that geographical funneling in this area is unlikely. No comparative data from Guatemala and Chiapas are available to compare feral nest densities, but the feral density in Chiapas was high, ca. 10 colonies per km^2, at this time which was approximately 18 months after Africanized bees first colonized the area (Ratnieks *et al.*, submitted). Resource gradient detection would presumably depend on some mechanism of long range scouting, which is so far unknown [or possibly anemotaxis (Fletcher, 1978)].

Although this is feasible on energetic grounds (Seeley, 1985), the navigational skills to enable a scout to return to her starting place are surely formidable, especially over the distances that might be necessary to detect such a gradient above local variation in food availability.

It is possible to discriminate experimentally between these competing hypotheses. In particular, the genetic hypothesis can be tested by looking for heritability of swarm orientation. Good evidence for this would be a correlation between dispersal orientations of related swarms, such as two swarms composed of worker offspring of the same mother queen, with suitable controls against the alternative explanations. These controls could be achieved by studying the related swarms in two widely separate locations well behind or in front of the colonizing front, thereby breaking any environmental correlation caused by geographic effects, and eliminating resource gradients caused by directional differences in the density of feral colonies. Additionally, the orientation of swarms taken from colonizing fronts moving in different directions (e.g., as will occur in Baja California when this peninsula is reached), or from behind the colonizing front could be compared in a single location. In areas behind the colonizing front, consistent migratory orientations would not be expected (see below).

SWARM ORIENTATION AND RATES OF COLONIZATION

If swarms move away from the colonizing front then the colonization of tropical America is the result of oriented, not random, movement of swarms into previously uncolonized areas. This may help to explain the rapid colonization of tropical America, which preceded at the remarkable rate of about 400 km per year in Central America, as evidenced by the first records of Africanized swarms in Darien, Panama in January 1982 (Buchmann, 1982) and Chiapas, Mexico in late 1986 (Moffett *et al.*, 1988).

Increase in colonization rate for oriented versus random swarm movement was investigated by computer simulation. A triangular grid representing suitable habitat was set up (see FIGURE 4) and each generation "colonies" located at grid points produced a number of "swarms." Swarms moved one "step" (the distance that swarms move before establishing a nest) to any adjacent grid point, either by equal movement in all six directions (i.e., 360°), or by oriented and equal movement in three directions (i.e., 180°). The number of swarms produced by each colony which "survived" to reproduce could be varied. Colony density was allowed to build up to 86,600 per grid square with no density dependence, after which density dependence maintained density at 86,600. This density is equivalent to 10 colonies per km^2 and a step length of 100 km between abandoning the parental nest and the establishment of a new nest, and are reasonable values (Ratnieks *et al.*, submitted; Rinderer and Hellmich, Chapter 5). However, the density used had almost no effect on the simulation

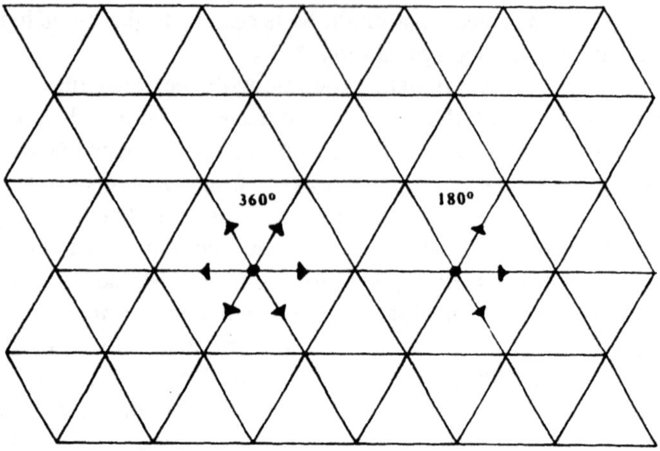

Figure 4. Part of the grid system used in computer simulation. Swarms could disperse randomly in all six directions (360^{0} degrees) or in three adjacent directions (180^{0} degrees). The number of generations for the swarm front to move 22 grid lengths in direction x was used to calculate the colonization rate (see text for details.)

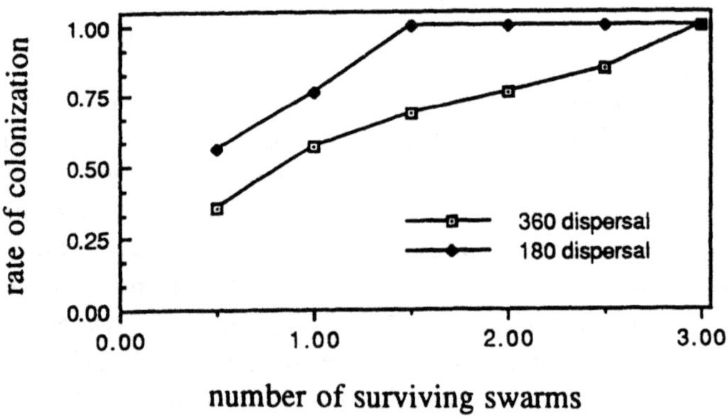

FIGURE 5. Rate of colonization as a proportion of maximum rate for 180^{0} and 360^{0} degree dispersal of swarms. A rate of one indicates that the colonizing front moves forward at the maximum rate allowed by swarm dispersal distances.

results. The simulation was started with 50 colonies on a central grid point, and was run until a grid point 22 steps away was reached.

The rate of colonization is defined as the number of generations taken for the "colonizing front" to travel the 22 steps divided by 22. The maximum colonization rate is one, and a rate of one half signifies that the colonizing front moves about half as far, per generation, as the distance moved by individual swarms. The results (FIGURE 5) show that the rate of colonization is less than one for random dispersal, unless colonies produce three surviving swarms per generation. For swarm dispersal oriented within 180° the rate of colonization reaches one for only 1.5 swarms per generation.

Data on swarm production by Africanized bees in French Guiana about two years after colonization (Otis, 1982) suggest that on average 2.8 reproductive swarms, one primary swarm and 1.8 afterswarms, are produced by Africanized bees housed in 21 l cavities in each generation. The number of swarms which go on to successfully establish nests and reproduce must be considerably lower than this. Otis (1982) records a 30% predation loss by ants each year, with the rate being much greater for smaller swarms, and an 8% loss of queens during mating flights. Loss during swarm dispersal is unknown. In addition, the absconding rate was 30% per year, and absconding results in a single swarm only. Thus, it is likely that each episode of swarming (i.e., both by reproduction and absconding) produces less than three swarms which go on to successfully found colonies. As a result, it seems likely that oriented movement of swarms away from the colonizing front will result in more rapid colonization. Within the range of 0.5 to 2 reproductively successful swarms per swarming episode, the rate of colonization is 30%-50% greater for the 180° versus the 360° strategy (FIGURE 5). More narrowly oriented swarm movement could be even more effective, particularly when few successful swarms are produced.

NATURAL SELECTION BEHIND THE COLONIZING FRONT

The above arguments suggest that natural selection at the colonizing front will favor traits, such as long distance swarm dispersal and short generation time, which enhance colonizing ability. How would selection differ in areas located behind the colonizing front, an area which now covers most of tropical America? Selection for some traits favoring colonizing ability would be relaxed or selected against, but others, such as short generation time, could still be favored, although regional differences in habitat quality and seasonality will probably lead to considerable diversity in reproductive and absconding strategies.

Swarm movement in a consistent direction and very long distance movement would no longer be universally favored by selection. Rather, selection would favor some combination of non-migration, opportunistic migration, and seasonal migration depending on local conditions. In other words, selection should probably cause movement behavior to revert to a state more similar to that which is thought to occur in Africa (Fletcher, 1978, Chapter 4; Seeley, 1985; Winston, 1987). In Brazil there is a suggestion (B. Viana, pers. comm.) that seasonal migration of swarms occurs between the San Francisco valley and adjoining cerrado areas of North-eastern Brazil.

Similarly, variation in local conditions may cause differences in swarming tendency. In regions with widely fluctuating population densities, "r-selected" strategies of early and multiple swarm production per swarming episode will contribute more to population growth, and therefore be favored over more conservative reproductive strategies. Fluctuating density could occur as a result of seasonal or unpredictable favorable seasons long enough for several swarming episodes, and separated by periods in which extensive colony mortality or absconding of colonies to other areas occurs. Nevertheless, reproduction may well be more conservative than at the colonizing front if colony densities behind the front are typically higher than at the colonizing front, and lead to competition for food which excessively penalizes small swarms and colonies. In areas where a relatively stable high population density is formed, perhaps because of more reliable year round food availability, density dependent competition for food resources may be more important leading to even more conservative reproduction. This idea is consistent with Michener (1975), who reported that beekeepers in Brazil claimed that swarming diminished a few years after Africanization, and Boreham and Roubik (1987) and Roubik and Boreham (1990) who showed that the number of swarms captured in the Panama canal zone declined after the second year following colonization. The existence of these differences in life history characteristics between Africanized bees in different areas could most effectively be looked for by comparing stocks from various places in a single apiary, using techniques such as those in Winston (1980).

COLONIZING ABILITY AND THE NON-INTROGRESSION OF EUROPEAN GENES

The data of Smith *et al.* (1989) and Hall and Muralidharan (1989) indicate that the mitochondrial DNA of Africanized bees at the colonizing front in Chiapas, Mexico and behind the colonizing front in Venezuela, Brazil, and Costa Rica, is entirely or almost entirely, of an African type. There are several processes which could limit the introgression of European genes. Perhaps the most important is that Africanized bees swamp the European population when large numbers of swarms move into new areas during colonization (Boreham and Roubik, 1987; Fierro *et al.*, 1988), and build up high density feral populations (Ratnieks *et al.*, submitted). In much of tropical America, European bees were never very abundant before Africanization. There were estimated to be 30,000 - 40,000 European colonies in Venezuela prior to Africanization, and 1-2 million afterward (Taylor, 1985). In French Guiana there were estimated to be 15

European colonies prior to Africanization (Winston and Taylor, 1980). However, it is unlikely that swamping is the full story. The evidence that only 1/49 (ca 2%) of the colonies studied by Smith *et al.* (1989) and Hall and Muralidharan (1989) in Mexico had European mitochondrial DNA works out at a per generation rate of introgression of 0.02%, assuming colonization has taken 100 generations and an equal rate of introgression occurred per generation. Such a low introgression rate would require about a 5000:1 numerical superiority of Africanized colonies, assuming random mating and equal drone production. Furthermore, the per generation rate is actually the geometric mean of the introgression per generation, so that generations with high introgression, as would be caused by pockets of European bees in areas where beekeeping occurred, have greater effect. FIGURE 6 shows the relationship between total introgression, and geometric mean introgression rate per generation.

Several processes additional to swamping have been proposed. Rinderer and Hellmich (Chapter 5) suggest that low introgression is largely a result of the parasitic behavior of Africanized bees, namely the replacement of queens in European race colonies by Africanized queens entering in small swarms, and the parasitism of European colonies by Africanized drones (Rinderer *et al.*, 1985). However, Hall and Muralidharan (1989) suggested that "direct takeover of colonies by invading African swarms is believed to have a minor role in the Africanization process." Further evidence for this process would be valuable, particularly for the takeover of queenright colonies. Queen parasitism would also seem to be an important facilitating factor for drone parasitism, by providing close proximity between Africanized and European colonies in apiaries. Assortative mating may also limit gene flow (Taylor 1985). Rinderer (1986) has also suggested that asymmetries in the mating biology of Africanized and European bees may favor one way gene flow of Africanized genes into European populations.

Another explanation is selection against European traits in hybrid bees (Smith *et al.*, 1989; M. L. Winston, pers. comm.). Hall and Muralidharan (1989) have suggested that "only non-neutral mtDNA could explain the failure of "Africanized" European maternal lines to enter the migrating feral population," and speculate that "European mitochondria may limit a metabolic capability necessary for long distance dispersal." The latter idea may be unlikely, given that European honey bee workers are capable of making repeated foraging trips of up to ca. 800 km in total length (Neukirch 1982) and make single foraging trips of up to 12 km (von Frisch, 1967), unless the queen is affected.

If, as argued above, the colonizing front of Africanized bees is composed of bees with characteristics that increase colonizing ability, then another selective process reducing introgression can come into play. European bees have much lower colonizing ability than Africanized bees at the colonizing front. One feral population of European honey bees in highland New Guinea colonized at about 15 km per year. This low rate of colonization probably results from more

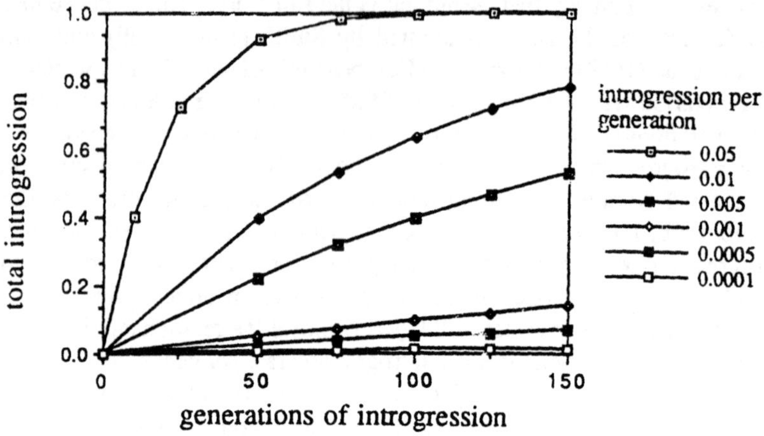

FIGURE 6. Introgression as a function of mean (geometric) rate of introgression per generation and number of generations.

limited swarm movement and more conservative reproduction. Swarms of European bees typically select a nest site less than a few kilometers, and frequently only a few hundred meters (Lindauer, 1961; Seeley and Morse, 1977; Gould, 1982), from the parental nest. European bees also produce fewer afterswarms per swarming episode. Winston (1980) measured a mean of 2.5 swarms per swarming episode (prime-swarm plus afterswarms) from European colonies housed in 42 l hives in Kansas, as against 2.8 swarms for Africanized colonies housed in 21 l hives in French Guiana (Otis, 1982). Winston (1980) also showed that the number of swarms was correlated with brood population, suggesting that European honey bees in equivalent 21 l hives may produce fewer swarms. Spivak (1989) measured the swarming rate of Africanized and European colonies for a year in 42 l hives at four elevations in Costa Rica a few years after colonization. The Africanized colonies swarmed on average 1.3 times and the European colonies 1.0 times per year. The number of afterswarms was not measured in this study.

If "hybrid" colonies between European and Africanized (i.e. colonization adapted) bees forming the colonizing front have some intermediate colonizing ability they will tend to be left behind the colonizing front, where the European genes will be swamped by the influx of large numbers of Africanized swarms. Although this process would only directly select out of the colonizing front those traits with lower colonizing ability, traits in gametic phase disequilibrium with non-colonizing traits would also be preferentially removed from the gene pool of the colonizing front. Initially at least, all European genes will be in disequilibrium with non-colonizing traits. Subsequent random matings of hybrid bees will rapidly reduce this disequilibrium, except for DNA tightly linked to "non-colonizing" genes.

The strength of this process in keeping European genes out of the colonizing front will depend very much on the colonizing ability of the various "hybrid" and backcross colonies. For example, if hybrid colonies composed of a European queen mated to Africanized drones at the colonizing front, and backcrosses of F_1 queens to Africanized drones rapidly regain the colonizing phenotype, then the "leaving-behind" of European genes will be weaker than if the colonizing phenotype is only slowly regained over several backcross generations.

Long distance swarm dispersal is presumably a complex adaptation composed of many interconnected traits (e.g., ability to perform and respond appropriately to long distance dancing, extreme engorgement with honey [Otis *et al.*, 1981] etc.). It is possible that "hybrid" swarms with workers deficient in one component may be unable to travel long distances. Similarly, swarms composed of a mixture of genotypes, such as a mix of Africanized and F_1 workers, may be unable to migrate. Experimentation (or possibly suitable analyses of mitochondrial and nuclear DNA) is necessary to determine the migratory ability of "hybrid" colonies, so that the possible strength of this process can be evaluated. Of particular interest would be hybrid colonies composed of a European queen mated to Africanized drones, and an Africanized queen mated to a mixture of both European and African drones.

An additional aspect of this potential mechanism for non-introgression is that it can only occur in the period when colonization is taking place. When Africanized bees reach the limits of their range, as will occur in the United States, it should be easier for European genes to introgress.

BEEKEEPING CONSIDERATIONS

These ideas may also provide some insight into the nature of Africanized bees as it affects commercial beekeeping, and efforts to control and improve Africanized bees. A corollary of natural selection for colonizing ability is that the Africanized honey bees forming the colonizing front are unlikely to be well suited for beekeeping. Modern beekeeping with movable frame hives requires stocks of bees that do not abscond, and that invest resources in high worker populations and honey production rather than in swarming (Fletcher, 1988). The irony of the situation is that Africanized bees with, on average, poor traits for beekeeping are the only ones available to beekeepers following colonization. Beekeepers must make a sudden transition from the use of stocks that for up to about one hundred generations (i.e., since the introduction of movable frame hives) have been selected for good beekeeping traits to a stock that for about the same number of generations has been selected for bad beekeeping traits.

In terms of control and improvement one obvious suggestion is that the potential of Africanized bees for beekeeping should not be judged by the first bees to arrive. In addition, the thought that selection within the American continent has had a significant effect on the nature of Africanized bees should lead to the expectation that selective breeding can also change them. The importation of African bees from Africa, or from behind the colonizing front in America, for incorporation into breeding stock in regions of colonization has been suggested by Fletcher (1988, Chapter 4). The arguments made above suggest that this is a reasonable idea, provided diseases are not also introduced.

CONCLUSIONS AND SUMMARY

The crux of this paper is the idea that the workings of natural selection make it inevitable that colonizing ability is selected for in the colonizing front of Africanized bees, with the essential condition that traits influencing colonizing ability have heritable variation, as is the case in other migratory insects. The colonization of a vast and favorable habitat, tropical America, has allowed directional selection for colonizing traits, such as rapid reproductive swarming, absconding, and long distance swarm migration, to proceed for approximately 100 generations. These traits are characteristic of African bees in Africa. In America they would have been further selected. An additional trait that would enhance colonizing ability is oriented swarm movement away from the swarming front. The conditions at the colonizing front would permit directional selection for a consistent innate preferred swarm movement orientation. However, it is not known whether genetic variation for this exists, thereby permitting evolution by natural selection to occur, although it has been shown to occur in other organisms.

Selection for colonizing ability in Africanized bees could also lower introgression of European genes into the colonizing front, if hybrids and backcrosses to Africanized bees have lower colonizing ability. In particular, it is possible that hybrid and backcross swarms may not be able to travel long distances successfully, because of the complex nature of this adaptation. The extent to which hybrids and backcrosses to colonization adapted Africanized bees recover migratory ability would determine the strength of this process in keeping selectively neutral European genes out of the colonizing front.

Another aspect of selection for colonizing ability is that the Africanized honey bees forming the colonizing front, which are also the first Africanized bees encountered by beekeepers, are likely to be especially unsuited to beekeeping, which requires non-absconding colonies which build up to high worker populations without swarming. Natural selection for Africanized bees behind the colonizing front may result in these traits being selected for under certain natural conditions, and also through selection by beekeepers, suggesting the value of importing carefully chosen stocks of Africanized bees from behind the colonizing front for use in stock improvement in areas of colonization.

Acknowledgements

The data reported in FIGURE 3 were collected while visiting the Centro de Investigaciones Ecologicas del Sureste (CIES), Tapachula, Chiapas, Mexico. P. Akers, H. Dingle, D. Fletcher, W. Getz, M. Spivak, and M. Winston made valuable comments. Funding was provided by a USDA grant to R. A. Morse and NSF grant 901347 BNS.

LITERATURE CITED

Berthold, P. J. 1988. Evolutionary aspects of migratory behavior in European warblers. *J. Evol. Biol.* 1:195-209.

Boreham, M. M., Roubik, D. W. 1987. Population change and control of Africanized honey bees (Hymenoptera: Apidae) in the Panama canal area. *Bull. Entomol. Soc. Am.* 33:34-39.

Brower, L. P. 1985. New perspectives on the migration biology of the Monarch butterfly, *Danaus plexippus* L. In *Migration: Mechanisms and Adaptive Significance*, ed. M. L. Rankin, *Contrib. Mar. Sci.* 27(suppl):748-785.

Buchmann, S. L. 1982. Africanized bees confirmed in Panama. *Am. Bee J.* 122: 322.

Carlson, D. A. 1988. Africanized and European honey-bee drones and comb waxes: analysis of hydrocarbon components of classification. In *Africanized Honey Bees and Bee Mites*, ed. G. R. Needham, R. E. Page, M. Delfinado-Baker, C. E. Bowman, pp. 264-274. Chichester, England: Ellis Horwood Limited.

Daly, H. V., Balling, S. S. 1978. Identification of Africanized honey bees in the western hemisphere by discriminant analysis. *J. Kans. Entomol. Soc.* 51:857-869.

Dingle, H. 1968. The influence of environment and heredity on flight activity in the milkweed bug *Oncopeltus*. *J. Exp. Biol.* 48:175-184.

Dingle, H. 1986. Evolution and genetics of insect migration. In *Insect Flight: Dispersal and Migration*, ed. W. Danthanarayana, pp. 11-26. Berlin: Springer-Verlag.

Dingle, H., Leslie, J. F., Palmer, J. O. 1986. Behavior genetics of flexible life histories in milkweed bugs. In *Evolutionary Genetics of Invertebrate Behavior*, ed. M. D. Huettel, pp. 7-18. New York: Plenum Press.

Fierro, M. M., Munoz, M. J., Lopez, A., Sumuano, X., Salcedo, H., Roblero, G. 1988. Detection and control of the Africanized bee in coastal Chiapas, Mexico. *Am. Bee J.* 128:272-275.

Fletcher, D. J. C. 1978. The African bee, *Apis mellifera adansonii*, in Africa. *Ann. Rev. Entomol.* 23:151-171.

Fletcher, D. J. C. 1988. Relevance of the behavioral ecology of African bees to the solution of the Africanized-bee problem. In *Africanized Honey Bees and Bee Mites*, ed. G. R. Needham, R. E. Page, M. Delfinado-Baker, C. E. Bowman, pp. 55-61. Chichester, England: Ellis Horwood Limited.

Gatehouse, A. G. 1986. Migration in the African armyworm *Spodoptera exempta*: genetic determination of migratory capacity and a new synthesis. In *Insect*

Flight: Dispersal and Migration, ed. W. Danthanarayana, pp. 128-144. Berlin: Springer-Verlag.

Gibo, D. L. 1986. Flight strategies of migrating monarch butterflies (*Danaus plexippus* L.) in southern Ontario. In *Insect Flight*, ed. W. Danthanarayana, pp. 172-184. Heidelberg, W. Germany: Springer Verlag.

Gilbert, L. E. 1985. Ecological factors which influence migratory behavior in two butterflies of the semi-arid shrublands of south Texas. In *Migration: Mechanisms and Adaptive Significance*, ed. M. L. Rankin, Contrib. Mar. Sci. 27(suppl):725-747.

Gould, J. L. 1982. Why do honeybees have dialects? *Behav. Ecol. Sociobiol.* 10:53-56.

Greenwood, J. J. D. 1990. Changing migration behavior. *Nature* 345:209-210.

Hall, H. G., Muralidharan, K. 1989. Evidence from mitochondrial DNA that Africanized honey bees spread as continuous maternal lineages. *Nature* 339:211-213.

Kigatiira, K. I. 1984. *Aspects of the ecology of the African honeybee.* PhD Dissertation. University of Cambridge, England.

Koeniger, N., Koeniger, G. 1980. Observations and experiments on migration and dance communication of *Apis dorsata* in Sri Lanka. *J. Apic. Res.* 19:21-34.

Lamb, R. J., Mackay, P. A. 1983. Micro-evolution of the migratory tendency, photoperiodic response and developmental threshold of the pea aphid, *Acyrthosiphon pisum*. In *Diapause and Life Cycle Strategies in Insects*, ed. P. K. Brown, I. Hodek, pp. 209-218. The Hague: Junk.

Leslie, J. F. 1990. Geographical and genetic structure of life-history variation in milkweed bugs (Hemiptera: Lygaeidae: *Oncopeltus*). *Evolution* 44:295-304.

Lindauer, M. 1961. *Communication Among Social Bees.* Cambridge, Mass: Harvard University Press.

McAnelley, M. L. 1985. The adaptive significance and control of migratory behavior in the grasshopper *Melanoplus sanguinipes*. In *Migration: Mechanisms and Adaptive Significance*, ed. M. L. Rankin, *Contrib. Mar. Sci.* 27(suppl):687-703.

Michener, C. D. 1975. The Brazilian bee problem. *Ann. Rev. Entomol.* 20:399-416.

Moffett, J. O., Maki, D. L., Andre, T., Fierro, M. M. 1987. The Africanized bee in Chiapas, Mexico. *Am. Bee. J.* 127:517-519,525.

Neukirch, A. 1982. Dependence of the life span of the honeybee (*Apis mellifera*) upon flight performance and energy consumption. *J. Comp. Physiol. B.* 146:35-40.

Otis, G. W. 1982. Population biology of the Africanized bee. In *Social Insects in the Tropics, Vol. 1*, ed. P. Jaisson, pp. 209-219. Paris: Université Paris-Nord.

Otis, G. W., Winston, M. L., Taylor, O. R. 1981. Engorgement and dispersal of Africanized honeybee swarms. *J. Apic. Res.* 20:3-12.

Perdeck, A. C. 1958. Two types of orientation in migratory starlings, *Sturnus vulgaris* L., and chaffinches, *Fringilla coelebs* L., as revealed by displacement experiments. *Ardea* 46:1-37

Perdeck, A. C. 1967. Orientation of starlings after displacement to Spain. *Ardea* 55:194-204.

Ratnieks, F. L. W., Piery, M., Cuadriello, I. The natural nest of the Africanized honey bee (Hymenoptera, Apidae) near Tapachula, Chiapas, Mexico. submitted.

Rinderer, T. E. 1986. Africanized bees: the Africanization process and potential range in the United States. *Bull. Entomol. Soc. Am.* 32:222-227.

Rinderer, T. E., Hellmich, R. L., Danka, R. G., Collins, A. M. 1985. Male reproductive parasitism: a factor in the Africanization of European honey-bee populations: *Science* 228:1119-1121.

Roubik, D. W. 1989. *Ecology and Natural History of Tropical Bees*. Cambridge, England: Cambridge University Press.

Roubik, D. W., Boreham, M. M. 1990. Learning to live with Africanized honey bees. *Interciencia* 15:146-153.

Schmidt-Koenig, K. 1985. Migration strategies of Monarch butterflies. In *Migration, Mechanisms and Adaptive Significance*, ed. M. L. Rankin, *Contrib. Mar. Sci.* 27(suppl):786-798.

Seeley, T. D. 1985. *Honeybee Ecology: A Study of Adaptation in Social Life*. New Jersey: Princeton Univ. Press.

Seeley, T. D., Morse, R. A. 1977. Dispersal behavior of honey bee swarms. *Psyche* 83:199-209.

Smith, R. J. F. 1985. *The Control of Fish Migration*. New York: Springer-Verlag.

Smith, R.-K. 1988. Identification of Africanization in honey bees based on extracted hydrocarbons assay. In *Africanized Honey Bees and Bee Mites*, ed. G. R. Needham, R. E. Page, M. Delfinado-Baker, C. E. Bowman, pp. 275-280. Chichester, England: Ellis Horwood Limited.

Smith, D. R., Taylor, O. R., Brown, W. M. 1989. Neotropical Africanized honey bees have African mitochondrial DNA. *Nature* 339:213-215.

Southwood, T. R. E. 1962. Migration of terrestrial arthropods in relation to habitat. *Biol. Rev.* 37:171-214.

Southwood, T. R. E. 1978. Escape in space and time. In *Evolution of Insect Migration and Diapause*. ed. H. Dingle, pp. 277-279. New York: Springer-Verlag.

Spivak, M. 1989. *Identification and relative success of Africanized and European honey bees in Costa Rica*. PhD Dissertation. Univ. of Kansas, Lawrence, Kansas.

Taylor, O. R. 1985. African Bees: potential impact in the United States. *Bull. Entomol. Soc. Am.* 31:15-24.

Vepsäläinen, K. 1978. Wing dimorphism and diapause in *Gerris*: determination and adaptive significance. In *Evolution of Insect Migration and Diapause*, ed. H. Dingle, pp. 218-253. New York: Springer-Verlag.

von Frisch, K. 1967. *The Dance Language and Orientation of Bees*. Cambridge, Mass: The Belknap Press of Harvard University Press.

Winston, M. L. 1979. Intra-colony demography and reproductive rate of the Africanized honey bee in South America. *Behav. Ecol. Sociobiol.* 4:279-292.

Winston, M. L. 1980. Swarming, afterswarming, and reproductive rate of unmanaged honey bee colonies (*Apis mellifera*). *Insectes Soc.* 27:391-398.

Winston, M. L. 1987. *The Biology of the Honey Bee*. Cambridge, Mass: Harvard University Press.

Winston, M. L., Otis, G. W., Taylor, O. R. 1979. Absconding behaviour of the Africanized honey bee in South America. *J. Apic. Res.* 18:85-94.

Winston, M. L., Taylor, O. R. 1980. Factors preceding queen rearing in the Africanized honey bee (*Apis mellifera*) in South America. *Insectes Soc.* 27:289-304.

Winston, M. L., Taylor, O. R., Otis, G. W. 1983. Some differences between temperate European and tropical African and South American honey bees. *Bee World* 64:12-21.

7

THE AFRICANIZATION PROCESS IN COSTA RICA

Marla Spivak[1]

As Africanized bees have migrated through the neotropics, they have hybridized with and ultimately displaced most of the resident European (*Apis mellifera mellifera* and *A. m. ligustica*) subspecies through a series of events now referred to as the "Africanization process" (Nogueira-Neto, 1964; Taylor, 1985; Rinderer, 1986). Originally, researchers considered Africanized bees to be a hybrid population which retained many characteristics of bees from Africa (Kerr, 1967; Kerr *et al.*, 1972; Gonçalves, 1974, 1975; Michener, 1975). In fact, various degrees of hybridization between European and African bees are evident in the Africanized population, as shown in studies of nuclear DNA (Severson *et al.*, 1988; Hall, 1990; W. S. Sheppard and D. Severson, pers. comm.), morphometrics (Buco *et al.*, 1987; T. E. Rinderer, pers. comm.), allozymes (Nunamaker and Wilson, 1981; Lobo *et al.*, 1989), and cuticular hydrocarbons (Carlson, 1988; Smith, 1988). Due to these various expressions of hybridization, Rinderer (1986) has described Africanized bees as a "hybrid swarm." However, recent studies of the mitochondrial DNA (mtDNA) of Africanized bees have indicated that an African (*A. m. scutellata*) genetic integrity has been maintained in colonies which comprise the migrating front and in areas where European bees have not maintained permanent populations or have not been continually reintroduced (Hall and Muralidharan, 1989; Smith *et al.*, 1989; Hall, 1990; Hall, Chapter 3). These researchers refer to the bees as neotropical African, rather than Africanized, to reflect the spread of an *A. m. scutellata* maternal lineage and the apparent lack of persistent hybridization. As Page (1989) points out, further research on co-occuring distributional patterns of mtDNA and nuclear DNA will probably clarify the contradictory views.

[1]Dr. Spivak is a Research Associate in the Center for Insect Science, University of Arizona, Tuscon, AZ, 85719, USA. Her address for correspondence is the USDA, Carl Hayden Bee Research Laboratory, 2000 E. Allen Rd., Tuscon, AZ 85719, USA.

Various mechanisms contribute to the rapid spread of Africanized bees and their displacement of European bees. Africanized bees have more rapid colony growth and have higher rates of swarming and absconding than European bees (Michener, 1975; Taylor, 1977; Winston, 1979; Otis, 1980, 1982; Winston *et al.*, 1983). These factors lead to the development of a rapidly growing and migrating feral population. Those colonies with high migratory tendencies may be continually selected for their migratory ability, forming the colonizing front (Fletcher, Chapter 4; Ratnieks, Chapter 6). In addition, Africanized colonies produce more drones than European colonies in the neotropics (Otis, 1980; Winston *et al.*, 1983; Rinderer *et al.*, 1987; Spivak, 1989) which may lead to a numerical advantage of Africanized drones in matings. Drone parasitism (Rinderer *et al.*, 1985; Rinderer and Hellmich, Chapter 5), and partial reproductive isolating mechanisms (Kerr and Bueno, 1970; Taylor, 1985; Taylor and Rowell, 1988) may also contribute, in varying degrees, to the rapid proliferation of Africanized bees.

How quickly does the Africanization process occur? From the point of view of a researcher or beekeeper in the field, what changes can be noted in the honey bee population as it becomes Africanized? One way to monitor the Africanization process, which differs from other studies that focus on gene flow or particular mechanisms of the process, is through an overview of the change in colony behavior and size of worker bee cells with time. Much attention has been given to the highly defensive behavior of Africanized bees (reviewed by Breed, Chapter 15; Collins and Rinderer, Chapter 16; Stort, Chapter 17), but this behavior may not always be distinct from that of European bees (Dietz *et al.*, 1985; Collins and Rinderer, 1986; Spivak *et al.*, 1988, 1989). Measurements of the diameter of worker cells may be the most accurate and easiest way to distinguish between Africanized and European colonies in the field (Cosenza and Batista, 1973; Rinderer *et al.*, 1986; Spivak *et al.*, 1988; Spivak, unpubl. data). The size of a bee is highly correlated with the size cell from which it emerges (Grout, 1937; Jagannadham and Goyal, 1983). Because Africanized bees are slightly smaller and construct slightly smaller cells than European bees (reviewed in Michener, 1975), changes in cell size over time help monitor the arrival of Africanized bees. Currently, the most commonly used laboratory method to identify Africanized colonies is a discriminant function analysis of morphometric characters (Daly and Balling, 1978; Daly, Chapter 2). Such an analysis gives maximum separation of characters, and has helped perpetuate the idea that the Africanized and European populations are phenotypically discrete groups (see Spivak *et al.*, 1988). Quite a different perspective emerges concerning the phenotype of Africanized bees when behavior and cell size are the only methods used to characterize colonies.

In this chapter I discuss in detail the first three years of Africanization in different areas of Costa Rica. From June 1984 to July 1986, I characterized many colonies in the field using behavioral assessments and cell size measurements and interviewed beekeepers in several regions. My aim was to answer the following questions: Can colonies be found with "intermediate" characteristics? In other words, are there colonies which are not typically "Africanized" or "European" but share characteristics of both types to varying degrees? If there are "intermediate" types, do they persist in areas where Africanized bees have established permanent populations without continual reintroduction of European stock? It should be pointed out that a colony with "intermediate" characteristics as determined in the field is not necessarily synonymous with a "hybrid" colony, as determined through controlled matings or genetic analysis.

I concentrate on three regions in this discussion: San Isidro del General and vicinity, the Meseta Central, and the western slope of Volcan Poás (FIGURE 1). These areas range from 700-2000 m in elevation and are characterized by different vegetative and climatic patterns (Holdridge, 1947; Hartshorn, 1983). The regions also differ in the degree of sophistication in management techniques practiced by beekeepers.

When the entire Africanization process is observed in the field, it is evident that an Africanized population, at any stage of the process, does not represent a well-defined, invariable phenotype. Rather, Africanized bees display a wide range of behavioral characteristics which span a continuum from relatively gentle and manageable to extremely defensive and difficult to manage (see also Collins *et al.*, 1988). The majority of Africanized colonies in Costa Rica during these years displayed characteristics intermediate between the two extremes. Manageable and productive Africanized colonies with intermediate characteristics were maintained by beekeepers who were willing to modify their management practices and requeen or kill the undesirable Africanized colonies.

METHODS USED TO IDENTIFY COLONIES

From July 1984 to July 1986, I collected samples of 100-200 bees from over 200 swarms and colonies. All samples and vouchers are in the Snow Entomological Museum, University of Kansas. Here I describe only those colonies for which the date when they were captured as swarms was known for certain. All colonies were maintained in standard 10-frame Langstroth equipment unless otherwise indicated. One or both of the following field techniques were used to characterize the 63 colonies described in this study:

Behavioral identifications: While performing colony inspections (i.e., noting the presence of a laying queen, presence of honey and pollen stores, and strength of the colony) I categorized individual colonies behaviorally as "strongly Africanized," "strongly European," or "intermediate." These categories are based solely on behavioral responses, and therefore do not necessarily reflect the genotype of the colonies. At least two people wearing full protective clothing

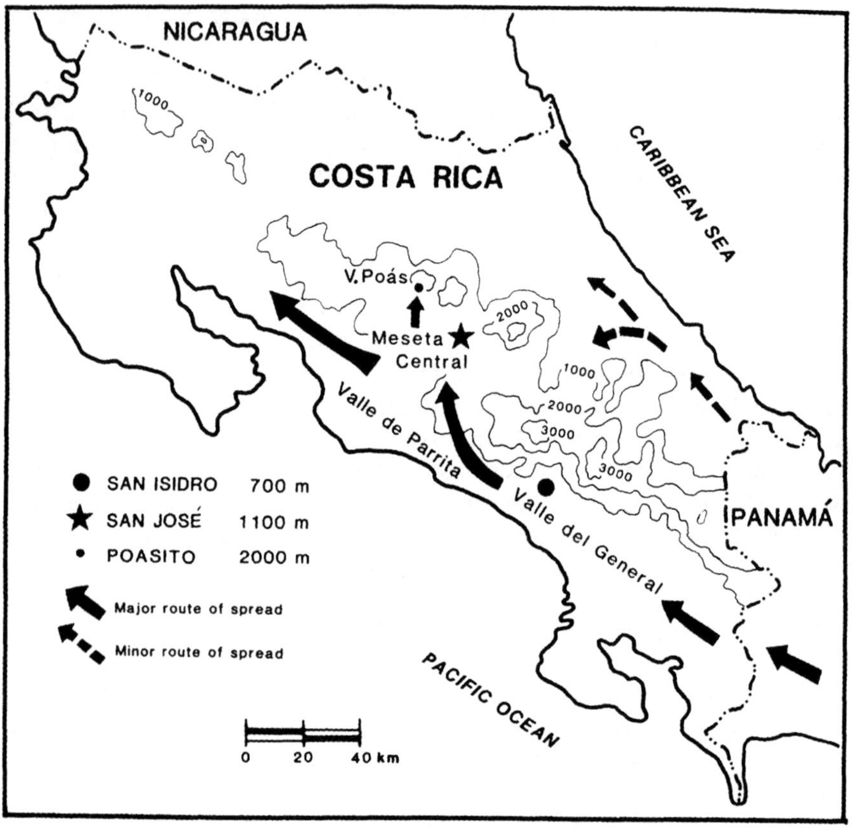

FIGURE 1. Major and minor routes of spread of Africanized honey bees through Costa Rica.

(veil, overalls, gloves, and boots) inspected each colony. If the colonies were found to be extremely defensive, it was always necessary to have one person apply smoke to the colonies while another removed the frames. All colony manipulations were made slowly and carefully, so as to create the least possible disturbance to the colony. Because climate and resource conditions may influence colony behavior (Gonçalves and Stort, 1978; Collins, 1981; Brandeburgo *et al.*, 1982; Villa, 1988), all assessments were based on repeated inspections spanning several months.

The criteria for classification into the "strongly Africanized" category were consistent and extreme defensiveness and "irritability." While inspecting these colonies, we would receive more than ten stings per visit through our protective clothing and gloves, and the bees did not remain calm on the combs. There seemed to be no management technique which would avoid eliciting a stinging response. In this category, the brood pattern often covered the entire face of a brood frame, and the queens were difficult to locate. "Strongly European" colonies were usually docile; we would not be stung more than twice, and the bees would remain relatively calm on the combs during the management period. Usually, the brood did not cover the entire face of a frame and queens were easily located. When managing "intermediate" colonies, we might receive between 0-10 stings; however, we were better able to control the stinging response through careful manipulations and proper use of smoke. The brood pattern in "intermediate" colonies more closely resembled that of "strongly Africanized" colonies.

Measurements of Cell Size: All measurements of cell size were made on worker comb which was built "naturally" (not based on foundation). If naturally built comb was not available, an empty frame was placed in the center of the brood nest of each colony to allow the bees to draw out comb. If a colony underwent a requeening event, at least 60 days were allowed to elapse before the empty frame was placed in the colony, so that all measured comb was built by the progeny of the queen present in the colony at the time. The mean diameter of the worker cells in the "naturally" built comb was recorded by measuring the linear distance spanned by ten cells. In all cases, all three diagonal rows of hexagonal cells were measured to average out any cell shape irregularities (Spivak *et al.*, 1988). No measurements were made of drone comb, comb filled with nectar or honey, or comb toward the edges of the brood nest, as these cells tend to be larger (Taber and Owens, 1970; Seeley and Morse, 1976).

European honey bees in the United States and Canada build worker brood cells (without the use of foundation) in which a series of ten cells measures from 5.2-5.4 cm (Taber and Owens, 1970; Michener, 1975). In the neotropics, European bee cell sizes range from 5.0-5.4 (Cosenza and Batista, 1973; Michener, 1975; Rinderer *et al.*, 1982, 1986). Measurements of ten cells of Africanized bees in the neotropics range from 4.6-5.0 cm (Barbosa da Silva and

Newton, 1967; Wiese, 1972; Cosenza and Batista, 1973; Rinderer *et al.*, 1982, 1986).

AFRICANIZATION OF THREE AREAS IN COSTA RICA

San Isidro del General

The town of San Isidro del General (elev. 700 m) is located in a valley (Valle del General) about 65 km north of the Panamanian border (FIGURE 1). This area is considered a secondary honey producing region (Kent, 1979). Yearly mean maximum temperatures range from 28-31°C, and mean minimum temperatures range from 17-19°C (Kent, 1979). The rainy season extends from April to December (precipitation from 200-450 mm/month), and the dry season from January to March (80-120 mm/month) (Kent, 1979). The land is mostly cultivated or is grazed by cattle, and it is intermixed with small strands of secondary forest. Before the arrival of Africanized bees the feral honey bee population was small. Before 1983, approximately 2,400 standard Langstroth colonies were operated by about 86 beekeepers (Kent, 1979; D. Quesada, pers. comm.). Honey yields per colony averaged from 13-19 kg (40-50 lbs), although up to double these yields were obtained by more experienced beekeepers (Kent, 1979 and references therein).

Movement of swarms into the area: The first Africanized swarm was detected on February 2, 1983 (dry season) by Fernando Sibaja near San Isidro. The identity was confirmed in the field by Jimmy Vargas from Ministerio de Agriculture y Ganaderia (MAG) of Costa Rica, and morphometrically by Dr. H. Daly in California. By July 1984 the area was considered to be "Africanized." Of the 86 beekeepers in the area prior to the arrival of Africanized bees only seven were managing their colonies, and virtually no one was producing honey. Numerous stinging incidents of both people and livestock were reported (D. Quesada, and J. Vargas, pers. comm.).

The Africanization process appeared to have happened in this area in the way most often described in the literature; a front of swarms, not barred by geographical obstacles, moved quickly into the lowland region mostly during the dry seasons. Many absconding and reproductive swarms issued throughout the rainy seasons. These occurrences resulted in the establishment of a high density feral Africanized population. Any new queens (by supersedure, or swarming) of the hived European colonies probably mated predominantly with the numerous Africanized drones from the feral population, resulting in rapid "Africanization."

On closer inspection, however, even by July 1986, not all of the swarms that were transferred to beekeeping equipment were extremely defensive, or "strongly Africanized." My observations every three months over two years show it was the minority of colonies which were consistently defensive and

TABLE 1. Behavioral assessments and cell sizes of swarms in chronological order of their arrival in the San Isidro area - 700 m. (str-Eur = "strongly European," str-Afr = "strongly Africanized," inter = "intermediate")

Date of swarm	Sample No.	Mean Cell Size (cm)	Behavioral Assessment
6/84	168	5.30	str-Eur
12/84	38	4.90	str-Afr
12/84	39	5.10	inter
1/85	34	4.90	inter
2/85	36	4.90	str-Afr
2/85	86	5.10	inter
3/85	87	4.90	inter
3/85	88	5.05	inter
3/85	89	5.10	inter
7/85	158	5.10	inter
7/85	169	4.90	str-Afr
8/85	157	5.00	inter
9/85	156	4.90	inter
9/85	167	4.90	str-Afr
1/86	191	5.00	inter
1/86	192	5.15	inter
2/86	193	5.10	inter
2/86	196	4.90	str-Afr

totally unmanageable. For example, of 18 colonies hived from new swarms for which I was able to make repeated behavioral assessments and obtain reliable cell measurements in this area between 1984 and 1986, one was "strongly "European" by behavior, 12 were "intermediates," and five were "strongly Africanized." The cell size of the "strongly European" colony was 5.30. All five "strongly Africanized" colonies had cell sizes of 4.90 cm. The mean cell size for the 12 "intermediate" colonies was 5.03 (s.d. = 0.09) (TABLE 1).

Beekeepers' Perspective. A common occurrence from the beekeepers' perspective was that in the first stages of Africanization, one to several colonies in an apiary became highly defensive relative to the behavior to which the beekeepers were accustomed. Upon entering an apiary and attempting to manage the colonies in the usual manner (i.e., without adequate protective clothing, using little smoke, and disregarding bumping or jarring of the colony), the beekeeper experienced extreme stinging responses from one or several of these more defensive colonies. The defensive behavior made it undesirable or impossible to inspect other colonies in the apiary. Since essentially all of the beekeepers in the area had access only to only low quality protective clothing (relative to the United States' standards), most beekeepers opted to give up their hobby or livelihood for something less hazardous. Most hived colonies were abandoned where they were previously located. Since they were near homes, penned animals, and roadsides, they caused numerous stinging incidents.

A specific example merits elaboration since it illustrates how beekeepers can control the Africanization problem in managed apiaries. One 23 year old beekeeper in this area, Ronald Montenegro, has persisted in managing Africanized honey bees successfully. In 1986, he produced an average of 40 kg/colony (90 lbs) from 90 colonies, which is equivalent to the highest yields recorded before 1983 (Kent, 1979). There are several reasons for his success: 1) He heeded advice from pamphlets and reports from other South and Central American countries on the importance of relocating apiaries 300-400 m from houses, animals and roadsides. He therefore put all his colonies on individual hive stands and distributed the apiaries into smaller groups before he noticed his colonies becoming Africanized. Consequently, he has had no stinging incidents or complaints from neighbors. 2) He obtained good protective equipment, including overalls, gloves, and boots which he uses consistently. 3) He pays careful attention to local resource conditions and anticipates dearths, or potential absconding conditions, by feeding colonies, and practices effective swarm control measures. 4) He is currently learning queen rearing techniques, which he will implement in his own apiaries. 5) He is familiar with those few colonies (3 in 90!, pers. obs.) which are consistently defensive. He either manages them last so as not to irritate the other colonies or plans to requeen them. Because he has modified his practices, is diligent and careful, he has monopolized the honey market in the area and plans to expand his operation slowly in the future.

Meseta Central (The Central Plateau)

The capital city of Costa Rica, San José, is located at 1100 m in the Meseta Central. This Central Plateau is also a secondary honey producing area (Kent, 1979). Annual mean maximum temperatures vary between 24-28°C, and mean minimum temperatures between 14-17°C (Kent, 1979). The dry season extends from December through April (precipitation, 5-25 mm/month), and the wet season from May through November (200-400 mm/month) (Kent, 1979). Below 1500 m, the land is largely cultivated. Most of the honey is produced in the coffee growing regions, with the principal nectar flow in March and April. Honey yields before the arrival of the Africanized bees averaged between 10-15 kg/colony (22-33 lbs), but more experienced beekeepers with modern equipment produced between 30-40 kg/colony (66-68 lbs) (Kent, 1979; F. Alvarado, J.M. Hidalgo, J.L. Montero, and W. Ramírez, pers. comm.).

TABLE 2. Behavioral assessments and cell sizes of swarms in chronological order of their arrival in the Meseta Central - 1100 m. (abbreviations as in TABLE 1)

Date of swarm	Sample No.	Mean Cell Size (cm)	Behavioral Assessment
4/84	9	4.90	str-Afr
10/84	7	4.90	inter
2/85	44	5.00	str-Afr
2/85	45	4.90	str-Afr
3/85	102	5.30	str-Eur
3/85	103	5.00	inter
4/85	106	4.90	str-Afr
4/85	107	4.90	str-Afr
4/85	108	4.90	str-Afr
4/85	144	5.10	inter
4/85	80	4.90	inter
4/85	54	4.90	str-Afr
6/85	120	4.90	inter
6/85	121	4.70	str-Afr
6/85	122	4.90	inter
6/85	123	4.70	inter
6/85	124	5.00	str-Afr
1/86	176	5.05	inter
2/86	177	4.90	inter
2/86	178	5.10	inter
2/86	179	4.80	inter
2/86	180	4.80	str-Afr
2/86	182	4.80	inter
3/86	185	5.05	inter
3/86	186	5.05	str-Afr
3/86	187	4.95	inter
3/86	188	5.10	inter
3/86	189	4.80	str-Afr
3/86	190	4.90	inter

Movement of swarms into the area: The arrival and spread of Africanized swarms into the Meseta Central took much longer than did the Africanization of the San Isidro area. The main reason for the delay was geographical; in order for swarms to reach the plateau, they had to follow the Pacific coast northwest to the Valle de Parrita, then turn north, while climbing in elevation to the San José area (FIGURE 1). Completing this circuitous route took one to two years. By June 1984, beekeepers and fire-fighters trained in swarm capture had captured only a few suspected Africanized swarms, and their arrival remained essentially unnoticed except by the more interested and attentive beekeepers.

Between July 1984 and July 1986, I collected samples from 39 swarms which were hived in the San José area. I made behavioral assessments and obtained reliable cell size measurements for 29 of these colonies. I considered one as "strongly European" by behavior, 16 as "intermediates" and 12 as "strongly Africanized." The cell size for the "strongly European" colony was 5.30. The mean cell size for the "intermediate" colonies was 4.94 cm (s.d. = 0.12), and for the "strongly Africanized" was 4.90 (s.d. = 0.10) (TABLE 2).

Beekeepers' perspective. At the end of the swarming and honey production season in 1984, I interviewed local beekeepers and asked if they thought their apiaries had any Africanized bees yet, and all responded that they did not (F. Alvarado, J. M. Hidalgo, and J.L. Montero, pers. comm.). Only Dr. Ramírez at the Universidad de Costa Rica had Africanized colonies which he had captured from swarms and was interested in maintaining for research purposes.

By the next dry season (January-April 1985), two years after the arrival of Africanized bees into Costa Rica, beekeepers definitely noticed that Africanized swarms were entering the San José area. Beekeeping equipment which was stored next to or inside homes or garages attracted many incoming swarms, and beekeepers were generally delighted at being able to capture many swarms with little effort. Both the Ministerio de Agricultura y Ganaderia (MAG) and Dr. Ramírez warned beekeepers through television and newspaper articles to relocate apiaries, and numerous articles on the potential dangers of collecting swarms were published (e.g., La Nación, 1 October 1985, 30 March 1986, 8 April 1986). A swarm capturing system was initiated in and around San José, and many swarms entering the city were removed at once, helping to minimize the danger to the public. Also, as the newspaper articles described Africanized bees in an educational manner, public hysteria and misinformation were kept to a minimum.

Many beekeepers heeded the advice of MAG and Dr. Ramírez. Serious attempts to requeen and maintain more gentle colonies were made by some more attentive and financially able beekeepers (e.g., see Cobey and Locke, 1986; Clarke, 1989). Attentive beekeepers were able to maintain "intermediate" types by consistently destroying or requeening the most "strongly Africanized" colonies (W. Ramírez, pers. comm.; pers. obs.). They also purchased or made better protective equipment (boots, better veils, gloves and overalls) and relocated their apiaries. For the first time, young beekeepers became interested in rearing and marking queens and then paid closer attention to general colony demographics in order to control excessive defensive behavior, swarming and absconding.

Honey yields dropped in the Meseta Central in 1984 and 1985. It is unclear whether the lower yields were due to unfavorable climatic and resource conditions, the arrival and increasing density of Africanized bees in the area, or a combination of these factors (F. Alvarado, W. Ramírez, pers. comm.). Skilled beekeepers with modern equipment claimed their honey production dropped from an average of 45 kg/colony (100 lbs) before the arrival of Africanized bees, to 30-34 kg/colony (65-75 lbs) in 1984, and 9-23 kg/colony (20-50 lbs) in 1985. In 1986, their yields increased again slightly, averaging 23-32 kg/colony (50-70 lbs) (F. Alvarado; J.M. Hidalgo; J.M. Montero, pers. comm.). Higher yields were obtained from both Africanized and European colonies placed in a large coffee plantation from March through April, 1986. The average for 15 colonies over these two months was 40 kg/colony (87 lbs) (Spivak *et al.*, 1989).

Volcan Poás (Western Slope)

Volcan Poás is an active volcano located about 40 km northwest of San José. Above 1500 m, on the western slope of the volcano, coffee is not grown, but some cold-tolerant crops are, and ornamental flowers are widely cultivated for export. Insecticides and herbicides are commonly used, and logging and grazing are changing the primary oak forests into eroding pasture lands, mixed with sparse patches of secondary forest. I surveyed colonies at 2000 m on a farm in Poasito. Mean annual daytime temperatures here range from 19-23°C, and nighttime temperatures range from 2-9°C, with occasional frosts. Precipitation levels are similar to those in the Meseta Central, or slightly higher (Kent, 1979).

Before the arrival of Africanized bees, local families routinely captured a few European swarms and hived them in rustic boxes, many times with immovable frames. Often these families did not own veils or smokers. They occasionally cut out combs of honey from the colonies, but basically the few colonies they kept were unmanaged. The lack of sophisticated beekeeping allowed me to observe a "feral" population closely. I inspected the largest apiary of such rustic colonies every three months from July 1984 to July 1986 to determine if and when Africanized bees would move up into the area, and what impact Africanized bees might have on the European bee population.

Movement of swarms into the area. From July 1984 to May 1985 all nine colonies (hived from swarms) in the apiary had cell sizes (built on naturally drawn comb) which measured 5.30 cm, a cell size suggestive of European bees. Behaviorally, all were considered "strongly European." In May 1985, combs from two newly captured swarms had mean cell sizes of 5.05 and 5.10 cm respectively. From the "irritable" nature of these new swarms, I suspected them to be slightly Africanized, but clearly in the "intermediate" category. In November 1985, the first "strongly Africanized" swarm entered an abandoned

TABLE 3. Behavioral assessments and cell sizes of swarms in chronological order of their arrival on the western slope of Volcan Poas - 2000 m. (abbreviations as in TABLE 1).

Date of swarm	Sample No.	Mean Cell Size (cm)	Behavioral Assessment
7/84	12	5.30	str-Eur
7/84	13	5.30	str-Eur
7/84	14	5.30	str-Eur
1/85	15	5.30	str-Eur
1/85	32	5.30	str-Eur
3/85	111	5.30	str-Eur
3/85	112	5.30	str-Eur
4/85	114	5.30	str-Eur
4/85	116	5.30	str-Eur
5/85	115	5.05	inter
5/85	117	5.10	inter
11/85	162	4.90	str-Afr
1/86	-	4.80	str-Afr
2/86	199	4.80	inter
2/86	198	5.10	str-Afr
3/86	200	5.05	str-Afr

box. The cells of this swarm measured 4.90 cm. November is usually the wettest month of the year in this area, and therefore a swarm was unusual, unless it was an absconding swarm. All of the original nine colonies in the apiary had dwindled and died by November, a fact which was also unusual as the majority of European bees in previous years had survived through the rainy season. Local people attributed these deaths to the indiscriminant use of insecticides with the recent cultivation of strawberries.

From January to July of 1986, four swarms entered abandoned equipment. One of the swarms had mean cell sizes of 4.80 cm, and it later absconded. Another, with cell sizes of 4.80 cm was manageable, and I considered it "intermediate" in behavior. The remaining two colonies had cell sizes of 5.00 cm and 5.10 cm respectively, and were consistently much more defensive although they had fewer bees than the one with smaller cell sizes. The family who owned the colonies killed one of the more defensive colonies because it was stinging their livestock. They finally moved the colonies farther from the house and then essentially abandoned their practice of gathering swarms. By July 1986, the only colony alive was the one which had cell sizes of 4.80 cm; the other three had dwindled and died. Of the four swarms considered by behavior to be "strongly Africanized," the mean cell size was 4.96 cm (s.d. = 0.14), and that of the "intermediates" was 4.98 cm (s.d. = 0.16) (TABLE 3).

COMPARISON OF BEHAVIOR, CELL SIZE, AND OTHER IDENTIFICATION TECHNIQUES

The main point of this comparison is to demonstrate that although measurements of cell sizes help monitor the arrival and spread of Africanized bees, they may not accurately predict the behavior of a colony. In this study, all colonies identified as "strongly European" in the field had cell sizes of 5.30 cm (n = 12). Colonies characterized as "intermediate" and "strongly-Africanized" had cell sizes which ranged from 4.7 cm to 5.1 cm. Although the mean cell sizes between the two behavioral groups (all locations combined) were significantly different (str-A: mean = 4.91 cm, s.d. = 0.09, n = 21; inter: mean = 4.98 cm, s.d. = 0.12, n = 31; t-test, $p < 0.05$), it is clear that a colony with a cell size of 4.7 cm, for example, may not always display extreme defensive behavior.

It is interesting to note that the Africanized colonies in this study were obtained from swarms just entering the country. These swarms comprised the migrating front and theoretically would have carried the *A. m. scutellata* mitochondrial DNA markers (Hall, Chapter 3). These results emphasize the need to judge the manageability and productivity of Africanized colonies based on general behavioral attributes, rather than measurements of size or degree of genetic affinity with *A. m. scutellata.* Mitochondrial DNA markers are very useful in population genetic studies to detect the flow of African genes into new areas, and, combined with other laboratory identification techniques, can be used to certify stocks as European where needed. Nuclear DNA markers will be more useful in determining levels of introgression and hybridization between the subspecies and may eventually be used to certify hybrid stocks if the need arises. However, in areas where Africanized bees establish permanent feral populations making it difficult and unrealistic to maintain strong European colonies, the only reasonable option for beekeepers and queen breeders is to select for and propagate the most gentle and productive Africanized colonies based on their behavior. The implications of the discrepancy between behavioral assessments and other identification techniques on control and regulatory procedures of Africanized bees is also discussed in Spivak *et al.* (1988, 1989).

CONCLUSIONS

The Africanization process in three areas of Costa Rica can be summarized as follows: In a lowland area such as San Isidro, there was a rapid establishment of a feral Africanized population, and there appeared to be a rapid turnover from a "strongly European" to a "strongly Africanized" population. In higher elevations and more geographically remote areas as in the Meseta Central and the western slope of Volcan Poás, there was a much slower establishment of a feral Africanized population, and the characteristics of the population did not appear to change as abruptly.

In all regions, there were beekeepers who were unwilling to modify their practices to adapt to the new circumstances. They soon experienced extreme stinging responses and high incidences of swarming and absconding. Ultimately, these beekeepers abandoned their colonies. Based on these occurrences, the idea was erroneously perpetuated that the entire population displayed uniform characteristics and that all bees were both dangerous to the public and undesirable from a management standpoint.

When swarms and colonies from all areas were observed and assessed on an individual basis, however, they clearly displayed a wide range of behavioral characteristics. It was the minority of colonies which were consistently unmanageable and extremely defensive. Those beekeepers who were willing to requeen or kill such colonies and modify their management practices were able to work Africanized colonies profitably and with minimal danger to the beekeeper or public (see also Kent, 1989 and Chapter 19).

The establishment of a feral Africanized population in recreational areas and cities and the stinging incidents which stem from these colonies have been the most pressing problems in the neotropics. Stinging incidents have been kept to a minimum in the San José area, however, because of the initiation of a swarm capturing program. The public also has been educated about Africanized bees through informative and non-alarmist newspaper articles.

These observations of the Africanization process in three different regions of Costa Rica emphasize how beekeepers can reduce the impact of Africanized bees by employing good management and selection practices. The introduction of a highly variable strain of honey bees in fact provides a source of variation from which to breed more productive, disease-resistant and mite-resistant bees. The answer to the question of how long the "intermediate" types persist in different areas without reintroduction of European stock depends on the collective cooperation of beekeepers and queen breeders (see Erickson *et al.*, 1986; Laidlaw, 1988; Spivak *et al.*, 1989). In areas where Africanized bees will or already have established a permanent population, it is not necessary to specify the degree of Africanization (through morphometric, allozyme or DNA analyses) to determine whether certain colonies should be propagated or eliminated. When the unmanageable Africanized colonies are consistently killed or requeened with Africanized stock which is gentle and productive, more manageable intermediate types can be maintained. This point has been demonstrated repeatedly by beekeepers and researchers in southern Brazil where they are currently working with productive, manageable and relatively disease-resistant Africanized bees (Kerr, 1966/67; Wiese, 1972, 1977; De Jong, 1984; De Jong *et al.*, 1984; Bradbear and De Jong, 1985; Camazine, 1986; Camazine and Morse, 1988;

Gonçalves *et al.*, Chapter 18; Page and Kerr, Chapter 8; Stort and Gonçalves, Chapter 17). If beekeepers are attentive and willing to follow advice from other beekeepers and researchers who have had positive experiences with Africanized bees, it is entirely feasible that the Africanization process can have positive consequences.

Acknowledgements

I extend thanks to all those in Costa Rica who patiently answered my questions and allowed me to repeatedly inspect colonies in their apiaries. Special thanks to Ana Lorena Castro, Fernando Segreda and Dr. William Ramírez for their help with colony inspections and sampling, and to Mary Jane West Eberhard, William Eberhard for their support. S. Cobey, E. Erickson, C.D. Michener, P. Payne, O.R. Taylor and W.T. Wcislo offered helpful criticisms of the manuscript. This work was funded by an NSF Dissertation Improvement Grant BSR-8313554, a Jessie Noyes Fellowship through the Organization of Tropical Studies, and a Fellowship from the Organization of American States 594908.

LITERATURE CITED

Barbosa da Silva, R. M., Newton, S. W. 1967. Sobre a bionomia da *Apis* Afro-europei do Brasil. *Bol. de Ind. Animal, 24 N.S. Unico-São Paulo, Brasil.* pp. 199-208.

Bradbear N., De Jong, D. 1985. The management of Africanized honeybees. *Information for beekeepers in tropical and subtropical countries. Leaflet 2.* London: Int. Bee Res. Assoc.

Brandeburgo, M. S., Gonçalves, L. S., Kerr, W. E. 1982. Effects of Brazilian climatic conditions upon the aggressiveness of Africanized colonies of honeybees. In *Social Insects in the Tropics,* ed. P. Jaisson, pp. 255-280. Paris: Université Paris-Nord.

Buco, S. M., Rinderer, T. E., Sylvester, H. A., Collins, A. M., Lancaster, V. A., Crewe, R. M. 1987. Morphometric differences between South American Africanized and South African *(Apis mellifera scutellata)* honey bees. *Apidologie* 18:217-222.

Camazine, S. 1986. Differential reproduction of the mite, *Varroa jacobsoni* (Mesostigmata: Varroidae), on Africanized and European honey bees (Hymenoptera: Apidae). *Ann. Entomol. Soc. Am.* 79:801-803.

Camazine, S., Morse, R. A. 1988. The Africanized honeybee. *Am. Sci.* 76:465-471.

Carlson, D. A. 1988. Africanized and European honey-bee drones and comb waxes: analysis of hydrocarbon components for identification. In *Africanized Honey Bees and Bee Mites,* ed. G. R. Needham, R. E. Page, M. Delfinado-Baker, C. E. Bowman, pp. 264-274. Chichester, England: Ellis Horwood Limited.

Clarke, W. W. 1989. Bill Clarke says: "We can work with the Africanized bee." *The Speedy Bee*. January, 1989. pp. 7-8.

Cobey, S., Locke, S. 1986. The Africanized bee: A tour of Central America. *Am. Bee J.* 126:434-440.

Collins, A. M. 1981. Effects of temperature and humidity on honeybee response to alarm pheromones. *J. Apic. Res.* 20:13-18.

Collins, A. M., Rinderer, T. E. 1986. The defensive behavior of the Africanized bee. *Am. Bee J.* 126:623-627.

Collins, A. M., Rinderer, T. E., Tucker, K. W. 1988. Colony defence of two honeybee types and their hybrid. 1. Naturally mated queens. *J. Apic. Res.* 27:137-140.

Cosenza, G. W., Batista, J. S. 1973. Morfometría da *Apis mellifera adansonii* (abelha africanizada) da *Apis mellifera caucasica* (abelha caucasiana) e sua híbridas. 2° *Cong. Bras. Apic. (Sete Lagoas, 1972)*, ed. G. W. Cosenza, pp. 53-56. Minas Gerais, Brazil. (Also, 1974, *Ciênc. Cult.* 26:864-866.)

Daly, H. V., Balling, S. S. 1978. Identification of Africanized honey bees in the western hemisphere by discriminant analysis. *J. Kans. Entomol. Soc.* 51:857-869.

De Jong, D. 1984. Africanized bees now preferred by Brazilian beekeepers. *Am. Bee J.* 124:116-118.

De Jong, D., Gonçalves, L. S., Morse, R. A. 1984. Dependence on climate of the virulence of *Varroa jacobsoni*. *Bee World* 65:117-121.

Dietz, A., Krell, R., Eischen, F.A. 1985. Preliminary investigation on the distribution of Africanized honey bees in Argentina. *Apidologie* 16:99-108.

Erickson, E. H. Jr., Erickson, B. J. Young, A. M. 1986. Management strategies for "Africanized" honey bees: Concepts strengthened by our experiences in Costa Rica. *Gleanings Bee Cult.* Part I. 114:456-457, 459. Part II. 114:506-507, 534.

Gonçalves, L. S. 1974. The introduction of the African bees (*Apis mellifera adansonii*) into Brazil and some comments on their spread in South America. *Am. Bee J.* 114:414-415, 419.

Gonçalves, L. S. 1975. Do the Africanized bees of Brazil only sting? *Am. Bee J.* 115:8-10, 24.

Gonçalves, L. S., Stort, A. C. 1978. Honey bee improvement through behavioral genetics. *Ann. Rev. Entomol.* 31:197-213.

Grout, R. A. 1937. The influence of size of brood cell upon the size and variability of the honeybee (*Apis mellifera* L.). *Res. Bull. Agric. Exp. Stat., Iowa State College, Ames, Iowa.* 218: 257-280.

Hall, H. G. 1990. Parental analysis of introgressive hybridization between African and European honeybees using nuclear DNA RFLP's. *Genetics* 125:611-621.

Hall, H. G., Muralidharan, K. 1989. Evidence from mitochondrial DNA that African honey bees spread as continuous maternal lineages. *Nature* 339:211-213.

Hartshorn, G. S. 1983. Plants. In *Costa Rican Natural History*, ed. D. H. Janzen, pp. 118-157. Chicago: University of Chicago Press.

Holdridge, L. R. 1947. Determination of world plant formations from simple climatic data. *Science* 105:367-368.

Jagannadham, B., Goyal, N. P. 1983. Morphological and behavioural characteristics of honeybees workers reared in combs with larger cells. *2nd Int. Conf. Apic. Trop. Climates, New Dehli, 1980*. Dehli: Yagantar Press.

Kent, R. B. 1979. Diversidad ecológica y las regiones apícolas de Costa Rica. *Rev. Geográfica* 90:65-95.

Kent, R. B. 1989. The African honeybee in Peru: an insect invader and its impact on beekeeping. *Appl. Geog.* 9:237-257.

Kerr, W. E. 1966/67. Solução e criar uma raça nova. *Guia Rural* 67:20-22. (Also, 1968. *O apicultor (Porto Alegre)* 1:7-10.)

Kerr, W. E. 1967. The history of the introduction of African bees to Brazil. *S. Afr. Bee J.* 39:3-5.

Kerr, W. E., Bueno E. 1970. Natural crossing between *Apis mellifera adansonii* and *Apis mellifera ligustica. Evolution* 24:145-155.

Kerr, W. E., Gonçalves, L. S., Blotta, L. F., Maciel, H. B. 1972. Biología comparada entre as abelhas Italianas (*Apis mellifera liguistica)*, Africana (*Apis mellifera adansonii*) e suas híbridas. *1° Congr. Bras. Apic. (Florianópolis, 1970)*, ed. H. Weise, pp. 151-185. Santa Catarina, Brazil.

Laidlaw, H. H. 1988. Thoughts on countering the Africanized bee threat. In *Africanized Honey Bees and Bee Mites*, ed. G. R. Needham, R. E. Page, Jr., M. Delfinado-Baker, C. E. Bowman, pp. 209-213. Chichester, England: Ellis Horwood Limited.

La Nación, 1 Oct, 1985. "El humo y la abeja africanizada." Costa Rica. p.1C.

La Nación, 30 March 1986. "Abeja africanizada: Como eliminar los enjambres." Costa Rica p.2C.

La Nación, 8 April, 1986. "Enjambres africanizados constituyen un peligro." Costa Rica p.2C.

Lobo, J. A., Del Lama, M. A., Mestriner, M. A. 1989. Population differentiation and racial admixture in the Africanized honeybee (*Apis mellifera* L.). *Evolution* 43:794-802.

Michener, C. D. 1975. The Brazilian bee problem. *Ann. Rev. Entomol.* 20:399-416.

Nogueira-Neto, P. 1964. The spread of a fierce African bee in Brazil. *Bee World* 45:119-121.

Nunamaker, R. A., Wilson, W. T. 1981. Comparison of MDH allozyme patterns in the African honey bee (*Apis mellifera adansonii* L.) and the Africanized population in Brazil. *J. Kans. Entomol. Soc.* 54:704-710.

Otis, G. W. 1980. *The swarming biology and population dynamics of the Africanized honey bee.* PhD Dissertation. Univ. Kansas, Lawrence. 197pp.

Otis, G. W. 1982. Population biology of the Africanized honey bee. In *Social Insects in the Tropics, Vol. 1*, ed. P. Jaisson, pp. 209-219. Paris: Université Paris Nord.

Page, R. E. Jr. 1989. Neotropical African bees. *Nature* 339:181-182.

Rinderer, T. E. 1986. Africanized bees: the Africanization process and potential range in the United States. *Bull. Entomol. Soc. Am.* 32:222-227.

Rinderer, T. E., Collins, A. M., Hellmich, R. L., Danka, R. G. 1987. Differential drone production by Africanized and European honey bee colonies. *Apidologie* 18:61-68.

Rinderer, T. E., Hellmich, R. L. II., Danka, R. G., Collins, A. M. 1985. Male reproductive parasitism: a factor in the Africanization of European honey-bee populations. *Science* 228:1119-1121.

Rinderer, T. E., Sylvester, H. A., Brown, M. A., Villa, J. D., Pesante, D., Collins, A. M. 1986. Field and simplified techniques for identifying Africanized European honey bees. *Apidologie* 17:33-48.

Rinderer, T. E., Tucker, K. W., and Collins, A. M. 1982. Nest cavity selection by swarms of European and Africanized honeybees *J. Apic. Res.* 21:98-103.

Seeley, T. D., Morse, R. A. 1976. The nest of the honey bee (*Apis mellifera* L.) *Insectes Soc.* 23:495-512.

Severson, D. W., Aiken, J. M., Marsh, R. F. 1988. Molecular analysis of North American and Africanized honey bees. In *Africanized Honey Bees and Bee Mites,* ed. G. R. Needham, R. E. Page, M. Delfinado-Baker, C. E. Bowman, pp. 294-302. Chichester, England: Ellis Horwood Limited.

Smith, D. R., Taylor, O. R., Brown, W. M. 1989. Neotropical Africanized honey bees have African mitochondrial DNA. *Nature* 339:213-215.

Smith, R.-K. 1988. Identification of Africanization in honey bees based on extracted hydrocarbons assay. In *Africanized Honey Bees and Bee Mites,* ed. G. R. Needham, R. E. Page, M. Delfinado-Baker, C. E. Bowman, pp. 275-280. Chichester, England: Ellis Horwood Limited.

Spivak, M. 1989. *Identification and relative success of Africanized and European honey bees in Costa Rica.* PhD Dissertation. Univ. Kansas, Lawrence.

Spivak, M., Ranker, T., Taylor, O. R., Taylor, W. Davis, L. 1988. Discrimination of Africanized honey bees using behavior, cell size, morphometrics, and a newly discovered isozyme polymorphism. In *Africanized Honey Bees and Bee Mites,* ed. G. R. Needham, R. E. Page, M. Delfinado-Baker, C.E. Bowman, pp. 313-324. Chichester, England: Ellis Horwood Limited.

Spivak, M., Batra, S., Segreda, F., Castro, A. L., Ramírez, W. 1989. Honey production by Africanized and European honey bees in Costa Rica. *Apidologie* 20:207-220.

Taber, S., Owens, C. D. 1970. Colony founding and initial nest design of honey bees. *Anim. Behav.* 18:625-632.

Taylor, O. R. 1977. The past and possible future spread of Africanized honeybees in the Americas. *Bee World* 58:19-30.

Taylor, O. R. 1985. African bees: potential impact in the United States. *Bull. Entomol. Soc. Am.* 31:15-24.

Taylor, O. R., Jr., Rowell, G. A., 1988. Drone abundance, queen flight distance, and the neutral mating model for the honey bees, *Apis mellifera.* In *Africanized Honey Bees and Bee Mites,* ed. G. R. Needham, R. E. Page, M. Delfinado-Baker, C. E. Bowman, pp. 173-183. Chichester, England: Ellis Horwood, Limited.

Villa, J. D. 1988. Defensive behaviour of Africanized and European honeybees at two elevations in Colombia. *J. Apic. Res.* 27:141-145.

Wiese, H. 1972. Abehlas Africanizas, suas carcterísticas e tecnologia de manejo. *1° Cong. Bras. Apic. (Florianópolis, 1970),* ed. H. Wiese, pp. 95-108. Santa Catarina, Brazil

Wiese, H. 1977. Apiculture with Africanized bees in Brazil. In *African Bees: Taxonomy, Biology and Economic Use,* ed. D. J. C. Fletcher, pp. 67-76. Pretoria: Apimondia.

Winston, M. L. 1979. Intra-colony demography and reproductive rate of the Africanized honeybee in South America. *Behav. Ecol. Sociobiol.* 4:279-292.

Winston, M. L., Taylor, O. R., Otis, G. W. 1983. Some differences between temperate European and tropical African and South American honeybees. *Bee World* 64:12-21.

8

HONEY BEE GENETICS AND BREEDING

Robert E. Page, Jr.,[1] and Warwick E. Kerr[2]

A process of "Africanization" of feral and commercial honey bee populations has taken place throughout South and Central America since 1957 (See Kerr, 1969; Kerr *et al.*, 1982; Lobo *et al.*, 1989). Associated with this process have been major adjustments in attitudes of the public toward beekeeping and adjustments within the beekeeping industry itself. The beekeeping industry of South America was not as developed nor as extensive as it is in North America, particularly the queen and package bee industries. Therefore, South American beekeepers lacked a fundamental mechanism for dealing with Africanized bees. Since then, some areas of Brazil have developed a queen rearing industry and through selective breeding have made good progress toward producing superior commercial stocks of Africanized bees and have made several changes in mangagemnt.

Africanization of feral and commercial populations began in Mexico in 1986 and will probably begin soon in the United States. In order to avoid the numerous problems observed in South and Central America, it is imperative that the queen-rearing industry of the United States remains strong both in production and in scientific knowledge. With Africanization, it is likely that queen breeding will become an essential component of queen production; therefore, a knowledge of honey bee genetics and breeding will be important.

In this chapter we discuss the genetic structure of breeding populations of honey bees, a knowledge of which is essential for constructing useful bee breeding programs. We then present different mating designs that can be used with different breeding programs. We discuss at length closed population breeding programs and present different methods for selecting superior stock and

[1]Professor Page is in the Department of Entomology, University of California, Davis, CA, 95616, USA.

[2]Professor Kerr is in the Department of Biology, Federal University of Uberlandia, 38400 Uberlandia, Minas Gerais, Brazil.

for predicting the response of a breeding population to selection. We finish by discussing the role of selective breeding programs in South American apiculture where Africanized bees have been a reality for over 30 years.

GENETIC STRUCTURE OF POPULATIONS

Because of arrhenotokous parthenogenesis, all functional gametes have their origin in the females, primarily queens, but occasionally also in uninseminated, egg-laying workers. Males do not produce unique genomes, barring mutations, but only replicate them. This haplodiploid system of reproduction has certain peculiarities that can be used to the benefit of breeders, and others that provide serious difficulties for breeding.

Population Genetics

The population genetics of honey bees, like all haplodiploid organisms, is analogous to sex-linked traits. The fundamental differences in haplodiploid and diploid population genetics involve the maintenance of genetic variability, rates of evolution, and asymmetries of genetic relationships among individuals.

A reduction in genetic diversity is expected with haplodiploidy because recessive genes are expressed in males and are subject to direct selection. This reduction will be greatest for genes that are expressed similarly in males and females. However, levels of genetic diversity of female haplodiploids should approach those of diploids as the correlation of gene expression between the sexes decreases (Pamilo and Crozier, 1981). Traits expressed only in queens and workers should be more variable than those expressed in only drones, or in both sexes.

Enzyme studies have shown a reduced amount of variability among Hymenoptera in general and honey bees in particular (Sylvester, 1986). However, selection studies have yielded high heritabilities of some traits, suggesting a large amount of genetic variability (see Collins, 1986). It is possible that more variability exists for those traits because they are expressed only in workers and queens and are not under strong natural selection.

The amount of genetic variation within a population is important in breeding. More variation results in more potential for a significant selection response (see discussions on selection and heritability below) and in a more rapid selective improvement. Natural populations of haplodiploid organisms are expected to evolve one-third faster than equivalent diploids (Hartl, 1972), a characteristic that may be important for understanding the process of Africanization in South and Central America.

Asymmetries in genetic relationships among nestmates of honey bees exist because of haplodiploidy (Crozier and Pamilo, 1980). Drones are derived directly from genomes of females and, therefore, share all of their genes in common by

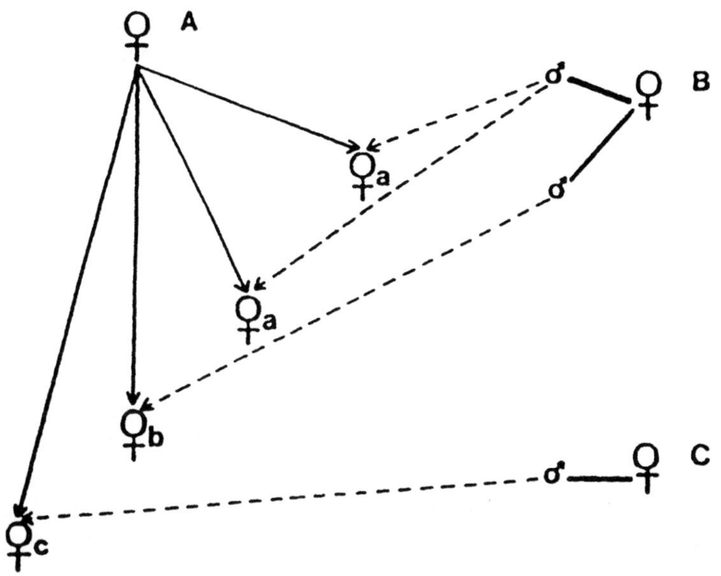

FIGURE 1. Subfamily relationships of females (workers and queens) of a colony. Each drone sires a separate subfamily. a-a, super sisters (G = 0.75); a-b, full sisters (G = 0.50); a-c, half sisters (G = 0.25). Solid lines represent egg-gamete paths while dashed lines represent sperm paths. Uncrossed female symbols are workers. From Laidlaw and Page (1986) by permission. Copyright 1986 by Academic Press.

descent with their "mothers." On the other hand a queen, or laying worker, shares only 1/2 of her genes in common with her own "sons."

Asymmetries also exist between drone and worker progeny of a queen. Drones share 1/2 of their genes in common with worker and drone siblings but workers share only 1/4 of their genes in common with their sibling drones and usually either 1/4 or 3/4 of their genes in common with sisters, depending upon whether they have different, unrelated fathers, or the same father, respectively (see FIGURE 1). These asymmetrical relationships are important for understanding the genetic structure of colonies and populations of honey bees, and are discussed in more detail below.

Terminology

There are two different sets of terminology found in the literature with respect to identifying familial relationships of haplodiploid female siblings

(Crow and Roberts, 1950; Laidlaw and Eckert, 1950; Polhemus *et al.*, 1950; Laidlaw, 1974; Page and Laidlaw, 1988). One set is based on the physical pairing of individuals and calls diploid female progeny that have the same queen mother and the same drone father full sisters. The other terminology is based on the genetic pairing of individuals and calls these diploid females super sisters and reserves the term full sister for those sisters that have the same mother but different drone fathers that are themselves brothers. With this genetic pairing model, queens can be considered as both male and female parents. They function as male parents by donating the male genomes to zygotes via their arrhenotokous sons. Genetic pairing terminology is more precise and is used throughout this text.

Polyandry and Sperm Utilization

Queens mate on the average with 7-17 different males, depending upon the population measured and, perhaps, the method of estimation (Taber and Wendel, 1958; Woyke, 1960; Adams *et al.*, 1977). The highest estimate of 17 matings was made for Africanized honey bees in Brazil. Matings take place while drones and queens are in flight at specific congregating areas within a few days after queens emerge as adults (Fyg, 1952; Ruttner, 1956a; Woyke, 1960; Ruttner and Ruttner, 1966). Queens frequently take more than one mating flight where, on any given flight, they may mate with up to 17 males (Woyke, 1960; 1964). During copulation males ejaculate approximately 5-10 million spermatozoa into the oviducts of the queen (Woyke, 1960; Kerr *et al.*, 1970, 1972). Kerr *et al.* (1970) estimated that Italian, Africanized and Italian X Africanized hybrid drones produce 5.46, 7.12, and 6.43 million sperm, respectively. Copulations last several seconds during which time the male undergoes an explosive ejaculation. The male then falls paralyzed to the ground and dies (Gary, 1963). The chitinous plates of the endophallus and a large quantity of mucous secreted by the male are left in the sting chamber of the queen and constitute a mating sign (Bishop, 1920; Laidlaw, 1944). The mating sign has little effect on preventing subsequent males from mating with the queen because males have an elaborate genital mechanism that results in the removal of the sign of the previous mate (Koeniger, 1984). The mating sign probably serves as a plug to prevent the backflow of semen into the sting chamber (Bishop, 1920).

After a mating flight, the queen returns to the colony where workers help remove the mating sign of the last mate. Over a period of up to 40 hours the spermatozoa migrate by both active and passive processes (mostly active) from the oviducts into the spermatheca (Ruttner, 1956b; Ruttner and Koeniger, 1971; Gessner and Ruttner, 1977; Woyke, 1983). Of the approximately 180 million spermatozoa received by a queen, each retains about 5-6 million in the spermatheca (Mackensen and Roberts, 1948; Woyke, 1960; Kerr *et al.*, 1962) which is roughly equivalent to the contribution of a single male.

It is likely that all of the mates of a queen contribute sperm to the spermatheca. Studies that used genetic markers have shown the representation of semen from all males used for instrumental insemination of queens (Moritz, 1983; Laidlaw and Page, 1984). The estimates of Taber and Wendel (1958) and of Adams *et al.* (1977) were estimates of the effective number of matings of naturally-mated queens, and show the representation of the semen of a large number of males (8-17). As estimates of effective number, they assumed equal contribution of each male to the spermathecae of queens and as such underestimated the actual number (see Page, 1986).

The sperm of the many different males are not completely mixed within the spermatheca (Taber, 1955; Kerr *et al.*, 1980; Page and Metcalf, 1982; Laidlaw and Page, 1984); however, for any moderately large batch of eggs laid by the queen, sperm of all males are used to fertilize some eggs. Patterns of sperm utilization by queens show consistency over time with unequal, but similar, contributions of each male (Page and Metcalf, 1982, Laidlaw and Page, 1984). There have been reports of large fluctuations in sperm use for queens of both European (Taber, 1955) and Africanized (Kerr *et al.*, 1980) descent; however, these large fluctuations were short in duration and occurred primarily at the beginning of egg laying of recently mated queens. Patterns of egg laying of queens in both of these studies fluctuated near the phenotypic proportions expected if each mate was contributing equally to egg fertilization.

Population Structure

The systems of mating and sperm utilization described above affect the genetic structure of honey bee populations and, consequently, honey bee breeding. Only by understanding the genetic structure of breeding populations can we breed intelligently.

The result of extreme polyandry and sperm mixing is that colonies consist of a large number of worker subfamilies. Members of each subfamily are super sisters of each other and half sisters of members of different subfamilies. The average genetic relationship among workers within a colony, measured as the pedigree coefficient of relationship, G, (Crozier, 1970; Pamilo and Crozier, 1982), approaches 0.25. The genetic variance within a population that is structured into families such as honey bee colonies is expected to be distributed as

$$V_W = (1 - r)V_T \qquad 1.$$

where V_T = the total additive genetic variance in the population, V_W = the variance between individuals within colonies, and r = the coefficient of genetic relationship between individuals within colonies (Crow and Kimura, 1970, p. 141). Replacing r in Formula 1 with G from above (a justifiable substitution), it is clear that close to 75% of the additive genetic variance in a population of honey bees is distributed within colonies compared with only 25% of the variance between colonies. Knowledge of how this variance is distributed becomes very important in predicting the outcome of selection.

SEX DETERMINATION

With the development of instrumental insemination in the 1930s came the ability to control matings. However, soon after its development a serious problem became apparent: inbreeding leads to severe inviability of workers. Over the next 20 years (1943-1963) a major cause of the inviability was determined to be the genic mechanism of sex determination (Mackensen, 1951, 1955; Rothenbuhler, 1957; Woyke, 1962, 1963a, 1963b, 1963c).

Primary sexual characteristics of honey bees are determined principally by a single gene locus, or a series of closely linked genes that segregate in Mendelian ratios, that has a series of different alleles, Xa, Xb, Xc, ..., Xi. Individuals that are heterozygous (have two different alleles at this X locus) develop phenotypically into females, either workers or queens. Individuals that are hemizygous (haploid) or homozygous at this locus develop phenotypically into males. Homozygous diploid males are consumed by the workers after they hatch from the egg and are, under normal circumstances, inviable (Woyke, 1963b, 1963c). As a consequence, with controlled mating, viability of brood in colonies can vary between 50% and 100%.

Any loss of viability can reduce colony performance. The effects of inviability of brood due to production of diploid males is complex. Woyke (1980, 1981) determined that differences in brood production between colonies with high (100%) and low (50%) viability in his study area were small during the spring but increased during the summer. Colonies that produced 50% diploid drones had 68% of the brood area of colonies with no diploid males. Adult worker populations of colonies were also affected by brood inviability. Colonies with 50% viability had 79%, 35%, and 65% of the number of adult workers in 100% viable colonies in the spring, summer, and autumn, respectively. Honey production of 50% viable colonies averaged between 8% and 75% of the production of 100% viable colonies for different honey flows in different seasons and different years.

The variance among colonies as a consequence of diploid drone production can affect the proper evaluation of colonies in a breeding program. For example, two colonies may show dramatic differences in honey production because one has 100% viable brood compared to the other with 50% viable brood. This type of variation would make selection for superior honey producing genotypes very difficult.

The variance can be controlled with instrumental insemination by controlling the number of males used for inseminating each queen. With a

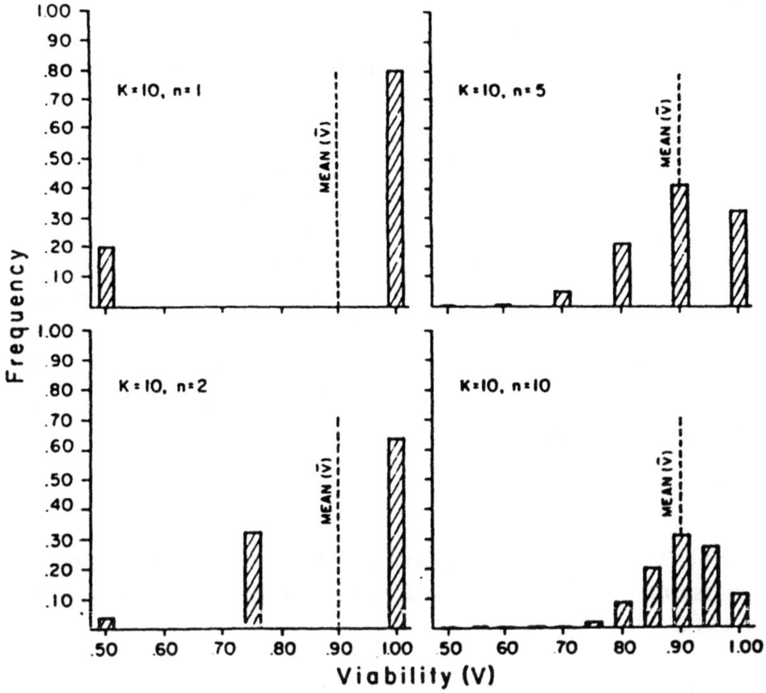

FIGURE 2. The distribution of brood viability among colonies within populations with K = 10 sex alleles as a function of the number of times queens mate, n. V is the mean viability for each population. It is assumed that all sex alleles are equally frequent and all n males contribute equal amounts of equally viable sperm for all inseminations. From Page and Metcalf (1982) with permission. Copyright 1982 by The University of Chicago.

single drone mating, two classifications of colonies are possible: those with 50% diploid drones and those with none. As the number of matings increases, the variance in diploid male production within the breeding population becomes redistributed (FIGURE 2; Page and Marks, 1982; Page and Metcalf, 1982). This relationship corresponds to the discussion above on the distribution of the variance between and within colonies. With a large number of matings, assuming the queens and drones are selected at random from a large panmictic population, each colony will have very close to the same viability of brood. Relatively equal viability can be achieved by collecting drones, pooling and homogenizing the semen of a large number of them, then inseminating all queens with the same mixture (Laidlaw, 1981, Page and Laidlaw, 1982b). The methodology for this technique has been established (Kaftangolu and Peng, 1980; Moritz, 1983).

The method of semen pooling and homogenization increases within family variance, which aids within family selection but decreases between family selection (Falconer, 1981 pp. 214-215). For breeding programs designed to select for colony performance characteristics, it results in selection for general combining ability of queen genotypes. It also establishes with higher certainty the genetic relationships among workers within colonies which is important for estimating heritabilities of traits (Oldroyd and Moran, 1983). It further results in higher overall brood viability in a closed breeding population by reducing the rate of loss of X alleles due to genetic drift by increasing the effective population size, N_e (Page and Laidlaw, 1982a; Moran, 1984). However, the effects of the increased number of mates on N_e are small above about 10 (Page and Laidlaw, 1982a). The use of this technique should be determined by the specific objectives of the program.

MATING DESIGNS

Useful systems of breeding honey bees employ three distinct mating designs: assortative, disassortative, and random mating. Each of these methods can be useful with different objectives.

Assortative Mating

Assortative mating can be achieved in two ways: (1) mating phenotypes that have similar characteristics, e.g. colonies that are good producers, and (2) mating genotypes that are similar, e.g. relatives (see discussion on heritability below for explanations of genotype and phenotype). Assortative mating of the first type is practiced for most breeding programs where deliberate inbreeding is not an objective.

Breeding of the second type is deliberate inbreeding. The purpose for deliberate inbreeding is to increase the probability of homozygosity at specific gene loci that are involved in determining specific characteristics. Inbred individuals are then crossed with other, unrelated inbred individuals, and in some crosses desirable heterosis will occur. Unfortunately, as inbreeding proceeds, homozygosity can occur at all loci of the genotype, including those that are not of particular interest to the breeder, and can result in severe inbreeding depression, as with the X locus. The proportion of initially heterozygous gene loci remaining heterozygous in any given generation, t, can be estimated by

$$H_t = H_0 (1 - F_t) \qquad 2.$$

where H_t is the proportion of heterozygous loci, H_0 is the proportion of loci heterozygous prior to inbreeding, usually assumed to be 0.50 (Crow and Kimura, 1970, p. 66). The coefficient of inbreeding, F, is the probability that both alleles at any gene locus are identical by descent from a common ancestor (see Crow and Kimura, 1970, p. 72) and can be determined by

$$F_X = \Sigma(1/2)^n(1 + F_A) \qquad 3.$$

where F_X is the inbreeding coefficient of individual X, and F_A is the inbreeding coefficient of the common ancestor in any given gamete path. Laidlaw and Page, (1986) describe a useful technique for determining inbreeding coefficients for individuals from established pedigrees. They also provide recursive formulae for determining the inbreeding coefficients for individuals when regular systems of inbreeding are used (see TABLES 1 and 2).

The number of different inbreeding designs are limited only by the imagination of the breeder. However, many would prove to be slow in terms of generation time, or difficult to conduct in terms of manipulation of colonies for drone and queen production. Laidlaw and Page (1986) discuss eight different systems of inbreeding, some of which have additional variations. Using the genetic pairing terminology introduced above, the inbreeding systems are as follows:

1...Self fertilization (the physical pairing of mother and son).
2...Parent-offspring mating (either mother-daughter or father-daughter).
3...Super-sister-super-sister mating (physical pairing of aunt and nephew).
4...Full-sister-full-sister mating (again physical pairing of aunt and nephew, hence the physical pairing terminology fails to distinguish between 3 and 4).
5...Half-sister-half-sister mating (these could be either paternal or maternal half sisters).
6...Cousin mating.
7...Aunt-niece mating.
8...Grandmother-granddaughter mating.

These different systems result in different rates of inbreeding for two reasons: (1) the level of consanguinity between parents is different, and (2) the generation times are different. Polhemus and Park (1951) pointed out the importance of the generation time for how fast inbreeding can take place within lines using different systems of mating. Mother-daughter mating is the most efficient system and has been used most extensively.

TABLE 1. Recursion Equations for Inbreeding Coefficients of Generation t(F_t) using Different Regular, Closed Systems of Inbreeding[a]

Mating	Equation
Selfing	$F_t = 1/2\ (1 + F_{t-1})$
Gametic backcross	$F_t = 1/2\ (1 + F_{t-1})$
Maternal mother-daughter	$F_t = 1/4\ (1 + 2F_{t-1} + F_{t-2})$
Paternal mother-daughter	$F_t = 1/2\ (1/2 + F_{t-1})$
Supersister-supersister	$F_t = 1/8\ (3 + 4F_{t-1} + F_{t-2})$
Full sister-full sister	$F_t = 1/4\ (1 + 2F_{t-1} + F_{t-2})$
Half sisters (drone mothers super sibs)	$F_t = 1/32\ (7 + 16F_{t-1} + 8F_{t-2} + F_{t-3})$
Half sisters (drone mothers full sibs)	$F_t = 1/16\ (3 + 8F_{t-1} + 4F_{t-2} + F_{t-3})$
Half sister\ (drone mothers half sibs)	$F_t = 1/8\ (1 + 6F_{t-1} + F_{t-2})$
Aunt-niece (single drone)	$F_t = 1/16\ (3 + 8F_{t-1} + 4F_{t-2} + F_{t-3})$
Aunt-niece (multiple drones)	$F_t = 1/8\ (1 + 4F_{t-1} + 2F_{t-2} + F_{t-3})$
Grandmother-granddaughter	$F_t = 1/8\ (2 + 2F_{t-2} + 3F_{t-3} + F_{t-4})$

[a] From Laidlaw and Page (1986), by permission. Copyright 1986 by Academic Press.

Disassortative Mating

Disassortative mating schemes are usually of two types for two different purposes: circular breeding schemes and interpopulational crosses.

Circular-breeding schemes have been employed when selection for a given trait is conducted on a small, genetically-closed population over several generations. The purpose of this type of design is to minimize inbreeding (Roberts, 1974; Rothenbuhler *et al.*, 1979; Hellmich *et al.*, 1985).

Interpopulational crosses are conducted to take advantage of heterosis. Different populations are likely to be homozygous (have only a single allele) for different genes. When individuals from different populations are crossed, heterosis can occur as a consequence of an increase in heterozygosity. Heterosis occurs as a consequence of an increase in proportion of loci that show dominance and is the opposite effect of inbreeding (Falconer, 1981, pp. 230-238). Inter-

TABLE 2. Inbreeding coefficients for generations of regular, closed systems of inbreeding. [a, b]

	Generation																	
	1	2	3	4	5	6	7	8	9	10	15	20	25	30	35	40	45	50
Selfing	.500	.750	.875	.948	.969	.984	.992	.996	.998	.999	1.00	1.00	1.00	1.00	1.00	1.00	1.00	1.00
Gametic backcross	.500	.750	.875	.938	.969	.984	.992	.996	.998	.999	1.00	1.00	1.00	1.00	1.00	1.00	1.00	1.00
Maternal mother-daughter	.250	.375	.500	.594	.672	.734	.785	.826	.859	.886	.961	.986	.995	.998	.999	1.00	1.00	1.00
Paternal mother-daughter	.250	.375	.438	.469	.484	.492	.496	.498	.499	.500	.500	.500	.500	.500	.500	.500	.500	.500
Super sister-super sister	.375	.562	.703	.797	.861	.905	.935	.956	.970	.979	1.00	1.00	1.00	1.00	1.00	1.00	1.00	1.00
Full sister-full sister	.250	.375	.500	.594	.672	.734	.785	.826	.859	.886	.961	.986	.995	.998	.999	1.00	1.00	1.00
Half sisters (drone mothers super sibs)	.219	.328	.438	.526	.602	.665	.718	.763	.800	.832	.929	.970	.987	.995	.998	.999	1.00	1.00
Half sisters (full sibs)	.188	.281	.375	.457	.527	.589	.642	.689	.729	.764	.883	.942	.971	.985	.993	.996	.998	.999
Half sister (drone mothers half sisters)	.125	.219	.305	.381	.449	.509	.563	.661	.654	.692	.827	.903	.946	.970	.983	.991	.995	.997
Aunt-niece (single drone)	.188	.281	.375	.475	.527	.589	.642	.689	.729	.764	.883	.942	.971	.985	.993	.996	.998	.999
Aunt-niece (multiple drones)	.125	.188	.250	.312	.367	.418	.465	.508	.547	.584	.726	.820	.882	.922	.948	.966	.978	.985
Grandmother-granddaughter	.125	.250	.281	.359	.430	.476	.527	.575	.614	.651	.789	.873	.923	.953	.972	.983	.990	.994

[a]Values presented were obtained from recursion equations given in Table 1.
[b]From Laidlaw and Page (1986) by permission. Copyright 1986 by Academic Press, Inc.

racial crosses have been used frequently, especially in Europe (see Ruttner, 1975, p. 36; cf. Oldroyd *et al.*, 1985).

Another type of interpopulational cross was suggested for closed population breeding programs (Page and Laidlaw, 1982b). Breeding populations that are maintained in isolation either by instrumental insemination or by geographic isolation can be divided into smaller subpopulations. Selection and maintenance take place within each breeding population, independently. Matings of drones and queens used for commercial colonies take place only between individuals belonging to different populations. The purpose of this proposed scheme is to reduce the amount of resources necessary to maintain breeding populations by instrumental insemination and, at the same time, to take advantage of interpopulational variance in X allele frequencies.

Random Mating

Open populations. A breeding program where the queens and drones are both taken randomly for mating is not a breeding program at all. However, it is possible to select only one sex and then allow random mating with unselected members of the other. This kind of open population breeding makes the most sense for honey bee breeding if the virgin queens are raised from selected queen mothers. However, a system could be designed where the drone mothers are selected and the queen mothers chosen at random. Such a system may be useful for changing the genetic composition of a feral population.

If queens are selected as breeders on the basis of predetermined individual or colony characteristics, then their daughter queens, or the performance of their colonies, can be expected to show a selection response. However, continual increase in desired characters is unlikely unless the unselected population that supplies the drones for the matings is sufficiently small to undergo significant gene frequency change as a consequence of migration of selected genomes from the breeding population. Most likely this would occur through drone production by the selected colonies.

Closed populations. Closed populations are free from the uncontrolled introduction of genetic material because both the queen and drone mothers are selected from within the population each generation. These populations must be maintained by either instrumental insemination or by complete geographical isolation. Closed population breeding programs have been responsible for most of the commercial successes in plant and animal breeding.

Honey bee breeders have been severely restricted in their use of closed populations for breeding because of a lack of understanding of the population genetics of sex determination. However, recent work using mathematical and computer simulation models has resulted in useful guidelines for this potentially powerful method of breeding (Kubasek, 1980; Page and Laidlaw, 1982a, 1982b;

Page and Marks, 1982; Page *et al.*, 1983, 1985; Moran, 1984; Moritz, 1984; Page and Laidlaw, 1985; Pamilo, 1986).

The following discussion is based primarily on the results of computer simulations. The assumptions for these simulations were uniform: A value of 85% (this equals 15% diploid drones, the expected value when there are six or seven X alleles, each at equal frequency in the population) was chosen as a minimum acceptable brood viability. Simulated programs were initiated by randomly selecting N (N = 10, 15, 25, 35, or 50) number of queens that had mated at random with ten different drones each from a very large base population that contained twelve (a likely number) sex alleles at equal frequency. Each generation, a very large number of virgin queens and drones were raised from each breeder queen in the closed population. Equal numbers of drones from each queen were pooled from which ten different drones were randomly selected from the drone pool as mates for each virgin queen. A new set of N breeder queens was then selected to be parents for the next generation according to one of two designs: random selection, or queen supersedure.

Random selection of queens. With this design, breeder queens were selected at random each generation from the very large pool of inseminated daughter queens. Selection was random with respect to the sex locus; this system does not disallow selection on the basis of other characters. Results of simulation studies show that 50 breeder queens are needed to guarantee that 95% of all such closed populations will maintain brood viability at or above 85% for 20 generations.

Queen supersedure. With this design, each breeder queen was replaced by one of her own superior-performing, mated daughters each generation. This selection scheme increases the effective breeding population (Page and Laidlaw, 1982b; Page *et al.*, 1983; Moran, 1984; Pamilo, 1986). About 35 queens yielded the same result as 50 queens with random selection of queens.

It is expected that selection will proceed more slowly with queen supersedure. However, the gains in brood viability and cost of maintaining fewer breeding colonies more than offset the decreased rate of selective improvement.

Top crossing. Top crossing is a method commonly employed by plant and animal breeders to rapidly increase the frequency of desirable traits in a breeding population. This method can also be applied to honey bees by selecting a superior performing queen, from either within or outside of the closed population, to be the drone mother for up to 100% of all daughter queen inseminations (Page *et al.*, 1985). Drones from the designated top-cross parent can be used in one of two ways: (1) Up to 100% of all daughter queens of breeders can be inseminated with the semen of drones only from the top-cross parent, and (2) Drones from the top-cross parent can be pooled with the other drones from the population in a proportion up to 100%. Drones can then be selected at random from this pool of drones for inseminations.

Top crossing results in a disruption of the frequencies of different X alleles, leads to a lower effective population size and, consequently, causes the more rapid loss of sex alleles and brood viability within the population. Computer simulations show the following: (1) Pooling drones results in slower loss of sex alleles. (2) Less than 75% of the drones should come from the top-cross parent. (3) Top crossing should not be performed on a closed population with less than 50 breeder queens. (4) Any given population should not be top crossed more than 3 times (assuming a breeding program life expectancy of 20 generations). (5) Top crossing should not occur in consecutive generations. (6) Queen supersedure designs result in slower rates of loss of brood viability. (7) Selecting the top-cross parent from outside the closed population is preferred (however, the population is then no longer technically closed).

Occasionally, it may become necessary to introduce new X alleles into a population in order to improve brood viability. This should be done in a way that minimizes the introduction of undesirable genes. Laidlaw and Page (1986) made specific recommendations on how introduction should be done. Novel sex alleles should be sought only in superior stock. Queens can be tested for novel sex alleles by inseminating them with pooled, homogenized semen from drones of all of the breeder queens in the population. Queens having novel sex alleles will have brood viability close to 100%. As a simple rule, virgin queens and drones raised from the selected, tested queen should be introduced into the population at an approximate frequency of

$$2(1-V)/3-2V \qquad 4.$$

where V is the average brood viability of the closed population.

Kubasek (1980) showed by computer simulation that the loss of brood viability in closed populations is negligible when queens are selected as breeders on the basis of brood viability. He also demonstrated that selection for other traits, such as honey production does not accelerate the loss of alleles when brood viability is also used as a criterion. His simulations used different assumptions from those discussed above and are not directly comparable.

Severson *et al.* (1986) discuss the organization and execution of a closed population breeding program in Madison, Wisconsin. Moritz (1984) discusses selection in small, closed populations.

METHODS OF SELECTION

Selection of superior stock is essential for the success of any breeding program. Breeders must develop clear objectives and have sufficient methods for quantifying and evaluating desirable traits. Once these criteria are defined, the breeder must then determine which of many different systems of selection will be used. Although there are innumerable possible selection schemes, they can be divided into three broad categories: gametic selection, individual selection, and progeny selection. Inbred-hybrid breeding and selection for multiple traits are also discussed in this section.

Gametic Selection

Gametic selection is a method of selection that could be of some use in honey bee breeding. Drones are derived directly fiom the gametes (eggs) of the queens and replicate the genome that they contain. Therefore, selected drones represent selected gametes. This system could result in very powerful, rapid selection for traits of drones and for traits of workers and queens that are highly correlated with drones, because there is no recombination in the males and, therefore, no variance (except for mutations) associated with gamete production.

Individual Selection

With individual selection, parents for the next generation are selected purely on their own performance or characteristics. With honey bees, queens can be selected on the basis of a specific trait, like size of thorax, or on the basis of colony performance, like honey production. Selection on the basis of colony performance is actually a type of progeny selection since the colony performance of most traits is dependent on the performance of worker progeny. However, because the worker progeny are not normally reproductive progeny, we will consider this to be a form of individual selection and consider progeny selection to be that which is based on the performance of daughter colonies.

Mass selection is a method of individual selection that has been very successful in plant and animal breeding programs. With mass selection, a proportion of superior queens is selected as parents from a population each generation. These selected queens then produce virgin queens and drones for the next generation. Matings can be assortative, disassortative, or random. Individual selected queens may contribute both virgin queens and drones, or virgin-queen and drone mothers may be maintained separately.

Each generation, a greater number of queens are raised, mated, and tested than the number selected as parents. The amount of selection response is dependent on the selection differential, S (the difference between the average of the population of potential parents for a trait and the average of the selected parents as measured in standard deviation units), and the heritability (discussed below) of the trait.

In very large closed populations, mass selection should result in a shift of gene frequencies for those genes under selection and result in a change in the mean value of the character. There should also be a reduction in the variance among individuals that are selected as parents, a logical result of the selection process. However, a reduction in the variance of offspring (daughter colonies) is not expected (see Allard, 1960).

In small closed populations, a change in gene frequencies and means for selected characters are also expected; however, an increase in uniformity of offspring is expected due to the fixation of some alleles at selected loci. Random fixation of alleles due to genetic drift will also take place. In very small populations the effects of genetic drift can be greater than those of artificial selection, in which case selection will not change the value of the selected character. Falconer (1981, p. 74) recommends that breeding populations have an effective size, Ne, of at least 1/4s, where s is the coefficient of selection at a given gene locus.

Breeding programs have been implemented in the states of So Paulo and Paraná, Brazil, that are based on a variation of the mass selection scheme presented above. Several different apiaries were established in different areas (environments). Queens from an equal number of superior performing colonies from each apiary location were selected to be queen and drone mothers for the next generation. These were then collected into one breeding population, crosses were made and new mated, daughter queens were installed in all colonies in the different apiaries and the process was repeated.

Progeny Selection

Progeny selection is frequently practiced in two forms: line breeding and recurrent selection. Again, there are as many different ways to organize programs as there are breeders. Selection is based on the performance of the progeny of a queen (her queen daughters or their colonies).

Line breeding. With line breeding, selection is on the basis of the performance of a given line, usually derived from a single queen. A number of daughter queens are raised from each of these lines and mated to drones from the same or different lines. These daughter colonies are then tested and breeder queens for the next generation are selected on the basis of the performance of the entire line.

Recurrent selection. Recurrent selection is a cyclical breeding system with three phases:

1. The development of progeny lines.
2. The evaluation of progenies in repeated trials.
3. Recombination of superior progenies to constitute the source population for the next cycle.

Cale and Rothenbuhler (1975) suggest a system of recurrent selection for honey bees. For additional discussion on recurrent selection consult Allard (1960).

Inbred-Hybrid Breeding

Inbred lines can be developed with any number of mating designs discussed above. Inbreeding can be rapid, using mating designs with high consanguinity and short generation times, or it can be done more slowly and coupled with selection to improve the lines for the desired characters.

Once lines are developed and sufficiently inbred so there is high uniformity among individuals within lines and between generations, they are crossed (combined) and tested. Two types of combining abilities are possible: general and specific. Lines that show good results in combination with many other lines demonstrate good general combining ability. Lines that are superior only in combination with other specific lines show specific combining ability.

The advantages of inbred-hybrid breeding are: (1) stocks are very uniform among individuals and between generations, (2) different desirable characteristics of different lines can sometimes be combined into one superior performing hybrid, and (3) hybrid bees often demonstrate heterosis (Cale and Gowen, 1956; Oldroyd *et al.*, 1985; Kulincevic, 1986).

The disadvantages of inbred-hybrid breeding are: (1) lines loose sex alleles and soon have 50% or less viable brood, (2) inbreeding depression on characters other than sex determination further reduce the viability of lines, and (3) lines are expensive to develop and maintain and are easily lost.

Selection for Multiple Traits

Sometimes a breeder wants to select for more than one trait (see Rinderer, 1986 for a more thorough discussion). There are three ways this can be accomplished: (1) selection of traits can be done in tandem, one after the other, (2) selection can be concurrent giving equal weight to each trait (independent, simultaneous selection), and (3) selection can be done concurrently using an index that is weighted for each trait.

With tandem selection a breeder selects for one trait at a time. Once that trait has improved sufficiently within the population, a second trait is selected, then a third, and so on. The major difficulty with this method of selection is that once artificial selection is relaxed on one trait, natural selection will move the mean of the selected trait back to its normal population value while others are being selected. This is particularly valid for traits that are under the influence of many variable loci.

TABLE 3. The relationship between the number of characters selected simultaneously and the number of colonies needed from which to select 100 breeder colonies that represent the top 20% of all colonies for all characters[a].

Number of characters	Number of colonies each generation
1	500
2	2,500
3	12,500
4	62,500
5	312,500
6	1,562,500

[a]From Rinderer, 1986

Tandem selection may be useful when traits are under the control of a single locus or a few loci. These traits may be rapidly fixed in the population and will not change while selection proceeds for other traits.

Independent, simultaneous selection is doomed to failure if very many traits are being selected because of the enormous amount of work required to maintain and test the necessary number of colonies (see TABLE 3). With this method, a breeder attempts to improve multiple characters at the same time by only selecting as parents those individuals that meet a specified minimum criterion for each trait. An enormous number of colonies are needed for this kind of selection even when predetermined selection differentials are small.

Selection index breeding is the best way to select for multiple traits (Falconer, 1981, pp. 294-300, Rinderer, 1986, pp. 317-319). A single selection index is assigned to potential parents on the basis of predetermined criteria. Each index is a function of the breeding value of the individual for each trait, the heritability of each trait, the relative economic value of each trait, and the genetic correlations (the degree to which the same genes affect different traits) between traits. Increasing positive genetic correlations improve concurrent selection.

PREDICTING RESPONSE TO SELECTION

The basis of plant and animal breeding lies in the principle that individual organisms vary with respect to appearance and performance. Every individual is unique. However, relatives tend to be more similar to each other than to nonrelatives. Breeders take advantage of this similarity by selecting those individuals that have the characteristics they desire and use them as parents with the expectation that their offspring will resemble them more in those characteristics than do individuals drawn at random from the population.

The source of variation among individuals can be genetic or environmental. The basis for similarity among related individuals can likewise be genetic or environmental. Genetic similarities are expected because relatives share common genes as a consequence of common descent. Environmental similarities are expected because relatives often share the same environment.

Breeders are primarily interested in the genetic differences and similarities. However, the genotype of an individual cannot, under most practical circumstances, be observed directly. Instead it is the phenotype of the individual that is observed, measured, and evaluated. The phenotype (P) is the sum of the genetic (G) and the environmental (E) effects (see Falconer, 1981):

$$P = G + E \qquad 5.$$

The genetic portion can be further divided into those components that are determined by the simple additive effects of genes (A), both within the same gene locus and between loci, the effects due to dominance of certain alleles at gene loci (D), and the interactions of different gene loci (I), called epistasis. The model becomes

$$P = A + D + I + E \qquad 6.$$

From this basic model, others can be developed that allow specific predictions about the probable success of breeding for specific traits. These models depend upon quantifying the observed phenotypic variance (V_P) in terms of the above components:

$$V_P = V_A + V_D + V_I + V_E \qquad 7.$$

To predict the response of a breeding population to selection, a measure is needed of the degree of genetic influence on the observed variability. There are two commonly used measures: (1) the degree of genetic determination, and (2) the heritability (h^2).

The degree of genetic determination considers all sources of genetic contribution and is defined as

$$V_G/V_P \qquad 8.$$

The measure of genetic determination often has a large amount of undetermined bias associated with it as an estimator and is, therefore, of little value in breeding (Falconer, 1981, p. 113).

Heritability is defined as

$$h^2 = V_A/V_P \qquad 9.$$

and is the proportion of the phenotypic variance that is a consequence of additive genetic effects. Only the additive component is a predictably useful estimator of selection response. Estimates of heritability can be used to predict selection response by

$$R = h^2 S, \qquad 10.$$

where R is the response to selection, the number of standard deviation units difference between the unselected population of the previous generation and the progeny of the selected parent, and S is the selection differential.

Heritabilities are usually estimated by comparing the phenotypes of relatives. They have specific properties and come with some precise assumptions of critical importance in honey bee breeding that must be met in order to be useful in Equation 10. Four assumptions are listed below:

1. Estimations of heritability can have a large, unknown amount of bias if comparisons are made between certain types of relatives. Full-sister and super-sister comparisons lead to an overestimation of heritability because the variability that is attributed to dominance is confounded with the variance attributed to the additive effects.

2. Estimates of heritability pertain to individuals, not groups. The underlying selection theory assumes that individual phenotypes are measured, or, with specific modifications, family averages are measured. If families are measured, then the genetic relationships of family members must also be known and the numbers of individuals per subfamily within a family known or at least equal. Individual performance of family members must be additive.

Often the traits that bee breeders want to improve are not expressed in individuals but only in groups, like honey production, pollen hoarding, or defensive behavior. The total numbers of individuals in a family (colony in this case), within subfamilies, and particularly the number belonging to the proper age or physiological state associated with the specific trait measured, are often unknown.

3. There should not be a correlation between genotypes and the environment. If a correlation exists, it can lead to an overestimation of heritability. The assumption is especially violated if genotypes influence the environment in a way that affects other genotypes (Falconer, 1981, p. 122). For example, bees that are genetically more predisposed to respond defensively to stimuli also alter the stimulus environment by releasing alarm pheromone. Bees of different genotypes, perhaps of different half-sister families, then respond to

the now higher level of stimulus. Or bees that are genotypically more predisposed to forage for pollen may reduce the colony stimulus for pollen collecting hence reducing pollen collecting behavior of members of some subfamilies. Calderone (1988) demonstrated a significant genotype x environment effect associated with different propensities among individuals belonging to different colony subgroups to forage for pollen.

4. There should not be an interaction between the genotypes and the environments in which they are measured. It is assumed that all environments have the same effect on all genotypes. However, cross-fostering experiments of Winston and Katz (1982) have shown that Africanized and European honey bees do not show uniform responses to different colony environments. In fact there are considerable genotype x environment interactions involved in foraging behavior. Oldroyd *et al.* (1985) showed a substantial genotype x environment interaction effect for colony weight gain. Violation of this assumption is usually not serious because the genotype x environment interaction variance is confounded with the environmental variance term.

Two final properties of heritability estimates need to be mentioned: (1) Heritabilities are gene frequency dependent. This means that two different populations, that are likely to have different gene frequencies, will yield two different estimates of heritability in the same environment. (2) Heritabilities are environment dependent. That is, the same population will likely yield two different estimates of heritability if measured in two different environments.

Based on the above discussion, it is apparent that heritability estimates of honey bees are difficult to obtain. Heritability estimates of group or social traits are not directly comparable with those obtained on individuals. Moritz (1986) discussed additional difficulties and proposed a method to estimate the degree of genetic determination for some social traits. A number of statistical designs have been employed to estimate components of variance used in estimating heritability of individual and group traits. For more information consult Rinderer (1977), Vencovsky and Kerr (1982), Oldroyd and Moran (1983) Collins *et al.* (1984), Milne and Friars (1984), Hellmich *et al.* (1985), Moritz and Hillesheim (1985), Oldroyd *et al.* (1985), Moritz (1986).

SELECTION IN BRAZIL

In this section, we discuss the effects of natural and artificial selection on commercial and feral honey bee populations in Brazil. Thirty years of Africanization and intensive bee breeding in some areas of Brazil have yielded obvious changes in characteristics of both feral and commercial populations.

Many of the changes have taken place through deliberate or accidental selection by commercial and hobby beekeepers. It is the nature of beekeepers to experiment with their bees and breed for traits that they believe are superior. Most population changes take place without empirical verification by scientists;

therefore, in this section some examples are given without scientific references, and instead reflect the experiences of one author (W. E. Kerr) during his 30 years of scientific investigation and public service work with Africanized honey bees and his personal communications with Dr. Lionel Gonçalves.

Defensive Behavior

In 1963-1964 an increase in the defensive behavior of honey bees was noted by beekeepers and other residents of Brazil. The number of serious stinging episodes had increased and honey production had declined because about 80% of the hobby beekeepers and about 10% of the professional beekeepers quit working with bees. As a consequence, in 1965 the Department of Genetics, University of São Paulo at Ribeirão Preto, began a breeding program in collaboration with the beekeepers in the states of São Paulo, Minas Gerais, Paraná, and Santa Catarina. The objective of this program was to increase honey production and decrease the defensive behavior of Africanized bees.

Initially, queens of excessively defensive colonies were replaced with Italian queens that were inseminated with Italian drones. This initial attempt failed because many of the best producers were also very defensive; therefore, selection resulted in lower honey production. The program was changed so that instead of eliminating queens on the basis of the defensive behavior of their colony, they were eliminated on the basis of productivity. Queens from the lowest 25% of colonies were replaced with an Italian queen, either virgin or inseminated with Italian drones. A drone comb was placed in each of these colonies. This program practiced mass selection on the queens and random mating with feral drones. The use of the drone comb resulted in the production of large numbers of drones with desirable genotypes that were available to mate with commercial and feral queens. Changes occurred in the feral population that allowed selective improvement of both the feral and commercial bees. A total of 20,000 virgin queens, daughters of Dadant Starline queen mothers, were distributed between 1963 and 1970 (Kerr, 1985). Lobo *et al.* (1989) recently demonstrated the admixture of genes of African and European descent in southern Brazil that likely occurred as a consequence of this program.

In order for these programs to continue, a source of gentle, productive stock is necessary. A closed population breeding program was implemented in October 1984 on Fernando de Noronha Island, 354 km off the coast of Brazil. This island did not have any honey bees prior to the introduction of 40 colonies of Italian bees used for this program (Kerr and Cabeda, 1985, Kerr, 1986; Malagodi *et al.*, 1986). This population provides inexpensive queens for Brazilian beekeepers.

Within Brazil today, bees in the northeast are becoming less defensive while bees in some areas of the State of São Paulo are becoming more defensive (Espenser Soares, pers. comm.). This difference in defensive behavior may be the result of an increase in agriculture, and artificial selection for more defensiveness in São Paulo. Some beekeepers in São Paulo were unhappy with colonies of gentle bees because gentle colonies were stolen readily, and they began selecting for more defensive behavior.

Swarming Behavior

Reproductive swarming is a behavioral response of successful colonies. The first reproductive swarm contains the old queen and approximately half of the workers. Additional reproductive swarms, or afterswarms, may be produced which contain a new queen and usually a smaller number of workers than the primary swarm (Winston, 1980). In Brazil, swarms of European bees contain a single queen; however, about 5.7% of swarms of Africanized bees contain two to four queens, all inseminated (Kerr *et al.*, 1970).

After swarming, a colony typically alights and clusters on a structure such as the branch of a tree and sends out scouts to find a new nest-site location. European bees usually select new nest sites within about 100-400 m of the old nest (Seeley, 1985, p. 73); however, Africanized bees may travel more than 40 km. If a colony does not find a suitable nest site, or fails to accept one of several optional sites (Lindauer, 1961, pp. 41-46) it usually initiates a nest at the location where it clustered. In Brazil, selection of exposed nest sites usually results in the death of European, but not Africanized colonies. Africanized bees begin construction of exposed combs and gradually build an envelope of wax and propolis around them, thereby providing protection. The behavior of building envelopes around exposed combs is increasing in the feral Africanized bee population of Brazil. This characteristic could be a consequence of inadvertent selection by beekeepers for a greater tendency to swarm. In some areas, beekeepers collect swarms as a method of increasing their numbers of colonies or the numbers of bees in their hives.

Production of Royal Jelly

In some areas of Brazil, "selection motivated" beekeepers have selected for bee colonies that produce large amounts of royal jelly. A test was conducted that compared five colonies that were selected for high honey production with five colonies that were selected for high royal jelly production by the Pascon Apiaries in Rio Claro. Workers from high royal jelly production colonies maintained fully developed hypopharyngeal glands five days longer than the workers from high honey production colonies.

Honey Production

Three different methods of mass selection for increased honey production have been employed throughout Brazil. Each system has been quantitatively modelled on the basis of results from selection programs and specific predictions of selection responses have been made for each of these models (Kerr and Vencovsky, 1982; Vencovsky and Kerr, 1982). The predictions assume that all beekeepers within a given region perform the same method of selection and breeding and that about 50,000 colonies are involved. The mass selection techniques and predictions are as follows:

1. Replace the queens in the 25% lowest honey producing colonies with queens raised from larvae of the 25% best colonies. A 20% improvement in honey production is expected each generation.
2. Place a single drone comb in the 25% highest producers. It is assumed that these drones will contribute about 50% of all drones in an area that mate with the new queens. An increase in honey production of 10% per generation is expected.
3. For beekeepers with larger numbers of colonies, select queens from the 25% best colonies from among all apiaries, as queen mothers. Each apiary should contain about 50 colonies. The 25% best colonies in each apiary each receive a drone comb. Virgin queens raised from the queen-mother colonies are then allowed to naturally mate at the individual apiary locations. This program should be applied only when there are more than 200 colonies available for selection. An increase of 30% honey production per generation is predicted.

Resistance to *Varroa jacobsoni*

Varroa jacobsoni was introduced into São Paulo in 1972 (De Jong and Gonçalves, 1981; De Jong *et al.* 1984). By 1980-1982 honey production in the State of São Paulo and in Northeastern Brazil had diminished 40-60%. Today, production is back to previous levels and researchers in Riberão Preto are trying to determine the mechanism of the apparent reduction in susceptibility of Africanized bees to *Varroa*. Evidently, natural selection, rather than artificial selection, was the cause.

Size of Brood Cells

Following Africanization, more than 50% of managed colonies rejected commercial wax foundation that was based on the larger cell sizes of Italian bees. Beekeepers in the state of São Paulo were organized into a program designed to select for an increase in the size of brood cells in combs constructed by Africanized honey bees. The Department of Genetics at the University of São

Paulo, Ribeirão Preto, recommended that beekeepers either kill the queens in colonies producing small-sized cells and allow the colonies to requeen themselves, or to substitute the Africanized queens of poor quality with Italians. Within three years, four or five generations, cell size had increased throughout the commercial population of São Paulo and was no longer a problem.

Nest Site Location

A survey conducted in December 1964, in Ribeirão Preto showed that nearly every termite hill within one kilometer of the bee laboratory (about 46 total) contained a colony of Africanized honey bees. Five years later another survey was conducted in a 40 km area around Ribeirão Preto. This survey failed to find a single colony of bees in a termite mound. Instead, colonies usually occupied holes in tree trunks and cavities in rocks, usually more than three meters above the ground.

Armadillos became very successful predators of ground-nesting colonies. Within about six generations, however, natural selection had eliminated ground nesting behavior from this bee population. The armadillos were the likely selective agent.

CONCLUSION

The process of Africanization of South and Central America has taken place during the lifetime of most of the bee researchers active today. It has been a process of great biological interest because of the potential demonstration of fundamental biological principles, many of which have proven intractable in laboratory studies and do not have a modern record in natural history. Field studies of community and behavioral ecology, population and quantitative genetics, and evolution of behavior have been possible, though underexploited.

The future holds even more promise for studies of basic biology of this natural phenomenon once it occurs in the United States populations. However, members of the bee industry in the United States do not hold Africanization in the same awe as biologists. Instead, those in the industry see it as a threat to their own livelihood. Genetics and breeding stand as the only means by which the bee industry can survive in its present state if Africanization of North America has similar effects to those in South and Central America. A strong queen-rearing industry will be essential and can only survive by combining good beekeeping with a knowledge of bee genetics. Hence, the joining of basic research in bee genetics and applied breeding is a clear necessity and, hopefully, will be achieved.

Acknowledgements

We thank the Tinker Foundation for a Foreign Research Travel Grant to R. Page that enabled us to work together. We also thank the Banco do Brasil, S. A. through its Fundo de Incentivo a Pesquisa Técnico Científica (FIPEC), the Conselho Nacional do Desenvolvimento Científico e Tecnológico (CNPq) for its grant to W. E. Kerr, and the Ohio State University Office of Research and Graduate Studies for funding various components of this work.

LITERATURE CITED

Adams, J., Rothman, E. D., Kerr, W. E., Paulino, Z. L. 1977. Estimation of the number of sex alleles and queen matings from diploid male frequencies in a population of *Apis mellifera*. *Genetics* 86:583-596.

Allard, R. W. 1960. *Principles of Plant Breeding*. New York: John Wiley and Sons.

Bishop, G. H. 1920. Fertilization in the honey-bee II. Disposal of the sexual fluids in the organs of the female. *J. Exp. Zool.* 31:267-286.

Calderone, N. W. 1988. *The genetic basis for the evolution of the organization of work in colonies of the honey bee, Apis mellifera (Hymenoptera: Apidae)*. PhD Dissertation. The Ohio State Univ., Columbus.

Cale, G. H., Gowen, J. W. 1956. Heterosis in the honey bee (*Apis mellifera* L.). *Genetics* 41:292-303.

Cale, G. H., Rothenbuhler, W. C. 1975. Genetics and breeding of the honey bee. In *The Hive and the Honey Bee,* ed. Dadant & Sons, pp. 157-84. Hamilton, Ill: Dadant & Sons.

Collins, A. M. 1986. Quantitative genetics. In *Bee Genetics and Breeding,* ed. T. E. Rinderer, pp. 283-304. Orlando, Fla: Academic Press.

Collins, A. M., Rinderer, T. E., Harbo, J. R., Brown, M. A. 1984. Heritabilities and correlations for several characters in the honey bee. *J. Hered.* 75:135-140.

Crow, J. F., Kimura, M. 1970. *An Introduction to Population Genetics Theory*. Minneapolis: Burgess.

Crow, J. F., Roberts, W. C. 1950. Inbreeding and homozygosis in bees. *Genetics* 345:613-621.

Crozier, R. H. 1970. Coefficients of relationship and the identity of genes by descent in the Hymenoptera. *Am. Nat.* 104:216-217.

Crozier, R. H., Pamilo, P. 1980. Asymmetry in relatedness: who is related to whom? *Nature* 283:60.

De Jong, D., Gonçalves, L. S. 1981. The Varroa problem in Brazil. *Am. Bee J.* 121:186-189.

De Jong, D., Gonçalves, L. S., Morse, R. A. 1984. Dependence on climate of the virulence of *Varroa jacobsoni*. *Bee World* 65:117-121.

Falconer, D. S. 1981. *Introduction to Quantitative Genetics*. London: Longman Group Limited. 365 pp. 2nd ed.

Fyg, W. 1952. The process of natural mating in the honey bee. *Bee World* 33:129-139.

Gary, N. E. 1963. Observations of mating behavior in the honey bee. *J. Apic. Res.* 2:3-9.

Gessner, B., Ruttner, F. 1977. Transfer der Spermatozoen in die Spermatheka der Bienenkönigen. *Apidologie* 8:1-18.

Hartl, D. L. 1972. A fundamental theorem of natural selection for sex linkage or arrhenotoky. *Am. Nat.* 106:516-524.

Hellmich, R. L., Kulincevic, J. M., Rothenbuhler, W. C. 1985. Selection for high and low pollen-hoarding honey bees. *J. Hered.* 76:155-158.

Kaftangolu, O., Peng, Y.-S. 1980. A washing technique for collection of honey bee semen. *J. Apic. Res.* 19:205-211.

Kerr, W. E. 1969. Some aspects of the evolution of social bees. *Evol. Biol.* 3:119-175.

Kerr, W. E. 1985. Africanized honey bees in Brazil. *Honey bee Sci.* 6:105-12. (In Japanese; English summary and translation)

Kerr, W. E. 1986. Introducão de abelhas no territorio Fernando de Noronha. 3. 30 anos antes e 30 depos. In *7° Congr. Bras. Apicul. (Salvador)*, ed. B. Viana, Bahia, Brazil. In press.

Kerr, W. E., Barbieri, M. R., Bueno, D. 1970. Reproducão de abelha Africana, Italiana e seus hibridos. In *1° Congr. Bras. Apic. (Florianópolis, 1970)*, ed. H. Weise, pp. 130-35. Santa Catarina, Brazil.

Kerr, W. E., Cabeda, M. 1985. Introdução de abelhas no território Federal de Fernando de Noronha. *Ciência e Cultura* 37:467-471.

Kerr, W. E., Martinho, M. R., Gonçalves, L. S. 1980. Kinship selection in bees. *Rev. Bras. Genet.* 3:339-343.

Kerr, W. E., Vencovsky, R. 1982. Melhoramento genético em abelhas. I. Efeito do número de colonias sobre o melhoramento. *Rev. Bras. Genet.* 5:279-285.

Kerr, W. E., del Rio, S. de L., Barrionuevo, M. D. 1982. The southern limits of the distribution of the Africanized bee in South America. *Am. Bee J.* 122:196-198.

Kerr, W. E., Zucchi, R., Nakadaira, J. T., Butolo, J. E. 1972. Reproduction in the social bees (Hymenoptera: Apidae). *J. N. Y. Entomol. Soc.* 70:265-276.

Koeniger, G. 1984. Funktionsmorphologische Befunde beider Kepolation der Honigbiene (*Apis mellifera* L.). *Apidologie* 15:189-204.

Kubasek, K. J. 1980. *Selection for increased number of sex alleles in closed populations of the honey bee. An investigation via computer simulation.* MS thesis. Louisiana State Univ., Baton Rouge.

Kulincevic, J. M. 1986. Breeding accomplishments with honey bees. In *Bee Genetics and Breeding*, ed. T. E. Rinderer, pp. 391-413. Orlando, Fla: Academic Press.

Laidlaw, H. H. 1944. Artificial insemination of the queen bee (*Apis mellifera* L.): morphological basis and results. *J. Morph.* 74:429-465.

Laidlaw, H. H. 1974. Relationships of bees within a colony. *Apiacta* 9:49-52.

Laidlaw, H. H. 1981. Honey bee genetics and its application to pollinator breeding. *Honey bee Sci.* 2:1-4. (in Japanese; English summary)

Laidlaw, H. H., Eckert, J. E. 1950. *Queen Rearing*. Hamilton, Ill: Dadant & Sons.

Laidlaw, H. H., Page, R. E. 1984. Polyandry in honey bees (*Apis mellifera L.*): sperm utilization and intracolony genetic relationships. *Genetics* 108:985-997.

Laidlaw, H. H., Page, R. E. 1986. Mating designs. In *Bee Genetics and Breeding*, ed. T. E. Rinderer, pp. 323-344. Orlando, Fla: Academic Press.

Lindauer, M. 1961. *Communication Among the Social Bees.* Cambridge, Mass: Harvard University Press.

Lobo, J. A., Del Lama, M. A., Mestriner, M. A. 1989. Population differentiation and racial admixture in the Africanized honey bee (*Apis mellifera* L.). *Evolution* 43:794-802.

Mackensen, O. 1951. Viability and sex determination in the honey bee (*Apis mellifera* L.). *Genetics* 36:500-509.

Mackensen, O. 1955. Further studies on a lethal series in the honey bee. *J. Hered.* 46:72-74.

Mackensen, O., Roberts, W. C. 1948. A manual for the artificial insemination of queen bees. *US Dep. Agric. Bur. Entomol. Plant Q. ET-250.*

Malagodi, M., Kerr, W. E., Soares, A. E. 1986. Introdução de abelhas na ilha de Fernando de Noronha. 2. População de *Apis mellifera ligustica. Ciênc. Cult.* 38:1700-1704.

Milne, C. P., Friars, G. W. 1984. An estimate of the heritability of the honey bee pupal weight. *J. Hered.* 75:509-510.

Moran, C. 1984. Sex-linked effective population size in control populations, with particular reference to honey bees (*Apis mellifera* L.). *Theor. Appl. Genet.* 67:317-322.

Moritz, R. F. A. 1983. Homogeneous mixing of honey bee semen by centrifugation. *J. Apic. Res.* 22:249-255.

Moritz, R. F. A. 1984. Selection in small populations of the honey bee (*Apis mellifera* L.). *Z. tierz. Zuchtungsbiol.* 101:394-400.

Moritz, R. F. A. 1986. Estimating the genetic variance of group characters: social behavior of honey bees (*Apis mellifera* L.) *Theor. Appl. Genet.* 72:513-517.

Moritz, R. F. A., Hillesheim, E. 1985. Inheritance of dominance in honey bees (*Apis mellifera capensis* Esch). *Behav. Ecol. Sociobiol.* 17:87-89.

Oldroyd, B., Moran, C. 1983. Heritability of worker characters in the honey bee *(Apis mellifera). Aust. J. Biol. Sci.* 36:323-332.

Oldroyd, B., Moran, C., Nicholas, F. W. 1985. Diallel crosses of honey bees. 1. A genetic analysis of honey production using a fixed effects model. *J. Apic. Res.* 24:243-249.

Page, R. E. 1986. Sperm utilization in social insects. *Ann. Rev. Entomol.* 31:297-320.

Page, R. E., Laidlaw, H. H. 1982a. Closed population honey bee breeding. I. Population genetics of sex determination. *J. Apic. Res.* 21:30-37.

Page, R. E., Laidlaw, H. H. 1982b. Closed population honey bee breeding. II. Comparative methods of stock maintenance and selective breeding. *J. Apic. Res.* 21:38-44.

Page, R. E., Laidlaw, H. H. 1985. Closed population honey bee breeding. *Bee World* 66:63-72.

Page, R. E., Laidlaw, H. H. 1988. Full sisters and super sisters: a terminological paradigm. *Anim. Behav.* 36:944-945.

Page, R. E., Laidlaw, H. H., Erickson, E. H. 1983. Closed population honey bee breeding. 3. The distribution of sex alleles with gyne supersedure. *J. Apic. Res.* 22:184-190.

Page, R. E., Laidlaw, H. H., Erickson, E. H. 1985. Closed population honey bee breeding. 4. The distribution of sex alleles with top crossing. *J. Apic Res.* 24:38-42.

Page, R. E., Marks, R. W. 1982. The population genetics of sex determination in honey bees: random mating in closed populations. *Heredity* 48:263-270.

Page, R. E., Metcalf, R. A. 1982. Multiple mating, sperm utilization, and social evolution. *Am. Nat.* 119:263-281.

Pamilo, P. 1986. Effect of supersedure breeding programs on the sex alleles of the honey bee. *J. Apic. Res.* 25:44-49.

Pamilo, P., Crozier, R. H. 1981. Genic variation in male haploids under deterministic selection. *Genetics* 98:199-214.

Pamilo, P., Crozier, R. H. 1982. Measuring genetic relatedness in natural populations: methodology. *Theor. Popul. Biol.* 21:171-193.

Polhemus, M. S., Lush, J. L., Rothenbuhler, W. C. 1950. Mating systems in honey bees. *J. Hered.* 41:151-155.

Polhemus, M. S., Park, O. W. 1951. Time factors in mating systems for honey bees. *J. Econ. Entomol.* 44:639-642.

Rinderer, T. E. 1977. Measuring the heritability of characters of honey bees. *J. Apic. Res.* 16:95-98.

Rinderer, T. E. 1986. Selection. In *Bee Genetics and Breeding,* ed. T. E. Rinderer, pp. 305-321. Orlando, Fla: Academic Press.

Roberts, W. C. 1974. A standard stock of honey bees. *J. Apic. Res.* 13:113-120.

Rothenbuhler, W. C. 1957. Diploid male tissue as new evidence on sex determination in honey bees. *J. Hered.* 48:160-168.

Rothenbuhler, W. C., Kulincevic, J. M., Thompson, V. C. 1979. Successful selection of honey bees for fast and slow hoarding of sugar syrup in the laboratory. *J. Apic. Res.* 18:272-278.

Ruttner, F. 1956a. The mating of the honey bee. *Bee World* 37:2-15, 23-24.

Ruttner, F. 1956b. Zur Frage der Spermaübertragung bei der Bienenkönigin. *Insectes Soc.* 3:351-359.

Ruttner, F. 1975. Races of bees. In *The Hive and the Honey Bee,* ed. Dadant & Sons, pp. 19-38. Hamilton, Ill.

Ruttner, F., Koeniger, N. 1971. Die Füllung der Spermatheka Bienenkönigin. Activ Wanderung oder passiver Transport der Spermatozen? *Z. Vergl. Physiol.* 72:411-422.

Ruttner, F., Ruttner, H. 1966. Untersuchungen uber die Flugaktivität und das Paarungsverhalten der Drohnen. *Z. Bienenforsch.* 8:332-354.

Seeley, T. D. 1985. *Honey bee Ecology: A Study of Adaptation in Social Life.* Princeton, NJ: Princeton University Press.

Severson, D. W., Page, R. E., Erickson, E. H. 1986. Closed population breeding in honey bees: a report on its practical application. *Am. Bee J.* 126:93-94.

Sylvester, H. A. 1986. Biochemical genetics. In *Bee Genetics and Breeding,* ed. T. E. Rinderer, pp. 177-203. Orlando, Fla: Academic Press.

Taber, S. 1955. Sperm distribution in the spermathecae of multiple-mated queen honey bees. *J. Econ. Entomol.* 48:522-525.

Taber, S., Wendel J. 1958. Concerning the number of times queen bees mate. *J. Econ. Entomol.* 51:786-789.

Vencovsky, R., Kerr, W. E. 1982. Melhoramento genético em abelhas. 2. Teoria e avaliação de alguns metodos de seleção. *Rev. Bras. Genet.* 3:493-502.

Winston, M. L. 1980. Swarming, afterswarming, and reproductive rate of unmanaged colonies *(Apis mellifera). Insectes Soc.* 27:391-398.

Winston, M. L., Katz, S. J. 1982. Foraging differences between cross-fostered honey bee workers *(Apis mellifera)* of European and Africanized races. *Behav. Ecol. Sociobiol.* 10:125-129.

Woyke, J. 1960. Naturalne i sztuczne unasienianie mateck pszczelich. [Natural and artificial insemination of queen honey bees.] *Pszczel. Zesz. Nauk.* 4:183-275. Summarized in *Bee World* (1962) 43:21-25.

Woyke, J. 1962. The hatchability of "lethal" eggs in a two sex-allele fraternity of honey bees. *J. Apic. Res.* 1:6-13.

Woyke, J. 1963a. Drone larvae from fertilized eggs of the honey bee. *J. Apic Res.* 2:19-24.

Woyke, J. 1963b. Rearing and viability of diploid drone larvae. *J. Apic. Res.* 2:77-84.

Woyke, J. 1963c. What happens to diploid drone larvae in a honey bee colony. *J. Apic. Res.* 2:73-76.

Woyke, J. 1964. Causes of repeated mating flights by queen honey bees. *J. Apic. Res.* 3:17-23.
Woyke, J. 1980. Effects of sex allele homo-heterozygosity on honey bee colony populations and on their honey production. I. Favourable development conditions and unrestricted queens. *J. Apic. Res.* 19:51-63.
Woyke, J. 1981. Effects of sex allele homo-heterozygosity on honey bee colony populations and on their honey production. II. Unfavourable development conditions and unrestricted queens. *J. Apic. Res.* 20:148-155.
Woyke, J. 1983. Dynamics of entry of spermatozoa into the spermatheca of instrumentally inseminated queen honey bees. *J. Apic. Res.* 22:150-154.

9

CONTINUING COMMERCIAL QUEEN PRODUCTION AFTER THE ARRIVAL OF AFRICANIZED HONEY BEES

Richard L. Hellmich II[1]

The Africanized honey bee (AHB) problem would be reduced significantly if beekeepers could control the mating activities of their queens and drones. Unfortunately, this control is not easily achieved. Queens and drones mate while in flight, sometimes several kilometers from their respective colonies (Peer, 1957; Szabo, 1986). Instrumental insemination (II) is the only known method to obtain matings with the genetics strictly controlled. This method is not practical on a commercial scale, however, because inseminated queens are generally inferior to naturally mated queens (Harbo and Szabo, 1984; Harbo, 1985). The limitations of II and the threat of Africanization have stimulated considerable research on the control of natural matings.

Some of the problems queen producers might encounter with AHB are identified in this chapter. Producers in areas of Africanization will have to consider whether or not they will be able to produce acceptable queens at reasonable costs. This ultimately will depend on mating control, or more specifically, on relative populations of managed and feral drones. Also discussed in this chapter is the thorny problem of defining acceptable levels of Africanization in commercial stock.

Additionally, this chapter considers methods to evaluate mating control. A study conducted with two queen producers in Texas is discussed which used mark-and-recapture procedures to estimate contributions managed and feral drones made to the natural matings of queens. A Mating Control Index is provided as a possible aid for bee breeders. And finally, various methods for improving mating control are discussed.

---

[1] Dr. Hellmich is with the USDA--ARS, Honey Bee Breeding, Genetics and Physiology Research Laboratory, 1157 Ben Hur Rd., Baton Rouge, Louisiana, 70820, USA.

ESTABLISHING REALISTIC GOALS

Most honey bee queens in the United States produced for sale come from Southern States and California, regions that are susceptible to Africanization. The options that queen producers in these regions have regarding an introgression of AHB genes range from complete acceptance to zero tolerance.

Africanized bees, besides being potential stinging hazards to laborers and customers, are not good candidates for standard queen production. Mating nuclei (5 l) stocked with Africanized bees produce fewer mated queens and abscond more often than those stocked with European bees (Hellmich *et al.*, 1986). Except for the occasional beekeeper who keeps defensive bees as a way to discourage thieves, most beekeepers agree that complete acceptance of AHB is not reasonable.

On the other hand, if a zero-tolerance policy is adopted toward Africanization; it would be difficult to produce saleable queens in Africanized areas. Queens mate with multiple drones that have the possibility of coming from numerous feral colonies (Taber and Wendel, 1958; Adams *et al.*, 1977). Consequently, even when there are small numbers of Africanized drones, in practice 100% production of European queens mated only with European drones is unlikely or impossible.

When some Africanization is tolerated, however, queen production in Africanized areas becomes more realistic. Mating control in the 90-95% range should be possible by many queen producers (Hellmich, 1988; Taylor and Rowell, 1988). Higher levels of control are prohibitive because the number of managed drones required to influence matings increases logarithmically compared to a linear increase in control percentages (Hellmich *et al.*, 1988). As an example, suppose a queen producer could control 50% of the matings with 10 drone source colonies. In order to control 90%, 95% and 99% of the matings that producer would need 90, 190 and 990 drone source colonies, respectively (Hellmich *et al.*, 1988). Thus, at some point higher levels of mating control become too expensive.

DEFINING ACCEPTABLE LEVELS OF MISMATING

A defense behavior experiment in Venezuela (Collins, 1987) suggests that low levels of Africanization might be tolerable. Colony defense does not appear to be dominated by a few workers in a colony which are Africanized-European hybrids. In fact, colonies with 90% European and 10% hybrid workers are manageable and virtually indistinguishable in behavior from pure European colonies. This is important because it rejects the idea that a small number of highly defensive bees can drastically influence colony defensiveness. These results suggest that beekeepers should be able to tolerate colonies headed by queens that are "mismated" with 10% or less Africanized drones.

While mating control might average 90%, this does not mean that every queen mates with 10% Africanized drones. Due to chance, and the fact that queens mate with many drones, some of the queens mate with less than 10% and some with more than 10% Africanized drones. TABLE 1 shows how the contribution from Africanized drones can vary among queens from the same apiary. Mating control in this table ranges from 70-99%. When there is 90% control (assuming queens mate with ten drones each) about a third (34.9%) of the queens mate only with European drones; however, approximately a quarter (26.41%) of them mate 20% or more with Africanized drones. At the extreme, about one out of 10,000 (0.01%) queens mate with 60% Africanized drones. When there is 95% mating control a majority of the queens (59.9%) mate only with European drones, but nearly one out of eleven (8.71%) of them mate 20% or more with Africanized drones.

Should all queens be rejected because of a few? One must consider that even some pure European colonies are unacceptable. Beekeepers commonly requeen European colonies that are too defensive. Perhaps when there is good mating control, most colonies headed by queens mated in areas with AHB will be indistinguishable from pure European colonies.

Establishing economic thresholds for tolerable defensiveness will not be easy. There must be a balance between business realities and social responsibilities. Much more information needs to be gathered, though, before some of these decisions can be made. Presently, there is a need to establish whether colonies with hybrid percentages greater than 10% are acceptable, or if they can be detected early and requeened.

EVALUATING MATING CONTROL

Until recently, queen producers had no method for determining the level of mating control. However, measuring Drone Equivalents (DE) of mating areas might be able to resolve this problem (Hellmich and Waller, 1990). The Drone Equivalent measures the effective population of drones within range of a queen mating yard. The method resembles the mark-and-recapture procedures that are commonly used by biologists to assess the size of animal populations. In such studies, animals are captured, marked and then released. Some of these animals are recaptured later along with non-marked animals. The number of animals in the population (X) is estimated by solving the following equation:

$$\frac{\text{\% non-marked animals in sample}}{\text{\% marked animals in sample}} = \frac{\text{\# animals in population (X)}}{\text{\# marked animals released}}$$

TABLE 1. Probability of a European queen mating with a designated percentage of Africanized drones when mating control in an apiary ranges from 70 to 99%. These values are based on the sampling distribution of the binomial. For these calculations queens are assumed to mate randomly with 10 drones. For example, when 95% of the matings are controlled, about 59.9% of the queens will mate with no AHB drones and about 31.5% will mate with 10% AHB drones.

	Mating Control Percentages				
Designated Percentage of AHB Drones	70	80	90	95	99
100	~0				
90	0.01	~0			
80	0.1	0.			
70	0.9	0.1	~0		
60	3.7	0.6	0.01	~0	
50	10.3	2.6	0.2	0.01	
40	20.0	8.8	1.1	0.1	~0
30	26.7	20.1	5.7	1.1	0.01
20	23.4	30.2	19.4	7.5	0.4
10	12.1	26.8	38.7	31.5	9.1
0	2.8	10.7	34.9	59.9	90.4

The underlying assumption is that the sample represents the actual ratio of marked:non-marked animals in the population.

To measure DE, a single-gene recessive trait for light-brown cuticle called cordovan functions as the marker. Cordovan queens are mated from an apiary that has a known number of cordovan drones. (This corresponds to number of marked animals released.) "Recapture" occurs when a queen mates with a cordovan drone which is detected by evaluating the worker progeny of the queen. Cordovan workers result from cordovan queens mating with cordovan drones; wild-type workers result from cordovan queens mating with wild-type drones.

Because most queens mate with several drones, cordovan queens usually produce both cordovan and wild-type workers.

If 10,000 cordovan drones are established in a mating apiary and 10% of the queens' progeny, on average, are cordovan then an estimate of the wild-type (non-cordovan) drone population is derived from:

$$\frac{\%\ \text{wild-type progeny in sample}}{\%\ \text{cordovan progeny in sample}} = \frac{\#\ \text{drones in population (X)}}{\#\ \text{cordovan drones in apiary}}$$

$$\frac{90\%}{10\%} = \frac{X}{10{,}000}, \qquad X = 90{,}000 \text{ drones}$$

The progeny ratio of this example suggests that cordovan queens, on average, mated with nine wild-type drones for every one cordovan drone. The influence of non-cordovan drones is equivalent to 90,000 cordovan drones that are centrally located, or 90,000 Drone Equivalents (DE).

This DE determination, however, does not distinguish between feral and managed wild-type drones. In a commercial apiary this would mean that a producer still could not evaluate mating control because the feral baseline is unknown. An estimate of the feral drone population in terms of Drone Equivalents (DE_{feral}) would solve this problem. Thus two types of DE determinations are necessary:

1) $DE_{commercial}$ (DE_c) -- conducted in commercial apiaries;
2) DE_{feral} (DE_f) -- conducted in remote apiaries.

Remote apiaries are located 8-10 km from managed colonies in areas that have terrain and flora similar to those near the commercial apiaries. Only mating colonies with cordovan virgins and drone source colonies that produce only cordovan drones are present in a remote apiary.

The DE_{feral} establishes a baseline of feral drone influence. A measure of the influence of managed drones, again in terms of DE, is calculated from:

$$DE_{managed} = DE_{commercial} - DE_{feral}$$

The $DE_{managed}$ measures the influence of drones from the mating apiary and drones from any other apiary that could mate with queens. These are all managed drones. (Note that the $DE_{commercial}$ measures the influence of all managed drones plus that of feral drones.) The ratio of $DE_{managed}$ to DE_{feral} determines whether it is possible to attain acceptable matings in a particular apiary.[2]

The DE_{feral} at one test area of Texas in July 1989 was 7,800 ± 3,700 (SD). Thus, an apiary with a $DE_{commercial}$ of 90,000 in this part of Texas during July would have yielded about 91% mating control. Actual estimates from two commercial queen producers during July were 93% in mating apiaries that were surrounded by several other apiaries ($DE_{commercial}$ = 108,600 ± 21,900) and 83% in more isolated apiaries ($DE_{commercial}$ = 46,600 ± 9,800).

Most of the commercial queens in Texas are produced in April and May. During this time the two producers had more drone source colonies near the mating apiaries than they did in July and August. Mating control estimates in Texas during April and May consequently are higher, 96-98% in the clustered apiaries and 93-96% in the isolated apiaries. FIGURE 1 shows mating control estimates for various levels of $DE_{commercial}$ and DE_{feral}.

Some assumptions must be made in such studies. One is that queens do not preferentially mate with cordovan or wild-type drones, that is, there is no assortative mating (positive or negative). This assumption is supported by tests in Kansas (O. R. Taylor, pers. comm.). Another assumption is that there is a low frequency of cordovan genes, preferably none, in the feral and managed populations. This assumption was true during the Texas test. Additionally, one must realize that the DE is not a static measure. It will vary yearly, seasonally and even daily as populations of drones fluctuate.

The mating control experiments suggest that some queen producers should be able to control 90-95% of the matings without substantially changing their procedures. Such levels of mating control should be acceptable even in areas of Africanized bees. Another point is that genetic control is probably higher than the mating control estimates. This is expected because the genetics of feral colonies near apiaries to some degree will be influenced by swarms and drones from managed colonies.

MATING CONTROL INDEX

The DE method can be used to derive a Mating Control Index (MCI) that might be useful to queen producers who want an easier way to predict mating control. Information necessary to derive a MCI includes a map showing all the apiaries within a 10-km radius and an estimate of the number of drones present at each apiary. The contribution managed drones have on matings at a particular apiary in terms of Drone Equivalents is calculated from:

---

[2] mating control = DE_m/DE_c or $(DE_c - DE_f)/DE_c$.

$$DE_{managed} = fd_1A_1 + fd_2A_2 + \dots fd_iA_i$$

where f is a function based on a dilution curve,[3] d_i = distance from designated mating apiary "i", and A_i= number of drones in the apiary "i". Distance is obtained from the maps. An estimate of the number of drones in an apiary that are old enough to fly is attained by trapping drones from a random sample of colonies. Flying drones from randomly selected colonies can be trapped with a queen-excluder trap (see Hellmich and Waller, 1990).

The $DE_{managed}$ for a mating apiary that is located near four outlying apiaries is determined by calculating five partial DE's as follows:

Apiary	distance(km)	weighted f	# of drones	partial DE
x	0	1.00	20,000	20,000
1	0.5	0.62	30,000	18,600
2	1.0	0.32	10,000	3,200
3	2.0	0.05	20,000	1,000
4	3.0	0.003	30,000	90

Total $DE_{managed}$ = 44,890

In this hypothetical apiary if the DE_{feral} = 10,000, then estimated mating control would be about 82% ($DE_{commercial}$ = 54,890; see FIGURE 1). The MCI weighting factors can be adjusted to improve the accuracy of this measure as more details about mating biology (e.g. directional effects) are learned.

INCREASING MATING CONTROL

Natural matings of queens can be influenced by either manipulating populations of drones (managed and unmanaged), or by using methods that isolate queens with desirable drones. Beekeepers with hundreds of managed colonies might be able to overwhelm feral drones through sheer numbers. Beekeepers with fewer colonies, however, might have to consider more inventive methods, such as modifying flight times of queens and drones, or using isolated mating yards. These methods and others that have been used to increase mating control of honey bees are considered in this section.

The obvious way for queen producers to increase managed drones is to move more drone source colonies into an area. When this is not possible, increasing managed drones in many instances can be accomplished by putting more drone

---

3 For these calculations a normal distribution with a standard deviation of 1 km is used. Rowell and Taylor (1988) suggest that the standard deviation estimates range from 800-1500 meters.

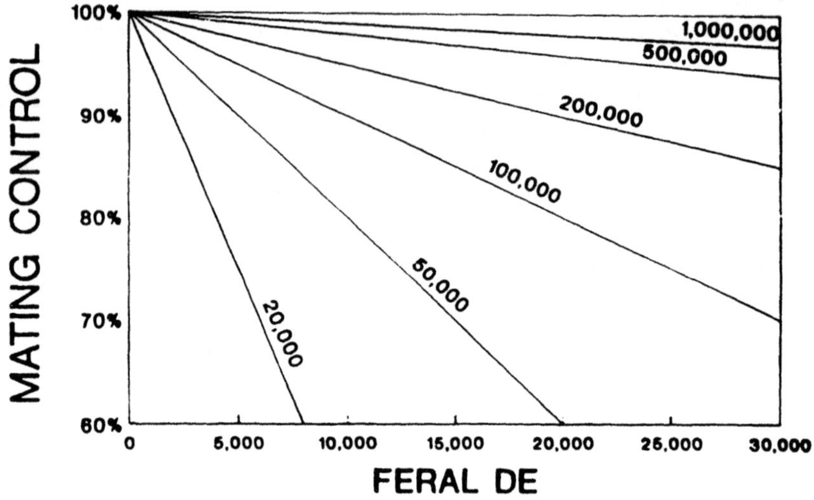

FIGURE 1. Estimated percentage of mating control when feral populations of drones (DE_{feral}) ranges from 0 to 30,000 and DE = 20,000, 50,000, 100,000, 200,000, 500,000 and 1,000,000.

comb in colonies. Queen producers commonly recommend a total area of drone comb equal one to three frames for each drone source colony. Clustering apiaries near mating yards also is an effective way to increase the proportion of managed drones in an area.

Decreasing feral drone populations for most queen producers probably is not a viable option because it takes considerable time to locate and destroy feral colonies. However, the baiting systems developed by Loper *et al.* (1987) and Williams *et al.* (1988) hypothetically could be used by some producers. The premise is that colonies for queen production could be established in an area after feral colonies and drones are reduced or eliminated with toxic baits. Limitations of such a system are being investigated.

Another method that queen producers might be able to use to increase mating efficiency is to establish drone source colonies at sites that will maximize control. Perhaps there is an optimal distance from a mating apiary that should be considered when locating drone source colonies. The problem with such a strategy, however, is that there are no clear recommendations. Many beekeepers and researchers believe that queens avoid inbreeding by not mating with drones near their colony. Perhaps queens prevent mating by flying low or by not opening the vaginal chamber. Until more is known, probably the best recommendation to queen producers is to try to move as many colonies as possible to within one or two kilometers of the mating apiary; or make no changes, if they have a system that is working.

Other methods of controlling natural matings involve geographic isolation, temporal isolation or a combination of the two. Geographic isolation has been achieved with varying degrees of success by locating mating apiaries on plains, deserts, islands, or high in the mountains (Laidlaw, 1979). Ferracane (1987) demonstrated that controlled natural matings are possible in the Adirondack Forest of New York. Undoubtedly, there are other locations in North America where isolated matings are possible. Some beekeepers are considering northern production of queens, whereas others either have established or are considering operations on islands.

Temporal isolation has been achieved by managing queens and drones to fly after normal mating flights have ceased. This has been accomplished by allowing colonies to fly after they had been confined in a cool basement (Dathe, 1867) or simply excluded (Bilash, 1955). One problem with this method, at least in the tropics, is that Africanized drones fly later than European drones (Hellmich *et al.*, 1990). Hellmich (1987) tried to modify flight activities and get European drones to fly earlier when fewer Africanized drones were flying by preventing them from flying for one or two days. European drones flew 20-30 minutes earlier than normal which was not considered sufficient to justify colony manipulations.

Waller *et al.* (1989) used both space and time to achieve isolated matings. They obtained pure matings during winter months at a desert location in southern Arizona. They moved drone-producing colonies from Tucson to a desert location where feral colonies had not yet started to produce drones. They obtained 100% mating control from December to February. Many versions of this system probably could be developed by innovative queen producers.

SUMMARY

Africanized honey bees probably will change the nature of queen production in the United States. The severity of this change will depend on whether some Africanization can be tolerated. If Africanization is absolutely not allowed, then no naturally mated queens will be produced for sale in most places that presently produce them. Such a position could weaken the bee industry and lead to an impaired defense against AHB. However, if low levels of Africanization are allowed, then many of the existing queen producers should be able to survive. Two queen producers in Texas and probably many others should be able to control 90-95% of the matings of their queens. Such levels of mating control will produce a few European queens mismated with 20-30% "unmanaged" drones, but the majority of them will be mated to 10% or fewer unmanaged drones. Queen producers, like all beekeepers, will need to be alert to identify acceptable and unacceptable colonies of bees, and will need to control their stock with frequent requeening. All these practices, including some carefully selected material from AHB, could result in improved honey bee stock.

Acknowledgements

In cooperation with Louisiana Agricultural Experiment Station. Mention of a proprietary product does not constitute an endorsement by the USDA.

LITERATURE CITED

Adams, J., Rothman, E. D., Kerr, W. E., Paulino, Z. L. 1977. Estimation of the number of sex alleles and queen matings from diploid male frequencies in a population of *Apis mellifera. Genetics* 86:583-596.

Bilash, G. D. 1955. An experiment in mating virgin queens and drones released for flight only at certain times. (Translated title.) Pchelovodstovo 32:23-29. Cited by Harbo, J. 1971. Annotated bibliography on attempts at mating honey bees in confinement. Bibliography No. 12. Int. Bee Res. Assoc.

Collins, A. M. 1987. Comparison of colony defense by European, hybrid (E x A), and mixed honey bee colonies. *Am. Bee J.* 127:842

Dathe, G. 1867. Anleitung zum Italisiren oder Züchtung der Italienischen Biene in Kästen und Körben. Nienburg, H. Bosendahl. Cited by Harbo, J. 1971. Annotated bibliography on attempts at mating honey bees in confinement. Bibliography No. 12. Int. Bee Res. Assoc.

Ferracane, M. S. 1987. A study of the feasibility of obtaining controlled natural-matings of honey bee queens in the Adirondack Forest of New York State *Am. Bee J.* 127:845.

Harbo, J. R. 1985. Instrumental insemination of queen bees -- 1985 (Part 2). *Am. Bee J.* 125:282-287.

Harbo, J. R., Szabo, T. I. 1984. A comparison of instrumentally inseminated and naturally mated queens. *J. Apic. Res.* 23: 31-36.

Hellmich, R. L. 1987. Flight-time differences between Africanized and European drones: implications for controlling matings. *Am. Bee J.* 127:846.

Hellmich, R. L. 1988. Influencing matings of honey bee queens with selected drones in Africanized areas. In *Africanized Honey Bees and Bee Mites,* ed. G. R. Needham, R. E. Page, M. Delfinado-Baker, C. E. Bowman, pp. 204-208. Chichester, England: Ellis Horwood Limited.

Hellmich, R. L., Collins, A. M., Danka, R. G., Rinderer, T. E. 1988. Influencing matings of European honey bees in areas with Africanized honey bees (Hymenoptera: Apidae). *J. Econ. Entomol.* 81:796-799.

Hellmich, R. L., Danka, R. G., Rinderer, T. E., Collins, A. M. 1986. Comparison of Africanized and European queen-mating colonies in Venezuela. *Apidologie* 17:217-226.

Hellmich, R. L., Rinderer, T. E., Danka, R. G., Collins, A. M., Boykin, D. L. 1990. Flight times of Africanized and European honey bee drones (Hymenoptera: Apidae). *J. Econ. Entomol.* In press.

Hellmich, R. L., Waller, G. D. 1990. Preparing for Africanized honey bees: evaluating control in mating apiaries. *Am. Bee J.* 130:537-542.

Laidlaw, H. H., Jr. 1979. *Contemporary Queen Rearing*. Hamilton, Ill: Dadant & Sons.

Loper, G. M., Williams, J. L., Taylor, O. R. 1987. Insecticide treatment of feral honey bee colonies by treating drones at DCA's. *Am. Bee J.* 127:847.

Peer, D. F. 1957. Further studies on the mating range of the honey bee, *Apis mellifera*. *Can. Entomol.* 89:108-110.

Rowell, G. A., Taylor, O. R. 1988. Some computer simulations using the neutral mating model for the honey bee, *Apis mellifera*. In *Africanized Honey Bees and Bee Mites,* ed. G. R. Needham, R. E. Page, M. Delfinado-Baker, C. E. Bowman, pp. 184-192. Chichester, England: Ellis Horwood Limited.

Szabo, T. I. 1986. Mating distances of the honey bee in North-Western Alberta Canada. *J. Apic. Res.* 25:227-233.

Taber, S., Wendel, J. 1958. Concerning the number of times queen bees mate. *J. Econ. Entomol.* 51:786-789.

Taylor, O. R., Jr., Rowell, G. A. 1988. Drone abundance, queen flight distance, and the neutral mating model for the honey bee, *Apis mellifera*. In *Africanized Honey Bees and Bee Mites,* ed. G. R. Needham, R. E. Page, M. Delfinado-Baker, C. E. Bowman, pp. 173-183. Chichester, England: Ellis Horwood Limited.

Waller, G. D., Hoopingarner, R. A., Martin, J. H., Loper, G. M., Meliton Fierro, M. M. 1989. Controlled natural matings of honey bee queens in Southern Arizona. *Am. Bee J.* 129:187-190.

Williams, J. L., Danka, R. G., Rinderer, T. E. 1988. Selective bait system for potential abatement of feral Africanized honey bees. *Am. Bee J.* 128:811.

Hajdu, [illegible] 1970. Probability [illegible] Advanced [illegible]

[illegible] control [illegible]

[illegible] [illegible]

[illegible]

PART THREE

Population Biology, Ecology, and Diseases

10

THE INSIDE STORY: INTERNAL COLONY DYNAMICS OF AFRICANIZED BEES

Mark L. Winston[1]

For someone with prior experience with temperate-evolved, European races of the honey bee *Apis mellifera*, the first impression of the tropical-evolved Africanized bee is one of frenetic activity. In comparison, bee races from the northern hemisphere are "temperate" in almost all respects, from their relatively docile defensive behavior to their calm, quiet demeanor in the hive. These first superficial impressions, when examined with the rigor of scientific research, rapidly deepen into profound differences in development, nesting biology, and behavior between temperate- and tropical-evolved races of honey bees. The key to the success of Africanized bees in South America lies largely within the hive, in their internal colony functions which are exquisitely crafted for success in tropical habitats. In this review I will examine the colony dynamics of Africanized bees by providing comparisons with their better-known cousins from temperate climates, and by this approach I hope to reveal the full range of remarkable adaptations which distinguish temperate and tropical honey bees (for other reviews, see Seeley, 1985; Winston, 1987). For clarity, "temperate-evolved" refers to the European races, and "tropical-evolved" includes Africanized and African bees.

THE DEVELOPMENT AND LIFE SPAN OF INDIVIDUAL BEES

The earliest portion of an Africanized worker's life already shows the more rapid pace which characterizes the life of tropical honey bees; their eggs hatch more quickly than those of European races. While egg hatching times can be enormously variable (reviewed by Jay, 1963), tropical-evolved bees have, on

---

[1]Professor Winston is in the Department of Biology, Simon Fraser University, Vancouver, British Columbia, V5A 1S6, Canada.

average, shorter time spans between egg laying and hatching, 70-71 hrs as compared to eggs of European-evolved bees which hatch in average times of 72-76 hrs (DuPraw, 1961; Tribe and Fletcher, 1977; Harbo *et al.*, 1981). The duration of the larval stages is also shorter for Africanized bees, averaging 4.2 days as unsealed larvae compared to 5.5 days for European bees (Jay, 1963; Tribe and Fletcher, 1977). The total development time from egg laying to emergence is, on average, 18.5 days for African-evolved workers and 21 days for European-evolved workers; queen development times are also shorter for African bees, 15 days vs 16 days for European bees (Jay, 1963; Kerr *et al.*, 1972; Tribe and Fletcher, 1977; Fletcher, 1978; Winston, 1979a). Drones develop in 24 days for both racial groups (Smith, 1960). Interestingly, hybrid workers between European and African races show intermediate development times, about 20 days from egg to adult, emphasizing the importance of a heritable factor in determining the duration of the brood stages (Garofalo, 1977).

Under normal colony conditions, brood mortality for both races is low, and colonies are able to rear most of the eggs laid by queens to the adult stage. One study which examined worker brood survival in colonies of Italian bees (*A. m. ligustica*) in the summer showed 85% of the eggs laid survived to the adult stage (Fukuda and Sakagami, 1968), and survival rates between 90 and 97% from egg to adult have been found for both European and Africanized races (Winston *et al.*, 1981). However, colonies founded by swarms show some dramatic differences in brood survival, with European bees maintaining their high survival rates but Africanized bees showing 32% brood mortality, mostly in the egg and young larval stages (Winston, 1979a). The reasons for this high brood mortality are not clear, but its impact is somewhat diminished by higher initial egg laying rates by Africanized queens. Nevertheless, new colonies of European bees are able to increase their worker populations more quickly than those of Africanized bees, at least for the first one or two brood cycles (Winston *et al.*, 1981).

Another major difference between temperate- and tropical-evolved honey bees is the shorter life span typical of tropical worker bees. Summer survival times for workers of European descent average 15-38 days in numerous studies (reviewed by Ribbands, 1953; Fukuda and Sekiguchi, 1966; Winston *et al.*, 1983), but only 12-18 days for Africanized bees during the equivalent South American dry season (Winston, 1979a; Winston *et al.*, 1981). Winter survival times also show dramatic differences; the average longevity of African bees during a Polish winter was only 90 days, as opposed to the 140 day average typical for European bees (Fukuda and Sekiguchi, 1966; Woyke, 1973). Longevity differences are also apparent during the wet season in South America, when Africanized and European workers show mean life spans of 22.7 and 26.3 days respectively (Winston and Katz, 1981).

A fourth major developmental difference between Africanized and European bees is in worker weight. Africanized bees are smaller than European bees, with average weights for unengorged and newly emerged workers of 62 and 93 mg respectively (Otis, 1982; Winston *et al.*, 1983 and references cited therein). These body size differences reflect size differences between cells built by workers of temperate- and tropical-evolved bees: 5.2-5.7 mm in European bees (width between opposite cell sides, means generally from 5.25-5.4 mm, Lee and Winston, 1985a and references cited therein); 4.8-4.9 mm in tropical African bees (Smith, 1960, 1961), and 4.6-5.1 mm for Africanized bees in South America (mean 4.8 mm, Rinderer *et al.*, 1982; Spivak *et al.*, 1988; and unpubl. obs.).

NEST ARCHITECTURE

The cell size differences cited above are only a small part of the differences in nest architecture between temperate- and tropical-evolved honey bee races. Feral nests of bees from these habitats differ in at least three other major attributes: nest size, amount of honey stored, and the building of nests outside of cavities. Feral European colonies are considerably larger than African or Africanized colonies, typically occupying cavities 45 l in volume (range 15-630), with comb areas of 23,400 cm^2 (range 13,000- 33,600, reviewed by Seeley and Morse, 1976 and Seeley, 1977). Africanized colonies have an average volume of 22 l (range 8.5-43.6) and comb area of 8,000 cm^2 in Peru (range 1,690-14,500) and 11,300 cm^2 in Venezuela (range 2,780- 21,190, reviewed by Winston *et al.*, 1983). In terms of Langstroth beekeeping equipment, this is the equivalent of one full brood box for European bees and less than half a box for Africanized bees. Almost no European colonies are found in cavities less than 22 l in volume, the average size for Africanized nests. Some African and Africanized colonies are extraordinarily small; we have found colonies, which had swarmed successfully, occupying the equivalent of 1.5 standard Langstroth deep frames.

There are also differences in honey storage between feral colonies of temperate and tropical *A. mellifera*. Unmanaged colonies of Africanized bees in 22 l hives store on average only one third as much honey as unmanaged European colonies in 40 l hives, 920 vs 2,810 cm^2 comb area (Winston *et al.*, 1981); this may be due in part to their smaller colony size. These observations agree with what is known about tropical African bees, where colonies often are small and store little honey (Fletcher 1978; Schneider and Blyther, 1988).

The frequency of exposed nests also differs between temperate and tropical bees. Such nests are uncommon with European bees (Byers, 1959; Avitabile, 1975; Sakai *et al.*, 1976; Seeley and Morse, 1976; Avitabile *et al.*, 1978), whereas many feral tropical African and Africanized bee colonies are found in the open, particularly in drier habitats (Smith, 1960; Fletcher, 1977, 1978). Such colonies may be suspended from tree limbs, rocks, or bunches of small branches, or under man-made structures such as overhanging eaves, open sewer manholes, or old tires. Four such nests were found on tree branches along one 500 m section of riparian forest in Venezuela, and 20% of all the nests measured in Venezuela were built outside of cavities (Winston *et al.*, 1983).

THE DYNAMICS OF DEMOGRAPHY

While the above review has highlighted specific attributes of development times, mortality schedules, worker size, and nesting biology in temperate and tropical honey bees, the most interesting level of colony dynamics involves the impact of these characteristics on colony demography, specifically colony growth rates, age structure, and caste ontogeny. When temperate and tropical bees are compared, some major differences in their patterns of colony demography are apparent which result in profound differences in colony dynamics.

The most readily apparent difference between temperate- and tropical-evolved bee races is in colony growth patterns. Newly hived swarms of European bees initially increase in population more rapidly than do Africanized bees, due to better brood survival and longer adult longevity. After about 45 days, European colonies have on average over 20,000 workers, as opposed to only 8,400 for Africanized colonies. However, the Africanized bees then experience an explosive period of population growth due to higher rates of brood rearing, and by 65 days after hiving are only about 4,000 workers lower in population than the European bees, 20,000 adults as compared to about 24,000 in European colonies. The high rate of brood rearing which Africanized bees are capable of is particularly impressive, with colonies often containing more than twice the brood of European colonies, although their comb area is generally less than half that of the temperate-evolved races (Winston *et al.*, 1981).

How can the physically smaller Africanized bee nests be used to rear so much more brood than their larger European relatives? Part of the answer lies in different patterns of comb utilization between the races. Africanized bees use over 80% of their comb throughout most of their colony cycle, while the larger European colonies generally use less than 65% of the available cells. Even at swarming, when colonies are at their largest, European bees are using only 72% of the available comb, while Africanized bees occupy 86% of their comb. The shorter development time of Africanized bees also allows for faster recycling of cells and thus increased brood rearing. Also, there is proportionally more comb area available for brood rearing in Africanized colonies since they store so much less honey than do temperate-evolved bees.

Another aspect of colony dynamics which differs between the races is worker age structure. The age distribution of workers in European colonies tends to be both more balanced and to show a higher mean age than that of Africanized bees (Winston, 1979a; Winston *et al.*, 1981). For example, the mean ages of workers in Africanized and European colonies at swarming are 10.0 and 15.7 days old respectively. The major consequence of this difference in age structure may be the higher brood mortality characteristic of newly founded Africanized bee colonies, and therefore their slower initial growth rate. The age distribution of adult workers in European colonies may be more favorable for brood rearing, since tasks performed by individual workers are largely dependent on age. Generally, young workers must feed larvae pollen and some nectar gathered by older workers in order for brood and emerging adults to survive (Haydak, 1937; Maurizio, 1950; Free, 1965). Differences in age-specific functions plus the high adult worker mortality rate in Africanized colonies may result in fewer nurse bees to feed brood, and thus poor survivorship.

We have also investigated age structure in swarms, with results that suggest one mechanism that Africanized bees may use to reduce their already high initial brood mortality in newly founded colonies, and to compensate for relatively short worker life spans (Winston and Otis, 1978). Swarms of European bees tend to have more young individuals (Butler, 1940; Meyer, 1956), but swarms of Africanized bees are even more heavily skewed towards young workers. Over 80% and up to 100% of 3-8 day old Africanized workers in colonies will issue with swarms. The advantage of this high, age-specific probability of young workers leaving in swarms is that it is mostly bees less than 10 days old which survive until the first emergence of new adult bees following colony founding; higher percentages of young workers in swarms increases the colony population at this critical point in the colony's survival.

Another aspect of colony demography which shows dramatic differences between temperate and tropical honey bee races is temporal division of labor. In honey bees, workers tend to perform in-hive tasks at younger ages, and then switch to external tasks such as foraging at older ages. The age at which European and Africanized bees make this important switch is lower for Africanized bees, which is consistent with the shorter life span which also characterizes tropical-evolved honey bees. The genetic basis of this difference in caste ontogeny was demonstrated in an experiment in which newly emerged workers of both races were marked and introduced into colonies of both races, and the ages the introduced workers initiated foraging were compared (Winston and Katz, 1982). The results were striking; in colonies of their own race, the tropical Africanized bees began to forage significantly earlier than did European bees in European colonies, 20 vs. 26 days. Even more impressive were the results from cross-fostered workers. Africanized bees in European colonies began foraging at older ages than they did in their own colonies, 23 days on average, but the European workers in Africanized colonies began foraging even earlier than the Africanized bees, at 14 days of age. Our hypothesis to explain these results is that the level of stimuli which induces foraging behavior is higher in Africanized colonies, and that Africanized workers have a higher threshold for these stimuli, which might include the age distribution of workers, worker longevity, colony size, amount of brood rearing, honey and pollen stores in the colony as well as availability of nectar and pollen in the field, general activity levels, and previous colony history. The full implications of this result for colony dynamics have yet to be explored, but an approach which examined complete ethograms of cross-fostered European and Africanized workers would undoubtedly yield much information concerning the role of ontogenetic caste differences in determining colony functioning.

DRONE PRODUCTION

Temperate- and tropical-evolved honey bee races differ in the percentage of drone comb constructed by colonies of similar sizes (Otis, 1980; Lee and Winston, 1985b). Africanized honey bees produce a larger percentage of drone comb at small colony sizes than temperate-evolved European races, and within the Africanized bee colonies studied, large colonies invest proportionately more energy in drone production in large than in small colonies. Both of these results may reflect different selective factors in temperate and tropical habitats (Fletcher, 1978; Winston *et al.*, 1983). In general, colony mortality is higher in the tropics, and the greater percentage of drone comb built by small tropical colonies may reflect an effort to produce some reproductives before the colony is lost to predation. In contrast, temperate-evolved honey bees suffer less predation but must have a large worker population and plentiful stores to survive the winter. Therefore, small colonies of temperate-evolved races do not initially allocate as many resources to drone comb production, but rather emphasize worker comb and worker brood rearing more than Africanized colonies.

SEASONALITY AND ABSCONDING

Seasonal patterns of brood rearing, swarming, absconding, nectar and pollen storage, and worker longevity are apparent for Africanized bee colonies in South America (Winston, 1980). Basically, during the wet season colonies show diminished brood rearing, slightly higher mean worker longevity, no swarming, a higher probability of absconding, and a shift in colony age structure towards older bees. However, these demographic responses to the dearth season are considerably less pronounced than those known for European bees under temperate and subtropical conditions during the winter, non-foraging period.

The most dramatic difference between temperate- and tropical-evolved bee races is in the frequency and pattern of absconding. In the tropics, colonies of African and Africanized bees abscond much more often than European-evolved colonies maintained in either temperate or tropical habitats. Absconding can be defined as the abandoning of a nest by a colony which forms a swarm and presumably re-establishes itself elsewhere. Absconding swarms differ from reproductive swarms in that few or no workers and no adult or viable immature queens are left behind in the original colony. The causes of absconding from feral nests fall roughly into two groups: disturbance-induced and seasonal or resource-induced absconding.

Disturbance-induced absconding is rare in temperate habitats, partly because there are few predators to induce such absconding but also due to temperate-evolved bees having a much lower tendency to abandon their nest. For tropical-evolved bees, absconding due to disturbance usually results from partial or total destruction of colonies by predators, destruction of comb by wax moths, fire in the proximity of the nest, heavy wasp or bird predation at the nest entrance, inability to regulate temperature due to cold or to excessive sunlight, and rain entering the nest (Fletcher, 1975-1976, 1978; Chandler, 1976; Woyke, 1976; Winston *et al.* 1979). In these cases, absconding generally occurs within hours or at most a few days following the disturbance.

Resource-induced absconding seems to result from a scarcity of nectar, pollen, and/or water, and occurs primarily during the dearth season in tropical habitats. Tropical African and Africanized colonies frequently abscond during these dearth periods; in French Guiana, 30% of the Africanized colonies abscond during the wet season, when there is relatively little flowering, presumably to search for areas with better resources (Winston *et al.*, 1979). Similarly, 79% of absconding swarms recorded in Brazil by Cosenza (1972) occurred during a dearth period. Absconding rates of tropical African bees are generally 15-30% a year, and can be as high as 100% in some areas. The highest incidence of African absconding occurs during the dry season, when there is less flowering, and also less water at a time when bees need it for regulating internal hive temperature (Smith, 1960; Fletcher, 1975-1976, 1978; Woyke, 1976). In contrast, resource-induced absconding by European bees in either temperate or tropical regions is infrequently recorded (Martin, 1963; Winston *et al.*, 1979; Robinson, 1982). Under wet season conditions in French Guiana, colonies of European bees maintained in the same manner as Africanized bees all dwindled and died instead of absconding, and interviews with African and South American beekeepers indicate that absconding among European bees is also infrequent under managed conditions (Winston *et al.*, 1979).

Resource-induced absconding differs from abscondings due to disturbance in the pattern of colony preparation prior to absconding. While disturbance can induce a colony to leave within hours or days, careful preparations are made by colonies for many weeks prior to abandoning the nest due to a resource shortage (Winston *et al.*, 1979). Colonies preparing to abscond begin reducing their brood rearing about 25 days before absconding, and rear no new larvae in the 10-15 days preceding absconding. The queen continues to lay a few eggs up to the time of absconding, although these are evidently consumed by the workers rather than being reared. Most of the stored pollen is also consumed prior to absconding, as is much of the stored honey. Interestingly, this type of seasonal absconding is timed so that the absconding swarm does not leave until the last sealed brood has emerged, so that absconding swarms have a reasonable population of young workers with which to initiate the new nest. Also, workers engorge with honey before absconding, so that the absconding colony only leaves their wax comb behind, taking all of its other resources with it in the form of newly emerged workers, honey, and protein from the recently consumed pollen. The protein is likely stored in the workers' fat bodies and hypopharyngeal glands.

Deteriorating resource conditions are not sufficient to explain absconding, however, since many colonies maintained at the same sites as absconding colonies persist throughout the dearth season. Resource-induced absconding is most likely induced by a combination of poor forage and internal colony conditions, particularly chronic low brood survival (Winston *et al.*, 1979; Winston, 1980). It is interesting that there are no differences in the amounts of honey, pollen, or brood areas between absconding and persisting colonies, but there are differences in the timing of swarming relative to the absconding season. Colonies which abscond are generally those which have swarmed within six weeks of the onset of a dearth period, which results in their having lower worker populations, more older workers, and higher brood mortality than colonies which swarmed more than six weeks before the absconding season.

Whatever the cause of absconding, colonies which abandon their nests may migrate long distances in search of better foraging conditions. Workers in absconding swarms carry almost twice the amount of honey as workers in reproductive swarms, providing fuel for longer dispersal distances (Otis *et al.*, 1981). The behavior of workers in absconding swarms also differs from reproductive swarms in that they frequently do not scout for new nest sites upon emergence, but travel cross-country for undetermined distances before scouting. In addition, absconding swarms will settle at interim sites and send out workers which search for floral resources, possibly scouting new areas for resource quality before searching for a nest site. Absconding swarms in Africa may travel as far as 160 km or more before constructing a new nest, migrating through areas of poor resources until they discover a better area with abundant but localized floral resources, such as *Eucalyptus* plantations (Nightingale, 1976; Fletcher, 1978).

EMERGENCY QUEEN REARING

Honey bee colonies are frequently able to replace their queens following sudden queen loss due to disease, predation, or other factors, and this replacement process has been described by Fletcher and Tribe (1977), Winston (1979b), Punnett and Winston (1983), and Fell and Morse (1984). However, there are some racial differences in responses to emergency queen situations between temperate- and tropical-evolved bees, most notably in the timing of new queen production and the frequency of brood movement and queen-loss swarming. For example, colonies of Africanized bees in South America tend to use older larvae to initiate queen rearing, which has the advantage of shortening the queenless period but the disadvantage of reduced queen viability resulting from the use of older larvae. The bees in Africa where the South American bees originated use larvae of ages similar to European races, suggesting that selection for a shorter queenless period for Africanized bees may have occurred in South America. The mean duration of queenlessness in African and Africanized colonies is only 23 days vs 29 days for European bees, due partly to the shorter queen rearing time for Africanized bees and partly to a shorter time between the establishment of the new colony queen and her initiation of egg laying. Also, Africanized workers move an average of 47% of brood reared as queens into empty cells, compared to about 4% for European bees. The advantage of this behavior is not clear, but it does result in more queen rearing at the periphery of colonies of Africanized bees during emergency queen rearing situations. Finally, a higher proportion of Africanized colonies swarm during emergency queen situations than European colonies, and usually produce two or three swarms as opposed to the one swarm more commonly issuing from European colonies. This higher frequency of swarming may be an effect of the higher tendency of tropically-evolved colonies to swarm under any conditions.

IMPLICATIONS FOR NORTH AMERICAN MANAGEMENT

In earlier articles, we have argued that the colony dynamics characteristics described here, as well as other attributes of Africanized bees such as their pronounced defensive behavior and frequent swarming, are all related traits which uniquely preadapted these bees for success in tropical South America (Winston *et al.*, 1983 and references cited therein). We also described two major selective pressures in tropical habitats, climate and heavy predation pressure, which are likely the major forces which have shaped the evolution of tropical bee societies. Unfortunately, the implications of these arguments for the pending arrival of Africanized bees in North America do not bode well for our beekeeping industry, as traits adapted for tropical habitats will have serious impact on bee management under temperate conditions.

The small colony size, diminished honey storage, and frequent swarming of Africanized bees will pose a major problem for beekeeping in more northerly environments, as colonies which move north on their own or are shipped north as packages will not grow to sufficient size to survive the temperate winter. Those few colonies which might grow to an overwintering size would likely be so aggressive as to be unmanageable. Adapting Africanized bees to the large cell size in European foundation will also make beekeeping with North American equipment more difficult.

Absconding will be another major problem. Many colonies will abscond prior to winter in response to the late summer and early fall dearth common in most parts of North America, which would also decrease overwintering success. In addition, Africanized bees abscond due to the slightest disturbance, and the migratory beekeeping common in our honey production and pollination industries today would not be possible with Africanized bees. These problems are serious ones, and will require innovation and resiliency on the part of the North American beekeeping community to overcome the problems caused by a tropical-evolved bee in a temperate habitat.

Acknowledgements

For my own research cited here, I am particularly grateful for the collaboration of G. W. Otis and O. R. Taylor. This research has been funded by the U.S. Department of Agriculture and National Science Foundation, Idaho State and Simon Fraser Universities, and the Natural Sciences and Engineering Research Council of Canada. Portions of this chapter have been taken from Winston (1987); the permission of Harvard University Press is gratefully acknowledged.

LITERATURE CITED

Avitable, A. 1975. Exposed combs of honey bees. *Am. Bee J.* 115:436-437.

Avitable, A., Sfafstrom, D., Donovan, K. J. 1978. Natural nest sites of honey bee colonies in trees in Connecticut USA. *J. Apic. Res.* 17:222-226.

Butler, C.G. 1940. The ages of the bees in a swarm. *Bee World* 21:9-10.

Byers, G.W. 1959. An unusual nest of the honey bee. *J. Kans. Entomol. Soc.* 32:46-48.

Chandler, M. T. 1976. The African honey bee *Apis mellifera adansonii*: the biological basis of its management. In *Apiculture in Tropical Climates*, ed. E. Crane, pp. 61-68. London: Int. Bee Res. Assoc.

Cosenza, G. W. 1972. Estudo dos enxames de migracão de abelhas africanas. In *1° Congr. Bras. Apicul. (Florianópolis)*, ed. H. Weise, pp. 128-129. Santa Catarina, Brazil.

DuPraw, E.J. 1961. A unique hatching process in the honey bee. *Trans. Am. Micros. Soc.* 80:185-191.

Fell, R. D., Morse, R. A. 1984. Emergency queen cell production in the honey bee colony. *Insectes Soc.* 31:221-227.

Fletcher, D. J. C. 1975-1976. New perspectives in the causes of absconding in the African bee (*Apis mellifera adansonii* L.). *S. Afr. Bee J. I & II.* 47:11-14; 48:6-9.

Fletcher, D. J. C. 1977. A preliminary analysis of rapid colony development in *Apis mellifera adansonii* L. *Proc. VIII Int. Congr. I.U.S.S.I.*, pp.144-45. Wageningen, Netherlands.

Fletcher, D. J. C. 1978. The Africanized honey bee, *Apis mellifera adansonii*, in Africa. *Ann. Rev. Entomol.* 23:151-171.

Fletcher, D. J. C, Tribe, G. D. 1977. Natural emergency queen rearing by the African bee *A. m. adansonii* and its relevance for successful queen protection by beekeepers. I and II. In *African bees: Taxonomy, Biology, and Economic Use,* ed. D. J. C. Fletcher, pp. 132-140, 161-168. Pretoria, South Africa: Apimondia.

Free, J.B. 1965. The allocation of duties among worker honey bees. *Symp. Zool. Soc. London* 14:39-59.

Fukuda, H., Sakagami, S. F. 1968. Worker brood survival in honey bees. *Res. Popul. Ecol.* 10:31-39.

Fukuda, H., Sekiguchi, K. 1966. Seasonal change of the honey bee worker longevity in Sapporo, North Japan, with notes on some factors affecting the life span. *Jap. J. Ecol.* 16:206-212.

Garofalo, C. A. 1977. Brood viability in normal colonies of *Apis mellifera*. *J. Apic. Res.* 16:3-13.

Harbo, J. R., Bolten, A. B., Rinderer, T. E., Collins, A. M. 1981. Development periods for eggs of Africanized and European honey bees. *J. Apic. Res.* 20:156-159.

Haydak, M. H. 1937. The influence of a pure carbohydrate diet on newly emerged honey bees. *Ann. Entomol. Soc. Am.* 30:258-262.

Jay, S. C. 1963. The development of honey bees in their cells. *J. Apic. Res.* 2:117-134.

Kerr, W. E., Gonçalves, L. S., Blotta, L. F., Maciel, H. B. 1972. Biología comparada entre as abelhas Italianas (*Apis mellifera liguistica*), Africana (*Apis mellifera adansonii*) e suas híbridas. In *1º Congr. Bras. Apicul. (Florianópolis)*, ed. H. Weise, pp. 151-185. Santa Catarina, Brazil.

Lee, P. C., Winston, M. L. 1985a. The influence of swarm population on brood production and emergent worker weight in newly founded honey bee colonies (*Apis mellifera*). *Insectes Soc.* 32:96-103.

Lee, P. C., Winston, M. L. 1985b. The effect of swarm size and date on comb construction in newly founded colonies of honey bees (*Apis mellifera* L.). *Can. J. Zool.* 63:524-527.

Martin, P. 1963. Die Steuerung der Volksteilung beim Schwarmen der Bienen. Zugleich ein Beitrag zum Problem der Wanderschwarme. *Insectes Soc.* 10:13-42. (I.B.R.A. translation E942).

Michener, C. D. 1974. *The Social Behavior of the Bees: A Comparative Study.* Cambridge, Mass: Harvard University Press.

Meyer, W. 1956. Arbeitsteilung im Bienenschwarm. *Insectes Soc.* 3:303-324.

Nightingale, J. A. 1976. Traditional beekeeping among Kenya tribes, and methods proposed for improvement and modernisation. In *Apiculture in Tropical Climates*, ed. E. E. Crane, pp. 15-22. London: Int. Bee Res. Assoc.

Otis, G. W. 1980. *The swarming biology and population dynamics of the Africanized honey bee.* PhD Dissertation. Univ. Kansas, Lawrence.

Otis, G. W. 1982. Population biology of the Africanized honey bee. In *Social Insects of the Tropics, Vol. I*, ed. P. Jaisson, pp. 209-219. Paris: Université Paris-Nord.

Otis, G. W., Winston, M. L., Taylor, O. R. 1981. Engorgement and dispersal of Africanized honey bee swarms. *J. Apic. Res.* 20:3-12.

Punnett, E. N., Winston, M. L. 1983. Events following queen removal in colonies of European-derived honey bee races. *Insectes Soc.* 30:376-383.

Ribbands, C. R. 1953. *The Behaviour and Social Life of Honey bees.* London: Bee Res. Assoc. Ltd. (Republished, 1964, New York: Dover Publ., Inc.).

Rinderer, T. E., Tucker, K. W., Collins, A. M. 1982. Nest cavity selection by swarms of European and Africanized honey bees. *J. Apic. Res.* 21:98-103.

Robinson, G. E. 1982. A unique beekeeping enterprise in Colombia. *Bee World* 63:43-46.

Sakai, T., Higg, K., Sasaki, M. 1976. Temperature constancy of a field-built natural comb of the European honey bee, *Apis mellifera* L. *Bull. Fac. Agric. Tamagawa Univ.* No. 16:55-63.

Schneider, S., Blyther, R. 1988. The habitat and nesting biology of the African honey bee *Apis mellifera scutellata* in the Okavanso river delta, Botswana, Africa. *Insectes Soc.* 35: 167-181.
Seeley, T. D. 1977. Measurement of nest cavity volume by the honey bee, *Apis mellifera. Behav. Ecol. Sociobiol.* 2:201-227.
Seeley, T. D. 1985. *Honey bee Ecology*. Princeton, NJ: Princeton University Press.
Seeley, T. D., Morse, R. A. 1976. The nest of the honey bees (*Apis mellifera* L.) *Insectes Soc.* 23:495-512.
Smith, F. G.. 1960. *Beekeeping in the Tropics*. London: Longmans.
Smith, F. G. 1961. The races of honey bee in Africa. *Bee World* 42:255-260.
Spivak, M., Ranker, T., Taylor, O., Jr., Taylor, W., Davis, L. 1988. Discrimination of Africanized honey bees using behavior, cell size, morphometrics, and a newly discovered isozyme polymorphism. In *Africanized Honey Bees and Bee Mites*, ed. G. R. Needham, R. E. Page, Jr., M. Delfinado-Baker, C. E. Bowman. pp. 313-324. Chichester, England: Ellis Horwood Limited.
Tribe, G. D., Fletcher, D. J. C. 1977. Rate of development of the workers of *Apis mellifera adansonii* L. In *African Bees: Their Taxonomy, Biology, and Economic Use*, ed. D.J.C. Fletcher, pp. 115-119. Pretoria, South Africa: Apimondia.
Winston, M. L. 1979a. Intra-colony demography and reproductive rate of the Africanized honey bee in South America. *Behav. Ecol Sociobiol.* 4:279-292.
Winston, M. L. 1979b. Events following queen removal in colonies of Africanized honey bees in South America. *Insectes Soc.* 26:373-381.
Winston, M. L. 1980. Seasonal patterns of brood rearing and worker longevity in colonies of the Africanized honey bee (Hymenoptera: Apidae) in South America. *J. Kans. Entomol. Soc.* 53:157-165.
Winston, M. L. 1987. *The Biology of the Honey Bee*. Cambridge, Mass: Harvard University Press.
Winston, M. L., Dropkin, J. A., Taylor, O. R. 1981. Demography and life history characteristics of two honey bee races (*Apis mellifera*). *Oecologia* 48:407-413.
Winston, M. L., Katz, S. J. 1981. Longevity of cross-fostered honey bee workers (*Apis mellifera*) of European and Africanized races. *Can. J. Zool.* 59:1571-1575.
Winston, M. L., Katz, S. J. 1982. Foraging differences between cross-forested honey bee workers (*Apis mellifera*) of European and Africanized races. *Behav. Ecol. Sociobiol.* 10:125-129.
Winston, M. L., Otis, G. W. 1978. Ages of bees in swarms and afterswarms of the Africanized honey bee. *J. Apic. Res.* 17:123-129.
Winston, M. L., Otis, G. W., Taylor, O. R. 1979. Absconding behaviour of the Africanized honey bees in South America. *J. Apic Res.* 18:85-94.
Winston, M. L., Taylor, O. R., Otis, G. W. 1983. Some differences between temperate European and tropical African and South American honey bees. *Bee World* 64:12-21.
Woyke, J. 1973. Experiences with *Apis mellifera adansonii* in Brazil and Poland. *Apiacta* 8:115-116.
Woyke, J. 1976. Brood-rearing efficiency and absconding in Indian honey bees. *J. Apic. Res.* 15:133-132.

11

POPULATION BIOLOGY OF THE AFRICANIZED HONEY BEE

Gard W. Otis[1]

The accidental "escape" of the African honey bee in South America in 1957 was the start of what has proven to be one of the most phenomenal biological "success stories" of an introduced species. Following their initial introduction near Rio Claro, São Paulo, Brazil, they are now found from Argentina (approx. 39°S) to northern Mexico (approx. 24°N), from sea level to 2900 m in mountains near the equator, and in a variety of habitats including tropical dry forest, savannas, deserts, rain forests, and cloud forests. Moreover, they expanded their range at the average rate of 500 km/year across northern South America and through Central America. Their ability to colonize areas and rapidly establish large populations in the tropics and subtropics is one of the most striking features of their biology.

The term "population biology" is used here in the broadest sense to indicate all those factors that affect the population size and demography of bee colonies, including colony birth and death rates, absconding, generation lengths, colony longevity, dispersal, and population growth rates. The details underlying the range expansion and population growth of the Africanized bee in the Americas have largely gone unrecorded, and therefore our understanding of these topics is still fragmentary. Consequently, this summary will not only attempt to synthesize the information we do have, but will also draw attention to those biological aspects we know little about that would benefit from additional research. Despite the general paucity of demographic data on Africanized bees (or any honey bee race or species for that matter), it will be evident to anyone familiar with European races of bees that unmanaged colonies of Africanized honey bees have strikingly different characteristics.

The most extensive research on demographic parameters of the Africanized bee was conducted in French Guiana from 1976-1977 by myself (Otis, 1980,

[1]Professor Otis is in the Department of Environmental Biology, University of Guelph, Guelph, Ontario, N1G 2W1, Canada.

1982) and co-workers Mark Winston and Orley Taylor. In order to facilitate reading of the remainder of this chapter, references to "French Guiana" will refer to Otis (1980) unless otherwise indicated. Fortunately, numerous studies by other researchers help to fill in gaps in our knowledge, providing us with a reasonable picture of the general population biology of Africanized bees.

COLONY REPRODUCTION: SWARMING

Honey bees reproduce by colony fission, or swarming. The swarming process begins when worker bees initiate queen rearing in response to a suite of demographic factors within the colony (Winston, 1979; Winston and Taylor, 1980; Winston, Chapter 10), a reduction in the effectiveness of queen pheromones that inhibit worker reproduction (Seeley, 1979; Baird and Seeley, 1983), successful foraging, and other factors (Severson, 1984). Usually, shortly after one or more queen cells that contain the oldest larvae have been sealed, a large proportion of the bees in the colony leave with the mated queen in what is called the prime swarm. The bees that remain with the nest continue to feed the remaining queen-destined larvae, resulting in immature queens that differ by as much as nine days in age. When the first virgin queen emerges from her cell, she can destroy all the remaining queen cells, thereby ending the swarming process. Often, however, the worker bees protect the remaining cells from being destroyed. After an interval of one to four days, an afterswarm issues from the colony. This afterswarming process can be repeated several times before one queen destroys the remaining queen cells with the aid of the workers, and mates, thereby becoming the resident queen of the colony. (see Otis, 1980 for a more detailed description of the swarming process).

Accurate data on the reproduction of honey bee colonies can only be obtained by compiling swarming histories of individual colonies. In French Guiana this was accomplished through detailed inspections conducted every day or two, of study colonies that were otherwise unmanaged. Colonies were maintained in 22 l hives. Large numbers of swarms were recorded. In addition to usually producing a relatively large prime swarm (FIGURE 1), from zero to four afterswarms (mean = 1.85 ± 0.186; n = 39) were also generated during each swarming sequence. A few atypical colonies failed to produce a prime swarm but subsequently swarmed with virgin queens in the characteristic pattern for afterswarming. Combining these data, the average "birth rate" was determined to be 2.81 swarms per colony per swarming sequence.

Reproductive rates typically have two major components: the number of offspring (in the case of honey bees, swarms) and generation lengths. For established colonies of honey bees the interval from one prime swarm to the next is a reasonable estimate of generation length. However, for swarms, the duration of the dispersal phase is unknown. Therefore, the interval from nest establishment to issuance of the first prime swarm only approximates generation

FIGURE 1. The high rates of colony reproduction and absconding by African bees result in the production of large numbers of swarms. These swarms are also the only mechanism for natural dispersal. This prime swarm contains approximately 16,000 bees.

length. The only quantitative data available on this parameter were obtained in French Guiana, and because of difficulties encountered in successfully hiving afterswarms with unmated queens, data from swarms weighing less than 1.0 kg are sparse (Otis, 1980). The information that is available, however, clearly indicates that Africanized bees have short swarm-to-swarm intervals. Over most of the year, prime swarms in their first reproductive cycle swarmed an average of 48.5 days (n = 18) after hiving. One colony produced a prime swarm only 31 days after nest establishment! The mean interval between prime swarms for established colonies that had already swarmed at least once was only slightly longer: 72.4 days (n = 19). In Brazil, colony growth is rapid (Anonymous, 1972; Gonçalves *et al.*, 1973), which undoubtedly indicates short swarm-to-swarm intervals there as well. Obviously, with such short intervals between reproductive events, individual colonies can undergo two to four episodes of swarming per year (Winston *et al.*, 1981; Otis, 1982). The exact number is probably determined by both genetic characteristics of the worker bees in a colony and by patterns of resource abundance (Seeley, 1985).

Prime swarms have higher relative fitness than afterswarms "at birth" as a consequence of numerous biological differences (Otis, 1980). Most important of these is size. Prime swarms (1.57 $\pm$ 0.088 kg, n = 25; 16,030 $\pm$ 1230 bees, n = 10) are nearly 2.5 times larger than first afterswarms (0.65 $\pm$ 0.040 kg, n = 28; 6910 $\pm$ 420 bees, n = 6), which in turn are larger than subsequent afterswarms (0.49 $\pm$ 0.043 kg, n = 24). The number of bees in the swarm influences the initial colony growth rate, the duration of the prereproductive period, and nest defense (see Colony Mortality, below) by the young colony. Prime swarms have somewhat greater nectar reserves because their workers have a higher average level of engorgement and they have few or no drones which require feeding (Otis, 1980). The age structure of prime swarms is more highly biased toward the very young bees (Winston and Otis, 1978), which should lead to proportionately less mortality of adult workers in the first few weeks after colony establishment and consequently enhanced larval survival (Winston, 1979). Finally, loss of queens during mating is not a problem for prime swarms as it is for afterswarms which have predominantly virgin queens. Unfortunately, no direct measurements of the relative fitness of prime swarms and afterswarms have been taken. Comparative data on growth rates, survival, and subsequent reproduction of colonies established from prime swarms and afterswarms of different sizes in unmanaged colonies in a natural setting are needed to elucidate the selective forces that have shaped the reproductive behavior of Africanized bees.

Recent data from southern Mexico emphasize the variability that can occur in sizes of swarms. J. Labougle (in prep.) trapped swarms at both Las Choapas on the Atlantic side of the Isthmus of Tehuantepec and Tapanatepec on the Pacific side. The sizes of the swarms caught in trap hives differed strikingly at the two sites. Swarms were large at Las Choapas (n = 68, mean = 2.09 $\pm$ 1.006 kg, mean = 24,160 bees), with 15 swarms weighing more than 3.0 kg. In contrast, the majority of swarms in Tapanatepec weighed less than 0.65 kg (n = 133, mean = 0.62 $\pm$ 0.417 kg, mean = 7,290 bees). The colonies from which these swarms issued were not studied, so it is impossible to know if they were prime swarms, afterswarms, or absconding swarms, the circumstances leading to these differences, or the selective value of swarm size in the two sites. In addition, the subsequent behavior of the colonies founded by these swarms of such different sizes was not followed, but there can be no doubt that such differences have a profound effect on many demographic characteristics of the two populations.

Temporal variation in colony reproduction presumably reflects seasonal climate and patterns of resource abundance. While tropical areas do not experience much annual variation in temperature, they do generally have discrete dry and wet seasons which influence the phenology of plants. In French Guiana during the major dry season (August-November), colonies produced relatively large numbers of afterswarms (mean = 2.3 per swarming episode, n = 18), swarm-to-swarm intervals were short (mean = 48.8 days for colonies swarming for the first time, n = 4) (Otis, 1980), and worker populations of colonies increased rapidly following swarming (Winston, 1980). In contrast, during December and January, the reproductive rate was lower. The mean number of afterswarms that issued per swarming cycle was only 1.3 (n = 10), and swarm-to-swarm intervals for colonies in their first reproductive cycle (mean = 64.5 days, n = 6) were considerably longer. Reproduction increased again briefly during February (mean = 1.9 afterswarms, n = 9; mean swarm-to-swarm interval during first cycle of 45.0 days, n = 8) before declining sharply in March. Only two of 48 colonies under observation in March swarmed, and neither produced any afterswarms. March also signaled the start of the major absconding season which peaked in April, apparently in response to poor foraging conditions (Winston *et al.*, 1979). Brood rearing, populations of worker bees, and mean age of worker bees also exhibited patterns similar to those mentioned above that reflect the influence of seasonal resource abundance on intracolony demography (Winston, 1980; Winston, Chapter 10). From other studies it is evident that swarming intensity varies seasonally throughout the neotropics (Kerr *et al.*, 1972; Boreham and Roubik, 1987) as well as in Africa (Smith, 1960; Ichikawa, 1981; Kigatiira, 1984), although it is not possible to differentiate reproductive swarming from absconding in most of these studies.

The majority of the data from French Guiana was obtained from colonies maintained in 22 l hives, although some hives were smaller (15 l) or larger (42 l), and feral nests were of a wide range of sizes. There is good evidence that colony size influences reproduction. Otis (1980) found that colony size, measured as total comb area at the time of swarming, was positively correlated with the number of sealed queen cells during swarming and the proportion of drone comb produced, which has an influence on the number of drones reared. Moreover, the number of afterswarms a colony produced was highly correlated with the number of queen cells sealed during the swarming cycle. Winston (1979) found a significant correlation between the area of sealed brood when the prime swarm issued and the subsequent number of afterswarms. Although colony size (comb area) was not significantly correlated with the number of afterswarms produced (Otis, 1980), the trend in these data (FIGURE 2) suggests a real influence of nest size on swarming.

Feral colonies of Africanized bees typically have small nests. Fortuitously, the mean comb area of nests built in 221 study colonies in French Guiana was almost identical to that of feral colonies in both French Guiana and Peru (Winston *et al.*, 1983), suggesting that the results from the research conducted on Africanized bees in French Guiana are representative of reproduction by feral colonies in general. The likelihood that nest size influences reproductive effort raises many unanswered questions. Colonies in their second and subsequent

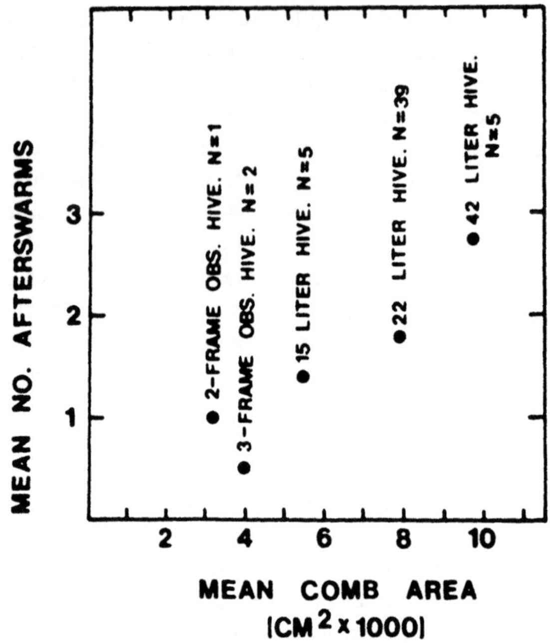

FIGURE 2. The larger a colony of Africanized bees is, the more afterswarms it is likely to produce during a swarming sequence. Large colonies also produce proportionately more drones.

episodes of swarming frequently add to the comb built in the previous swarming cycle. Does this affect the number and/or the size of the swarms produced? If so, is this an artifact of colony growth patterns or is it an adaptive strategy that has been selected for? Recent data on European honey bees in Ontario, Canada, indicate that colonies in 42 l hives that are similar to feral colonies in volume have higher net reproductive rates than do colonies in smaller or larger hives (Morales, 1986). Therefore, the choice exercised by a swarm of bees when colonizing a nest cavity may have profound effects on the fitness of a colony. In two separate studies, Rinderer *et al.* (1981, 1982) found that swarms of Africanized bees prefer to occupy relatively large cavities (20-120 l), although Seeley (1988) pointed out how little we know about nest-site preferences of swarms. In contrast, Africanized bees rarely build nests larger than 30 l in volume no matter how large the nesting cavity (Otis, 1980; Winston and Taylor, 1980). Beekeepers working with Africanized bees frequently indicate that the frequency of swarming is reduced when colonies are kept in large hives

(Michener, 1975; G.W. Otis, pers. obs.). What is the relationship between cavity volume and nest size in Africanized bees, and why do they appear to prefer such large cavities when it may actually reduce swarming? Finally, does the larger size of feral colonies in Venezuela than in French Guiana and Peru (Winston *et al.*, 1983) represent the operation of different selective pressures or is it a consequence of the interaction of intracolony demography, climate, resource abundance, and foraging success? At this time we are far from a comprehensive understanding of the ramifications of nest size on colony reproductive success.

COLONY MORTALITY

Predators frequently attack honey bee colonies in the tropics (Smith, 1960; Fletcher, 1978; Otis, 1982; Seeley *et al.*, 1982). Hives in French Guiana were placed on tin cans just above ground level, and were therefore easily accessible to predators such as army ants (*Eciton burchelli*), other ants, anteaters, and armadillos. Data from 73 colonies over 341 colony months indicated that the daily probability of being attacked by predators was 0.001. From that the annual rate of attack by predators can be estimated to be 30% (Otis, 1980, 1982). It was not possible to determine how many swarms were able to abscond from hives following predator attack, and therefore this predation rate I recorded is an upper estimate of mortality. In other parts of South America humans are important predators on honey bee nests (Posey, 1983), although they probably only occasionally cause the death of the bees themselves. Unfortunately, there are no other detailed data available on colony mortality of Africanized bees. Although many feral colonies of Africanized bees are located near ground level, those that nest higher in trees probably experience reduced predation pressure. It would be interesting to know how nest height, worker population, habitat characteristics, seasonal changes in predator abundance or behavior, and other variables influence predation and nest defense.

Newly founded colonies are particularly vulnerable to predation. When bees from swarms are scouting for a new domicile, they are able to assess many characteristics of potential nest sites (Seeley, 1982). However, they probably cannot determine the presence of predators such as predaceous ants in the vicinity prior to actual colonization of a domicile. They are consequently at relatively high risk initially. In French Guiana seven out of 30 colonies were attacked and killed by ants (predominantly *Camponotus* sp.) within the first two weeks after nest establishment. Swarm size greatly influenced the probability of survival. None of the swarms that weighed 1.0 kg or more were killed by ants. In contrast, 62.5% (n = 8) of the swarms smaller than 0.50 kg were attacked (Otis, 1982).

Another mortality factor of Africanized bees for which we have few data is the loss of queens on their mating flights. Nearly all afterswarms have virgin queens when they leave the parental colony (Otis, 1980) and the parental colony itself contains a single virgin queen upon completion of the afterswarming process. Before they can initiate egg-laying the queens must take one to three mating flights (Taylor *et al.*, in press), during which they are vulnerable to predation. Ruttner (1985) indicated that 20% of queens in temperate regions are lost during mating. A lower proportion (7.8%, n = 90) of queens in French Guiana disappeared on mating flights. This represents a major source of mortality, however, considering that Africanized bee colonies reproduce several times per year and that a large proportion of their reproductive effort is devoted to afterswarms. Polygyny during swarming and of the parental colony before queen mating could provide immunity from queen loss, as apparently occurs in several North African races of honey bees (Ruttner, 1985). However, polygynous afterswarms of Africanized bees at the time of issuance are relatively uncommon. Only one of 20 afterswarms examined in French Guiana had more than one queen. In addition, when swarming has been completed, Africanized bee colonies quickly return to a monogynous condition prior to mating by the new resident queen. The occurrence of monogyny during these times when colonies have a high risk of loosing their queens suggests that other selective pressures must favor monogyny (Ruttner, 1985). This topic has never been investigated.

ABSCONDING AND COLONY FUSION

Occasionally honey bees totally abandon their nests. This behavior, termed absconding, occurs regularly in tropical honey bees (Smith, 1960; Fletcher, 1975-1976; Woyke, 1976; Winston *et al.*, 1979. Absconding can be induced by disturbances, such as fire, infestation by wax moths, and severe manipulation of colonies (Fletcher, 1975-1976). As already mentioned, attack by a predator often results in the absconding of the surviving bees (Winston *et al.*, 1979). In contrast, during periods of resource dearth, high temperatures, or extensive rain, many colonies of Africanized bees initiate an orderly process of brood reduction some 20-25 days prior to actually absconding (Winston *et al.*, 1979). Chandler (1976) differentiated "absconding" that resulted from disturbance, from "migration" which is caused by lack of adequate resources. This choice of terms does not facilitate understanding of the absconding process (Fletcher, 1978). However, it may be useful to distinguish between swarms resulting from different underlying conditions, especially because their subsequent behavior may be different (Kigatiira, 1984).

Absconding is extremely difficult to incorporate into a model of honey bee population dynamics. The frequency of absconding varies both temporally and spatially, with regular annual absconding rates of 15-30% that can increase to 100% in some circumstances (Fletcher, 1975-1976). Because tropical climatic patterns vary from one year to the next, the exact timing and intensity of absconding cannot be predicted with accuracy. Even more problematic is determining the survival of swarms created during absconding. Absconding is not a mortality factor of colonies *per se*, but rather it increases the risk of mortality. Swarms produced during absconding should be subject to higher predation risks as a result of having to colonize new nest sites that may be near nests of predaceous ants. For a period of several days up to a few weeks while the bees are exposed to the elements in swarms, the constant attrition of workers as the older individuals die would slowly reduce the populations of swarms. The age structure of workers in swarms created during absconding is substantially older than in reproductive swarms (Winston *et al.*, 1979), which would influence the initial colony growth rates once they establish nests again, again influencing their ability to drive off predators. Unfortunately, because virtually all African and Africanized bee swarms disperse for some distance soon after leaving their nests (Otis, 1980), it is impossible to estimate the extent to which absconding contributes to colony mortality.

Africanized bee swarms, and particularly absconding swarms, often behave in a manner that further complicates the relationship between absconding and mortality. Swarms fuse with one another, forming "megaswarms." Under some situations, supernumerary queens are eliminated by the workers, apparently by forming a small ball of workers surrounding a queen, that moves to the bottom of the swarm and eventually falls to the ground (Kigatiira, 1984, 1988). When this happens the colony headed by the queen which is eliminated essentially dies, while the other colony experiences an instantaneous increase in worker bees, thereby enhancing its chances of surviving and reproducing successfully. In other instances, queens coexist within the swarm in an unknown manner, giving rise to a substantial proportion of polygynous swarms during the absconding season (Smith, 1960; Cosenza, 1972; Silberrad, 1976; Boreham and Roubik, 1987). Cosenza (1972) found that 16 out of 31 swarms examined during a period of food scarcity in Minas Gerais state, Brazil, had more than one queen, while during the majority of the year swarms were monogynous. In South America, up to 14 queens have been recorded in a single swarm (Anonymous, 1972), while in Africa, a few swarms have as many as 500 queens (OPIDA, 1977). Sometimes, after fusing, swarms fission off the megaswarm, resulting in a return to a monogynous condition (Michener, 1975).

Swarms of Africanized bees also join established colonies (Michener, 1975; Danka and Rinderer, 1988). In general this behavior goes unnoticed unless queen bees are marked and colony development is being closely monitored. During my research in French Guiana I observed this behavior on one occasion, when a swarm joined a colony which had become queenless. Danka and Rinderer (1988) also reported "queen parasitism" in colonies that were queenless, as well as in those that had failing queens or were weak. The advantage to the swarm in such an instance is great because it is able to take over a proven nest site with combs, food reserves, and a labor force of worker bees. If joining queenless colonies is a common strategy of Africanized bees, many colonies that have no queens and

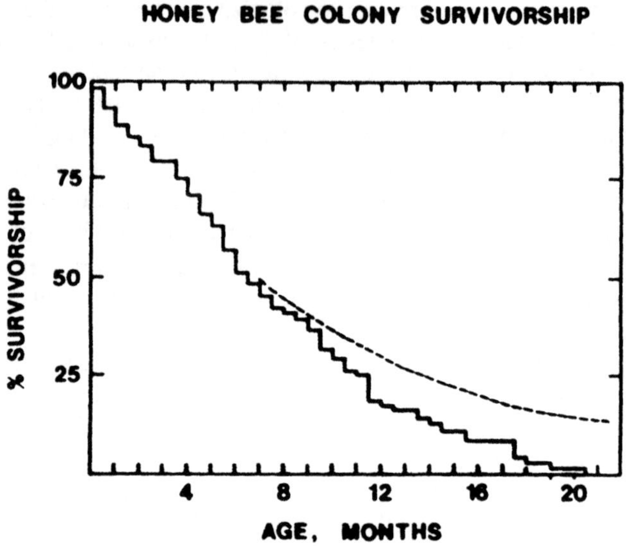

FIGURE 3. Colonies of Africanized bees rarely survive in the same nest for a long period of time because of attack by predators, absconding, and death of queens during mating at the end of swarming sequences. The high rate of colony turnover is evident in this survivorship curve (or more approximately, "residency" curve) for established colonies (Otis, 1982). (Reprinted with permission of Université Paris-Nord: Paris).

therefore are genetically "dead" (except for the drones they produce from worker-laid eggs) and approaching biological death may appear to survive because of swarm fusions. In other cases, swarms join queenright colonies. I have observed swarms of Africanized bees enter and take over small colonies of European bees in Venezuela on several occasions, and beekeepers frequently observe this behavior. Such take-overs may be common in queen-rearing operations during periods of reproductive swarming (Taylor, 1985).

Vergara *et al.* (1989) have recently studied usurpation of European bee colonies by swarms of Africanized bees in Mexico. There were 22 colony takeovers in 240 colonies over a 14 month period. Half of those were of queenless colonies. Surprisingly, the other invasions were of predominantly strong colonies (n = 8). The histories of the fusing swarms were unknown. This is the only detailed study of the swarm-fusing behavior of Africanized bees. Because most fusions probably involve unrelated colonies, inclusive fitness arguments cannot be invoked to explain the biological basis for the behavior. The advantage in individual fitness that accrues to a swarm that joins a queenless colony is clear. However, when swarms join queenright colonies or other swarms, the fates of the "joining" queens and the resident ("joined") queens are unclear but critical to understanding the evolution of this behavior. It is also unclear why fusion events are often peaceful, while on other occasions they are accompanied by fighting (Fletcher, 1978). Future research on this topic should prove to be extremely interesting.

SURVIVORSHIP

Honey bee colonies are affected by predation, loss of queens following swarming, and absconding. As pointed out above, these factors do not necessarily result in death of the bees themselves, but because absconding swarms cannot usually be followed after they have dispersed, it is not possible to quantify actual mortality. However, attack by predators and absconding do result in the disappearance of colonies from their nests. From the data I obtained from all study colonies in French Guiana, I constructed a "survivorship curve" for established colonies that had survived past the initial two-week, high risk period (FIGURE 3) (Otis, 1982). Because absconding and queenlessness were considered as colony losses, this might be more appropriately termed a colony "residency" curve. Some colonies were still resident at the end of my study, so consequently their contribution to residency was underestimated by the actual data; the dashed line in the figure corrects for that underestimation. The results of this exercise were surprising. Only 50% of the colonies were still present after seven months, and by the end of two years, only 10-20% of the original colonies would still have been alive and resident in their original nests. While FIGURE 3 is not exactly a "survivorship curve" in the standard sense, nevertheless it is helpful in pointing out the high rate of colony turnover and the vagility of a representative Africanized bee population. Actual survivorship curves for prime swarms and afterswarms of various sizes, from their production until their subsequent reproduction, would be helpful for understanding the evolution of the reproductive strategy of the Africanized bee.

COLONY DISPERSAL

Swarms of Africanized bees disperse extensively before colonizing new domiciles. Much of our understanding of this phenomenon comes from anecdotal information. For example, swarms of bees have naturally colonized several islands, including one 11 km off South Africa (Hannabus, 1939), l'Ile du Gran Connetable and Les Iles du Salut 12-20 km off French Guiana (Taylor, 1977), and Trinidad, which is 14 km from mainland Venezuela (Hallim, 1984). One swarm landed on a boat 10 km from land in the Amazon River (Michener,

1975). All of these instances necessarily resulted from single long distance flights. In addition, swarms frequently settle in a site for a few days before moving on again (Kigatiira, 1984). While they are in this interim site the worker bees forage for nectar and occasionally pollen. Apparently they can sustain themselves in this manner for at least a period of several days (O. R. Taylor and M. Peterson, pers. comm.). Foraging from swarms may actually allow bees from the swarm to assess the resource conditions before "deciding" whether to establish a nest or to move on again (Taylor and Otis, 1978).

The actual distances travelled by individual swarms are generally unknown. The average rate of range expansion of 300-500 km per year through much of South and Central America (Taylor, 1977, 1985) is only possible if each swarm travels an average distance of 50-80 km (assuming about six swarm movements per year; Taylor, 1977), or if some swarms fly very long distances. Long distance flights may occur regularly. Antonio Both (pers. comm. to O. R. Taylor) in Brazil reported following a single swarm by car, motorcycle, and bicycle for more than 100 km. Nightingale (1976) believed that swarms in Kenya regularly migrated as far as 160 km. In the only detailed study of swarm movements to date, Kigatiira (1984) and his coworkers in Kenya attempted to follow 250 swarms by foot and motorcycle. They lost contact with most of these before they settled in new sites. The longest recorded flight from an apiary to another location was 15.1 km. Several swarms took additional flights on subsequent days. Interestingly, after resting for a day or less, some swarms departed again in a very different direction, while others maintained a relatively constant flight direction (Kigatiira, 1984). The causes underlying this variability are not understood.

Beekeepers frequently report unidirectional movements of swarms. Kigatiira (1984) has presented a large data set on this phenomenon. Absconding and migrating colonies (*sensu* Chandler, 1976) within an apiary consistently departed in directions that fell within a 90° sector. The flight directions were unrelated to either wind direction or the cause of hive abandonment. Nonrandom swarm movements could have a large effect on population dynamics. For example, in parts of Africa large proportions of the colonies in some areas migrate seasonally, presumably to areas with better resources (Nightingale, 1976). In fact, during periods of flowering of *Eucalyptus grandis* in South Africa, huge numbers of swarms migrate into the tree plantations where they are captured in bait hives and managed for honey production (Fletcher, 1978). The plantations occur in areas that previously contained marginal bee forage, so it appears that the bees are responding to the food resources produced by *Eucalyptus* and choosing to move into the plantations. This phenomenon can result in populations that fluctuate greatly both as a result of colony reproduction and patterns of immigration and emigration. Unfortunately, so little is known about dispersal of Africanized bee swarms in South America that at present its possible importance in population dynamics can only be mentioned.

Natural movement of swarms is the primary factor that has contributed to the range expansion of the Africanized bee in the New World. Climate and/or habitat seem to have strongly influenced this dispersal. Taylor (1977) summarized information that suggested that range expansion is more rapid in tropical dry habitats that receive less than 2,000 mm of rainfall per year than in wetter habitats. For example, following their rapid movement (300-400 km/year) through the seasonally dry coastal region of the Guianas, Africanized bees expanded their range more slowly across the wet forested region of the Orinoco River delta, from central Guyana to Trinidad (160 km/year). In contrast, they pushed up through the Gran Sabana to northwestern Venezuela very quickly at a rate of 400-500 km/year, and have dispersed through the drier regions of Central America at this same rate. The strong seasonal climate that alternates between dry, resource-rich periods and wet, resource-poor periods, may facilitate long distance dispersal by swarms (Taylor, 1977).

DENSITY OF FERAL COLONIES

Population densities are a function of the amount of time that has elapsed subsequent to colonization of a region, the population growth rate, food resources, and competition with other floral visitors. There are almost no quantitative data on this subject. As swarms of Africanized bees move into new areas of South America, they usually do so at very low densities such that the first swarms go undetected. For example, in Costa Rica the first swarm was detected in the northwestern corner of the country (Santa Rosa National Park; D. H. Janzen, pers. comm to O. R. Taylor) at approximately the same time as Africanized bees were just being discovered near the eastern border with Panama. Elsewhere, as in southwestern Guyana and southeastern Venezuela, bees were abundant shortly after they first arrived there (Taylor, 1977). Differences in abundance so soon after colonization cannot be due to reproduction alone, and extensive immigration probably contributed extensively to the rapid population increase. In fact, people in that region reported seeing several swarms per day flying overhead in the same direction during certain seasons (Taylor, 1977), which is suggestive of large scale immigration.

Carrying capacities vary both seasonally and geographically, probably in large part due to the abundance of resources. Competition with other bees may also be important, but has never led to the exclusion of Africanized bees from any region of the New World. Population estimates from areas with moderately good foraging conditions, such as Maturín, Venezuela, and Pucallpa, Peru, are 10-20 feral colonies per km^2 only two to three years after colonization (Taylor, 1985). Roubik (1988) gave a general estimate of six colonies per km^2 for much of tropical America. In savanna ("cerrado") regions of Goias and Mato Grosso, Brazil, Kerr (1971) found an average of 107.5 colonies per km^2! Densities of honey bee nests are very low at the upper elevational limits of Africanized bee distribution in South America (J. Villa, pers. comm.). They are likewise scarce in most rain forest habitats. These densities may fluctuate widely between different seasons, first increasing when food resources are abundant and colonies are reproducing extensively, and then declining as they abscond and emigrate from the region.

Additional aspects of population density will be considered in the next section on population growth.

RATE OF POPULATION INCREASE

Africanized bees have several characteristics that hinder the analysis of population growth. For most animals, offspring are roughly equivalent in size and fitness at the time of birth, and individuals are usually distinct entities. With Africanized bees, swarms range in size from afterswarms with virgin queens and only approximately 1,000 worker bees to large prime swarms with 22,000 or more bees. The reproductive potential at "birth" of swarms so disparate in size is clearly very different. Colonies can abscond from relatively secure nests to form swarms, which substantially changes their risk of mortality. Moreover, because swarms can fuse with other swarms or colonies, the concept of an individual colony becomes blurred.

These complicating factors do not lend themselves well to analysis by standard demographic techniques. Because colonies cannot be aged accurately, and in fact are potentially immortal as long as their queens become mated after swarming, it is not possible to construct current (time-specific) life tables. On the other hand, there are rarely enough swarms available at the same time to allow for construction of cohort (age-specific) life tables, besides which the swarms would not be equivalent in size, age and mating status of their queens, and other characteristics. The use of population projection matrices is more feasible because they rely on probabilities of reproduction and mortality which could be changed at several points during the analysis. However, even with that difficulties arise. The variable fitness of swarms remains a problem as it was for life-table analyses. Moreover, probabilities of some colony-level events depend upon the previous history of the colony. For example, colonies in French Guiana that swarmed in February had a much higher probability of absconding in March-April than those that last swarmed in December or January (Winston *et al.*, 1979). No standard demographic techniques allow historical events to affect the analyses.

I chose to determine the population growth rate of the Africanized bee population I studied in French Guiana (Otis, 1980) by developing a stochastic simulation. The simulation begins with a single colony. The outcome of every event faced by a colony, such as survival from one day to the next, intervals between each swarm, loss of the queen after mating, and absconding, is determined randomly using the probabilities for the event that were generated from field data. For example, there were 29 experimental data points for the time interval between the start of the study and the dates on which colonies swarmed for the first time. With the use of a random number generator, one of these data points is randomly selected for the particular simulation. Each swarm that is produced during the remainder of the 318-day study period is entered into an array. When the original colony has either died or lived to the end of the study period, the simulation moves down the array to the first swarm that has been produced and follows its subsequent reproduction and survival. Once all offspring have been sequentially handled in this manner, the simulation program calculates the number of colonies alive on the first day of each month. The approximate population increase is the net number of colonies alive at the end of the study period, as determined by averaging the outcomes of a large number of simulations performed in this manner.

A simulation of this nature is limited only by the complexity of the model and the extent to which the input data, which are subject to sampling error, are representative of true population parameters. For some events in this simulation, the probabilities could be reliably estimated because the data sets that determined them were large. For example, the likelihood of a colony losing its queen during mating was determined to be 8% based on observation of 90 swarming sequences. The number of afterswarms produced during a swarming sequence, however, was determined from only 39 detailed swarming histories. This is an adequate sample for estimating the approximate probabilities for this parameter. However, when trying to incorporate seasonal differences into the simulation, there were never more than eight observations for this parameter within a given month. Therefore, the influence of seasonal factors could only be roughly estimated. Other probabilities lacked quantification and had to be guessed, such as the mortality rate of absconding swarms.

It is already evident from the discussion in this chapter that several assumptions had to be made concerning the biology of Africanized bees in order to perform the simulation. Mortality of absconding swarms was estimated to be 50%. The probability of mortality from one day to the next remained constant at 0.001 even though defensive behavior of colonies changed during the swarming cycle. It was assumed that the reproduction and survival of unmanaged study colonies in hives were the same as for feral colonies. In the absence of data, the effects of migration into and out of the population were ignored. As with any model, some assumptions were necessary. As additional details become available through further research, they could be incorporated into the simulation.

Despite these limitations, the results of the various simulations were instructive. Both a standard simulation and the simulation that incorporated seasonal differences in numbers of afterswarms and generation lengths were performed (FIGURE 4). These two simulations yielded equivalent results. Even though there were extensive colony losses, the population increased at a remarkably rapid rate. Growth rate was approximately geometric from August through February except for a slight decline in October due to a few colonies that absconded then. The reproductive rate slowed in March, and then declined during the major absconding season. Using the seasonal simulation model, the number of colonies increased during the 318-day study interval from one to 14.9, and the population was estimated to have increased 16-fold in a year. The growth patterns evident in FIGURE 4 are specific to the French Guiana population during the third year after the arrival of the Africanized bee. In its present form, the simulation could not be used to analyze honey bee demography elsewhere. However, if one had the appropriate data base and event probabilities, this approach could be used to generate appropriate simulations for other populations.

One way to further understand the details of population growth is to test the sensitivity of the simulation to changes in its components, a process generally known as sensitivity analysis. Input parameters were varied by a fixed percentage (e.g., 20%) and the impacts those changes had on the simulation were determined. The relative sensitivity values (S), which were calculated by dividing the percent change in the population growth rate by a predetermined (20%) increase in input parameters, proved to be illuminating. When all generation lengths were increased by 20%, the growth rate declined by 26.4% (S = -1.32). Increasing the number of afterswarms yielded a sensitivity of +1.02. These two parameters had the largest sensitivity values, and both are determined by factors that cannot be altered by human-designed control programs. In contrast, increasing the annual mortality of established colonies (S = -0.30) and the loss of queens on mating flights (S = -0.08), which are components of proposed control programs, have much less impact on population growth. (Eliminating queens on mating flights is somewhat akin to causing them to become hybridized by European drones, except the latter would have an even lower sensitivity). The interpretation of these sensitivity analyses is that it would take exceptionally effective control programs to have much impact on the feral population of honey bees, and that feral colonies will probably continue to reproduce at a very rapid rate as long as ecological conditions allow.

An aspect of Africanized bee populations that has never been adequately quantified is how populations change several years after the initial colonization period. Over the first two to four years, the number of feral colonies and swarms generally increases (Taylor, 1985). Then the numbers appear to decline again (Michener, 1975). I attempted to model the changes that would be likely to occur under conditions of increased competition and colony density by decreasing the number of afterswarms, and increasing generation lengths, annual mortality rates, and the proportion of colonies that abscond during the dearth period (Otis, 1980) but the results were not very satisfying because these input changes were wholly based on speculation.

Recent field data from Panama presented by Boreham and Roubik (1987) clearly demonstrate that a decline in population size occurs within only a few

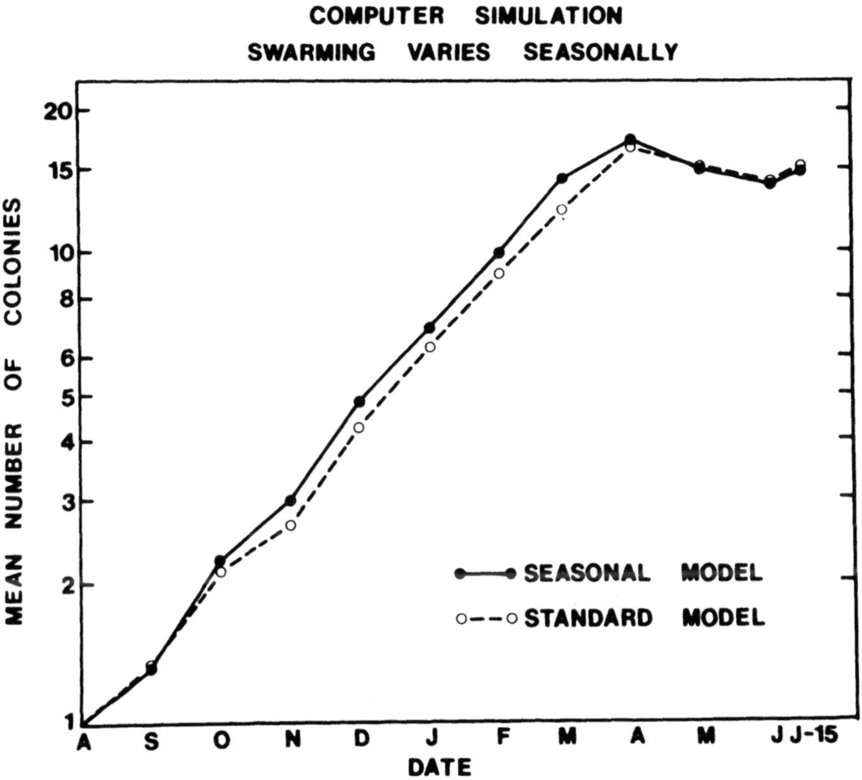

FIGURE 4. The results of a computer simulation of an Africanized bee population in French Guiana, South America (Otis, 1980), indicate how quickly populations can increase. This population was studied in the third year following colonization of the site. The seasonal model utilized slight monthly differences in generation lengths and number of afterswarms, whereas the "standard" model was based on values for these parameters averaged over the entire study period.

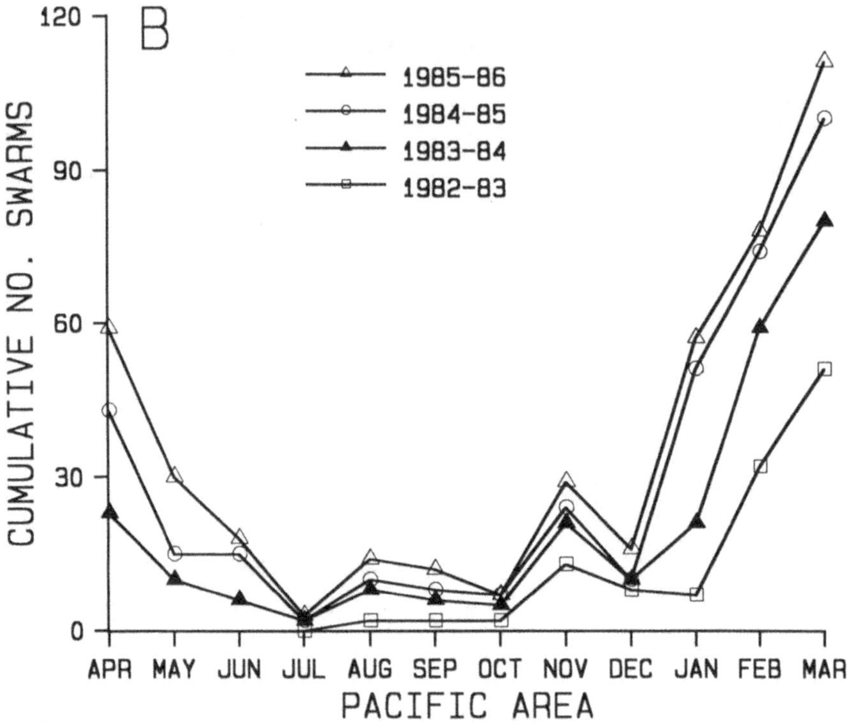

FIGURE 5. The changes in a population of Africanized bees during the first four years after initial colonization of an area are reflected in this graph of the cumulative number of swarms for each month of the year recorded from the Pacific side of the Panama Canal Area (Boreham and Roubik, 1987). Initially, the number of swarms increased very quickly, but for each succeeding year the number declined. Similar trends were observed in swarm data from the Atlantic side of the Panama Canal Area and on the number of nests discovered in both areas. (Reprinted by permission of the Entomological Society of America).

years. They recorded all swarms (FIGURE 5) and feral colonies within the Panama Canal Area. The first colonies appeared in the area in April 1982. In addition to a strong seasonal component to the data, the number of swarms increased greatly over the first two years. By the third year, however, the proportional increase in swarms was slightly less, and by 1985-1986, discoveries of both swarms and nests dropped substantially, suggesting that the population had declined.

Where did the colonies go? Competition may have become so intense that swarming was sharply curtailed, thereby giving the appearance that population density had declined. Alternatively, the population may have exceeded the carrying capacity within the short span of two years, following which many colonies died or emigrated. It is tempting to speculate that a population may undergo a dramatic shift in characteristics as competition increases. The characteristics such as extensive swarming and absconding that contribute heavily to individual fitness during the early colonization phase when the bees are moving into an "empty niche" may soon prove to be detrimental to colonies that must compete for what have become limited resources. With increased competition, other attributes may become important such as food hoarding, production of fewer swarms that have greater fitness (larger size), and larger colony size that allows better nest defense. Although this pattern has not been verified for Africanized bees, this general concept of demographic attributes of colonizing organisms as compared to populations that are near the carrying capacity of the environment forms the basis of r- and k-selection theory (Stearns, 1976).

CONCLUDING REMARKS

This discussion of the population biology of the Africanized bee has pointed out numerous topics that deserve further study. Several aspects of Africanized bee biology appear to be very different from the attributes of European honey bees. However, simultaneous comparative studies at a tropical study site are required to determine the extent to which these differences are genetic or environmental. Comparative demography of these two geographic types of bees could be particularly interesting at progressively more northern sites where the selective advantage should gradually shift from Africanized to European bees.

Africanized bees have colonized much of Mexico and their arrival at the United States border within the next year has been predicted (Taylor, 1977, 1985). Several aspects of their population biology make it improbable that their northward dispersal can be stopped or significantly slowed. The dry, tropical environments they will be moving through are similar to areas further to the south where they have expanded their range most rapidly and have quickly built up large populations. Although control schemes have been implemented in Mexico, the long-distance dispersal abilities of Africanized bees coupled with their ability to survive and reproduce in a feral setting make it likely that some swarms will move through zones of control virtually unaffected. The sensitivity analyses of my simulation emphasize the difficulties of being able to cause significant reductions in the growth rates of feral populations. At present there appear to be no major barriers to the movement of Africanized bee swarms into the United States. We should examine more critically the possible interactions between Africanized and European bees that could occur within the United States, and begin immediately to seek long-term solutions to the problem.

LITERATURE CITED

Anonymous. 1972. *Final report of the Committee on the African honey bee*, Washington, D. C.: Natl. Res. Counc. Natl. Acad. Sci. 95 pp.

Baird, D. H., Seeley, T. D. 1983. An equilibrium theory of queen production in honey bee colonies preparing to swarm. *Behav. Ecol. Sociobiol.* 13:221-228.

Boreham, M. M., Roubik, D. W. 1987. Population change and control of Africanized honey bees (Hymenoptera: Apidae) in the Panama Canal area. *Bull. Entomol. Soc. Am.* 33:34-39.

Chandler, M. T. 1976. The African honey bee -- *Apis mellifera adansonii*: the biological basis of its management. In *Apiculture in Tropical Climates*, ed. E. E. Crane, pp. 61-68. London: Int. Bee Res. Assoc.

Cosenza, G. W. 1972. Estudo dos enxames de migracão de abelhas Africanas. *1° Cong. Bras. Apic. (Florianópolis, 1970)*, ed. H. Weise, pp. 128-129. Santa, Catarina, Brazil.

Danka, R. G., Rinderer, T. E. 1988. Social reproductive parasitism by Africanized honey bees. In *Africanized Honey Bees and Bee Mites*, ed. G. R. Needham, R. E. Page, M. Delfinado-Baker, C. E. Bowman, pp. 214-222. Chichester, England: Ellis Horwood Limited.

Fletcher, D. J. C. 1975-1976. New perspectives in the causes of absconding in the African bee (*Apis mellifera adansonii* L.). *S. Afr. Bee J.* I & II. 47:11-14; 48:6-9.

Fletcher, D. J. C. 1978. The Africanized honey bee, *Apis mellifera adansonii*, in Africa. *Ann. Rev. Entomol.* 23:151-171.

Gonçalves, L. S., Kerr, W. E., Chaud Netto, J., Stort, A. C. 1973. Relatorio final do grupo de estudo americano sobre as abelhas africanas. *2° Cong. Bras. Apic. (Sete Lagoas, 1972)*, ed. G. W. Cosenza, pp. 211-268. Minas Gerais, Brazil.

Hallim, M. K. I. 1984. *Extension program suggestions for solving problems caused by Africanized bees in Trinidad and Tobago*. M.Sc. Thesis, University of Guelph, Guelph. 268 + 11 pp.

Hannabus, C.H. 1939. Swarm flies seven miles. *S. Afr. Bee J.* 13:8.

Ichikawa, M. 1981. Ecological and sociological importance of honey to the Mbuti net hunters, eastern Zaire. *African Study Monogr.* 1:55-68.

Kerr, W. E. 1971. Contribuiçao à ecogenetica da algumas espécias de abelhas. *Ciênc. Cult. São Paulo* 23 (Suppl):89-90.

Kerr, W. E., Gonçalves, L. S., Blotta, L. F., Maciel, H. B. 1972. Biología comparada entre as abelhas Italianas (*Apis mellifera liguistica)*, Africana (*Apis mellifera adansonii*) e suas hı́bridas. In *1° Congr. Bras. Apic. (Florianópolis, 1970)*, ed. H. Weise, pp. 151-185. Santa Catarina, Brazil.

Kigatiira, K. I. 1984. *The aspects of the ecology of the African honey bee*. PhD Dissertation, University of Cambridge, England. 229 + 27 pp.

Kigatiira, K. I. 1988. Amalgamation in tropical honey bees. In *Africanized Honey Bees and Bee Mites,* ed. G. R. Needham, R. E. Page, M. Delfinado-Baker, C. E. Bowman, pp. 62-71. Chichester, England: Ellis Horwood Limited.

Labougle, J. In prep. *Comparative studies of African honey bees in southern Mexico.* PhD Dissertation. University of Kansas, Lawrence.

Michener, C. D. 1975. The Brazilian bee problem. *Ann. Rev. Entomol.* 20:399-416.

Morales, G. 1986. *Effects of cavity size on demography of unmanaged colonies of honey bees (Apis mellifera L.).* M.Sc. Thesis, University of Guelph, Guelph. 81 + 5 pp.

Nightingale, J. A. 1976. Traditional beekeeping among Kenya tribes, and methods proposed for improvement and modernisation. In *Apiculture in Tropical Climates,* ed. E. E. Crane, pp. 15-22. London: Int. Bee Res. Assoc.

OPIDA. 1977. Bull. Tech. Apicole 4(2): cover.

Otis, G. W. 1980. *The swarming biology and population dynamics of the Africanized honey bee.* PhD Dissertation. University of Kansas, Lawrence. 197 pp.

Otis, G. W. 1982. Population biology of the Africanized honey bee. In *Social Insects of the Tropics, Vol. 1,* ed. P. Jaisson, pp. 209-219. Paris: Université Paris-Nord.

Posey, D. A. 1983. Folk apiculture of the Kayapó Indians of Brazil. *Biotropica* 15:154-158.

Rinderer, T. E., Collins, A. M., Bolten, A. B., Harbo, J. R. 1981. Size of nest cavities selected by swarms of Africanized honey bees in Venezuela. *J. Apic. Res.* 20:160-164.

Rinderer, T. E., Tucker, K. W., Collins, A. M. 1982. Nest cavity selection by swarms of European and Africanized honey bees *J. Apic. Res.* 21:98-103.

Roubik, D. W. 1988. An overview of Africanized honey-bee populations: reproduction, diet, and competition. In *Africanized Honey Bees and Bee Mites,* ed. G. R. Needham, R. E. Page, M. Delfinado-Baker, C. E. Bowman, pp. 45-54. Chichester, England: Ellis Horwood Limited.

Ruttner, F. 1985. Reproductive behavior in honey bees. In *Experimental Behavioral Ecology and Sociobiology,* ed. B. Holldobler and M. Lindauer, pp. 225-36. Sunderland: Sinauer Assoc.

Seeley, T. D. 1979. Queen substance dispersal by messenger workers in honey bee colonies. *Behav. Ecol. Sociobiol.* 5:391-415.

Seeley, T. D. 1982. How honey bees find a home. *Sci. Am.* 247:158-168.

Seeley, T. D. 1985. *Honey bee Ecology: A Study of Adaptation in Social Life.* Princeton, NJ: Princeton University Press.

Seeley, T. D. 1988. What we do and do not know about the nest-site preferences of African honey bees. In *Africanized Honey Bees and Bee Mites,* ed. G. R. Needham, R. E. Page, M. Delfinado-Baker, C. E. Bowman, pp. 87-90. Chichester, England: Ellis Horwood Limited.

Seeley, T. D., Seeley, R. H., Akratanakul, P. 1982. Colony defense strategies of the honey bees in Thailand, *Ecol. Monogr.* 52:43-63.

Severson, D. W. 1984. Swarming behavior of the honey bee. *Am. Bee. J.* 124:204-210; 230-232.

Silberrad, R. E. M. 1976. *Bee-keeping in Zambia.* Bucharest: Apimondia.

Smith, F. G. 1960. *Beekeeping in the Tropics.* London: Longmans.

Sterns, S. C. 1976. Life-history tactics: a review of the ideas. *Quart. Rev. Biol.* 51:3-47.

Taylor, O. R. 1977. Past and possible future spread of Africanized honey bees in the Americas. *Bee World* 58:19-30.

Taylor, O. R. 1985. African bees: potential impact in the United States. *Bull Entomol. Soc. Am.* 31:15-24.

Taylor, O. R., Otis, G. W. 1978. Swarm boxes and Africanized honey bees: some preliminary observations. *J. Kans. Entomol. Soc.* 51:807-817.

Taylor, O. R., Otis, G. W., Kukuk, P., Spivak, M. Timing of mating flights of queens of African and European honey bee races in South America (*Apis mellifera* L.). *J. Apic. Res.* In press.

Vergara, C., Dietz, A. Perez, A. 1989. Usurpation of managed honey bee colonies by migratory swarms in Tabasco, Mexico. *Am. Bee J.* 129:824-825.

Winston, M. L. 1979. Intra-colony demography and reproductive rate of the Africanized honey bee in South America. *Behav. Ecol. Sociobiol.* 4:279-292.

Winston, M. L. 1980. Seasonal patterns of brood rearing and worker longevity in colonies of the Africanized honey bee (Hymenoptera: Apidae) in South America. *J. Kans. Entomol. Soc.* 53:157-165.

Winston, M. L., Dropkin, J. A., Taylor, O. R. 1981. Demography and life history characteristics of two honey bee races (*Apis mellifera*). *Oecologia* 48:407-413.

Winston, M. L., Otis, G. W. 1978. Ages of bees in swarms and afterswarms of the Africanized honey bee. *J. Apic. Res.* 17:123-129.

Winston, M. L., Taylor, O. R. 1980. Factors preceding queen rearing in the Africanized honey bee (*Apis mellifera*) in South America. *Insectes Soc.* 27:289-304.

Wins:on, M. L., Otis, G. W., Taylor, O. R. 1979. Absconding behaviour of the Africanized honey bees in South America. *J. Apic Res.* 18:85-94.

Winston, M. L., Taylor, O. R., Otis, G. W. 1983. Some differences between temperate European and tropical African and South American honey bees. *Bee World* 64:12-21.

Woyke, J. 1976. Brood-rearing efficiency and absconding in Indian honey bees. *J. Apic. Res.* 15:133-132.

12

FORAGING BEHAVIOR AND HONEY PRODUCTION

Thomas E. Rinderer[1] and Anita M. Collins[2]

The indigenous honey bee of the East African dry savanna, *Apis mellifera scutellata* (Ruttner, 1986), was recommended by Smith (1953) as the bee of choice for importation to the tropics of India or the Far East. Although he recognized the "bad tempered" nature of the bee, he considered it to be improvable through selective breeding and "the best of all the honey producers in the tropics." Three years after Smith's evaluation of the Eastern African subspecies appeared, 63 living *A. m. scutellata* queens arrived in a shipment of 133 queens sent to Brazil from primarily South Africa (Kerr, 1957). Forty of these queens survived introduction procedures and produced colonies; 26 of which were considered to be "exceptionally prolific, productive and vigorous" (Kerr, 1957). Breeding plans for these colonies were interrupted when an accident enabled 26 colonies to abscond into a Eucalyptus forest (Kerr, 1967; Gonçalves, 1975).

The intention of this breeding program was to crossbreed the African stock with Italian (*A. m. ligustica*) stock. The crossbred colonies were to serve as a base population in a selection program designed to produce a stock which was both gentle and excellent in honey production (Kerr, 1957). The chief genetic source of productivity was to be the group of imported African queens. *A. m. scutellata* colonies were considered to be "very prolific, better than the Italian in honey production, colony development, and adaptability to climatic conditions, flora and places where colonies are established" (Kerr and Portugal-Araújo, 1958). In part, honey production was thought to be due to the lengths of times in which African bees foraged since "they start work earlier and finish later than

[1] Dr. Rinderer is with the USDA--ARS, Honey Bee Breeding, Genetics and Physiology Research Laboratory, 1157 Ben Hur Rd., Baton Rouge, Louisiana, 70820, USA.

[2] Dr. Collins is with the USDA--ARS, Honey Bee Research Laboratory, 2413 E. Highway 83, Weslaco, TX, 78596, USA.

Italians and black bees, sometimes before sunrise and up to some minutes before nightfall" (Kerr and Portugal-Araújo, 1958).

The view that honey bees of one subspecies invariably will store more honey than the bees of another subspecies is consistent with inaccurate but long held beliefs (or hopes) of beekeepers. Such beliefs tempt beekeepers with the thought that somewhere in the world there exists a bee stock which is vastly superior to the bees in their own apiaries.

Occasionally, action based on such beliefs has resulted in serious errors. Recently, Ukrainian beekeepers forgot that their grandfathers who settled the eastern Primor'e region of Russia near the end of the 19th century took Ukrainian *Apis mellifera* with them (the Primor'e area only had *Apis cerana).* The Ukrainian bees did well there and, due to very large honey crops secured from extensive lime forests, attained sufficient fame that the Latin trinomial *Apis mellifera acervorum* was proposed for them. In the 1960s this special "far Eastern bee" was "introduced" to Siberia, the western part of the Soviet Union and Bulgaria. These returning Ukrainian bees brought with them the parasitic *Varroa jacobsoni* and a resultant large-scale death of colonies (Alpatov, 1976).

The introduction of African bees to the Americas has had equally disruptive results. Although there are some reports from Brazil of increased honey production with Africanized bees, there are many more reports of decreased production, disrupted beekeeping practices and seriously depressed beekeeping industries. In part, these difficulties stem from the colony defense of Africanized bees (Collins *et al.*, 1982). Equally important, the expected production from Africanized bees has not been generally realized. While it is true that Brazilian honey production has increased in the past 30 years (Gonçalves, 1975) this is due to improved agricultural and apicultural practices rather than to Africanized bees (Wiese, 1977). Similar changes in a much smaller developing country with tropical and subtropical climates, Mexico, have caused that nation to increase its honey production several times more than the Brazilian increases (Labougle and Zozaya, 1986). This was accomplished without Africanized bees. Furthermore, where Africanized bees have entered countries with some degree of advanced apiculture, production has dramatically fallen (Rinderer, 1985; Cobey and Locke, 1986; Swezey, 1986).

The biological principles which underlie the disparate reports of Africanized and European honey bee honey production are rooted in the behavioral adaptations of these bees to the differing ecological characteristics of their evolutionary homes. Several behavioral differences relating to foraging and honey production by Africanized and European honey bees may be tested experimentally. How these different behavioral patterns are adaptive becomes clear when they are considered in the context of the floral environment in which they evolved.

In this chapter we will first examine a model of European honey bee foraging and supporting evidence. Using this model as a standard, we will then examine the comparative foraging and honey production of Africanized and European honey bees. Finally, we will offer an ecologically based interpretation accounting for the differences and provide a model of Africanized honey bee foraging.

EUROPEAN HONEY BEE FORAGING AND HONEY PRODUCTION

Foraging by European honey bees is highly responsive to seasonally related nectar flow conditions. Von Frisch (1967) experienced difficulties training bees to artificial feeding stations in spring but not in autumn. He attributed these training difficulties to competition from "abundant flowering and a good supply of food." The responses he saw were the result of an elegant regulating system which guides bees to the most productive foraging possible in the various seasons of the year.

The first hint that organized seasonal regulation of foraging occurs in a honey bee colony came from a project in which bees were being bred for increased honey production. Since production records of field colonies contain a substantial environmental variance component, the project involved selecting colonies for breeding based on the performance of a small group of bees from each colony in laboratory cages (FIGURE 1) (Kulincevic *et al.*, 1973). These cages were usually fitted with a small piece of comb and two feeders, one containing 50% (wt/wt) sucrose in water solution and the other containing water. Bees in such cages removed the sucrose solution from its feeder and stored or "hoarded" (Kulincevic and Rothenbuhler, 1973) it in the comb provided. To verify that the differences in the amount of comb in the cages were not a source of experimental variation, cages were fitted with one, two, or three pieces of comb having 47, 94 and 140 sq cm of exposed surface area. Surprisingly, cages with one, two and three pieces of comb had bees hoarding, respectively, 0.11, 0.14 and 0.19 ml of solution/bee/day (FIGURE 2A)(Rinderer and Baxter, 1978). A later experiment showed that increasing the amount of comb up to 280 cm^2 continued to increase rates of hoarding (Rinderer, 1982a).

The principles underlying these experimental designs were then applied to field colonies (Rinderer and Baxter, 1978). Of 20 equal-size field colonies in an apiary experiencing the season's major nectar flow, ten were given empty honey storage combs having 4.06 m^2 of comb surface area (CSA) and ten were given combs having 1.88 m^2 of CSA. After 15 days the colonies with more comb had stored an average of 51 kg of honey and nectar while the colonies that had less comb had stored a smaller average of 36 kg. The storage combs were removed and the colonies were then transported to a new location having an intense nectar flow. There, the CSA treatments were reversed using fresh storage combs: those colonies that had previously received 1.88 m^2 of CSA were given 4.06 m^2 and those that had previously been given 4.06 m^2 were given 1.88 m^2 of CSA. After ten days, the colonies with more CSA had stored an average of

FIGURE 1. A laboratory hoarding cage.

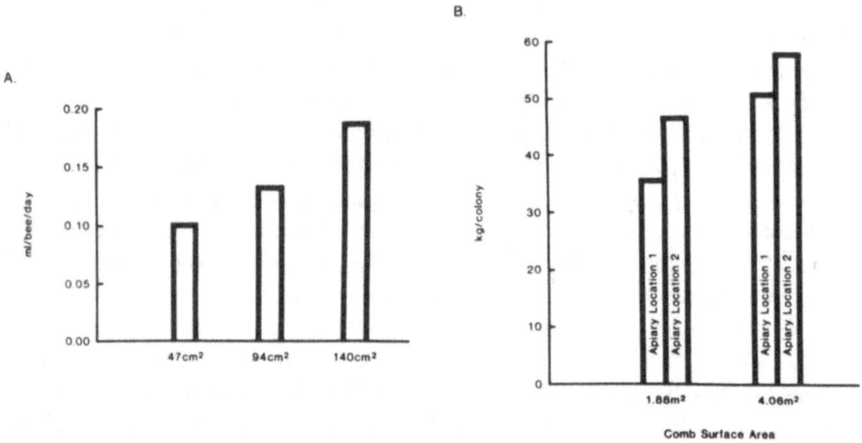

FIGURE 2. (A) Hoarding (ml) of sugar syrup by bees in hoarding cages supplied with three amounts of comb surface area. (B) Honey yields (kg) of colonies in two different apiary locations supplied with two amounts of comb surface area. (From data of Rinderer and Baxter, 1978. Copyright in public domain.)

58 kg of honey and nectar while those with less comb had stored an average of 47 kg. In no case was a colony allowed to fill completely all its storage combs before the experiment ended. Thus, the results of the laboratory experiment were supported by the experiment using field colonies (FIGURE 2B). At least under strong nectar flow conditions, large amounts of CSA resulted in increased nectar gathering by bees and consequently greater honey production.

To test the comb stimulation hypothesis further, an experiment was conducted which was based on the hypothesis that if empty comb does stimulate hoarding behavior, then bees transferred from one level of empty CSA to another would show predictable changes in their hoarding behavior (Rinderer and Baxter, 1979). Bees transferred to greater CSA should increase their hoarding while bees transferred to lesser CSA should decrease their hoarding. In order to accomplish such transfers, bees were first caged for three days with either one piece (47 cm^2) or three pieces (140 cm^2) of empty comb. The bees were then transferred to new cages: from cages with three combs to cages with three combs (3-3), from three combs to one comb (3-1) from one comb to three combs (1-3), and from one comb to one comb (1-1). The experiment continued for the following four days.

Before transfer, the bees with three pieces of comb hoarded more sucrose solution than those with one piece (1 piece, mean $\pm$ SEM = 5.58 $\pm$ 0.09; 3 pieces, 7.25 $\pm$ 0.11; $P < 0.01$). After transfer, the bees in treatment group 1-3 hoarded 15.0 $\pm$ 0.5 ml; more than any other treatment group ($P < 0.05$). Bees in group 3-3 continued a relatively high rate of hoarding (11.7 $\pm$ 0.56 ml). The hoarding of group 3-3 was numerically but not statistically higher than that of bees in group 3-1 (9.7 $\pm$ 0.45 ml) and it was significantly higher ($P < 0.05$) than that of group 1-1 bees (8.8 $\pm$ 0.34 ml). Group 3-1 bees hoarded numerically but not significantly more than group 1-1 bees.

These results were consistent with the comb stimulation hypothesis. The data collected before bees were transferred were similar to comparative data from other experiments: bees with more comb hoarded more sucrose solution or stored more honey. Most importantly, as predicted, the transfer of bees to greater or lesser amounts of comb resulted in a raising or lowering of the hoarding rate.

An unexpected aspect of the results was the suggestion that past experience with comb amounts affects the bees hoarding rate. Bees in group 3-1 tended to hoard more than bees in group 1-1. This indicated that the change to a less intense response, although reasonably rapid, was not immediate. Also, those bees in group 1-3 hoarded more than the comparable controls in group 3-3. This indicated that bees are additionally stimulated by large amounts of empty comb after limited exposure to empty comb.

One of the simplest explanations of these results is that the internal chemistry of bees responds to some type of stimulation from comb. Possibly, the relative abundance of some physiological compound varies with comb availability. An abundance of precursors in the bees might be available to be changed to a super-normal abundance of a regulating compound in situations similar to those of the 1-3 treatment group. The response of the 3-1 treatment group might be the result of a time lag required for the reduction in the titer of a regulating compound in the haemolymph or nervous system. Explorations of the neurochemistry of bees under various levels of comb stimulation may reveal compounds which are responsible for regulating the intensity of hoarding, foraging, and honey production.

Regardless of precisely how stimulation by comb is internally translated by bees and transformed into the behavioral result of more intensive foraging, the effect of comb on bees is itself known to be mediated chemically. This conclusion was suggested when bees in hoarding cages were only occasionally observed on comb not being used for immediate storage. Seemingly, additional comb had its effect even though bees only infrequently came into direct contact with it.

This hypothesis of chemical stimulation was tested with modified hoarding cages (Rinderer, 1981). The modifications permitted air to be pumped first through a 2 l plexiglass box filled with an experimental material and then into hoarding cages just above the comb. Three experiments were then used to evaluate the effects on hoarding of volatiles from empty comb held at 35°C (the temperature of a colony's brood area) from empty comb held at 5°C and from comb filled with stored honey at 35°C.

Only the volatiles from the warm empty comb increased hoarding rates. Thus, volatiles from empty comb at a temperature similar to that occurring in the brood area of bee nests increased the hoarding of bees. These volatiles were not given off in effective amounts by cold empty comb or warm comb that contained stored honey.

Probably, these volatiles are pheromones incorporated into comb by bees as it is built, repaired and maintained. In an experiment comparing new, light colored comb and older comb which had been used several times for brood rearing, both types of comb increased hoarding to similar levels (Rinderer and Baxter, 1980). Since the stimulatory mechanism of comb is volatile chemicals, the continued ability of older comb to induce increased hoarding likely results from the renewal of its stimulatory properties. Probably, renewal occurs as the bees clean and repair comb.

The performance of bees in hoarding cages is a good, although not a complete, predictor of the foraging and honey storage of bees in the field. Genetic differences between various stocks of bees identified in hoarding experiments were similar to differences discovered in field trials in some but not all cases (Kulincevic and Rothenbuhler, 1973; Kulincevic *et al.*, 1974). Also, hoarding experiments with comb impregnated with 2-heptanone showed that this chemical strongly increased the intensity of hoarding (Rinderer, 1982c). Yet, a similar field trial failed to result in differences between treated and untreated colonies.

While some effects that were apparent in laboratory experiments were less apparent in field experiments, the opposite was true in experiments using empty comb as a treatment variable. Often the differences found were greater in field experiments. For example, observation hives having three times as much empty comb as control hives stored almost ten times as much honey (Rinderer, 1982b). During this experiment, the recruitment dancing in the colonies reflected the differential honey storage. In colonies with more comb, more bees danced, the dances had greater average durations, and more recruit bees followed individual dances. Consequently, the combined effects of these variables resulted in a two-fold increase in the rate of recruitment.

Further insight into the nature of the influence of additional comb on recruitment dancing was provided by an experiment in which bees were trained to a feeding station (FIGURE 3) (Rinderer, 1982b). Only marked bees which had previously been trained to the feeding stations were permitted to forage at them during the experiment. The experiment using a descending sequence of sucrose solutions (2.0 M, 1.0 M, 0.5 M and 0.25 M) simultaneously at separate feeding stations for observation colonies given one or three empty storage combs.

The data from the experiment were the numbers of bees foraging at the feeding stations and the numbers of bees doing recruitment dances after foraging during equal time periods for each hive type (CSA treatment) and each sucrose concentration. Bees from the two hive types behaved quite differently. Comparatively, bees from hives having three storage combs foraged in significantly reduced numbers at the station when it contained less concentrated food. Yet, if they foraged at all, they had a significantly higher likelihood of doing recruitment dances after collecting sucrose concentrations below 2.0 M. Thus, the stimulation from additional empty comb increased the selectivity of the bees' food choice and also increased their tendency toward recruitment dances and group foraging. Bees that were stimulated by lesser amounts of comb were less selective in food choice, less likely to dance, and more likely to engage in individual foraging; that is, they tended toward gleaning.

Further support for the observation that additional CSA stimulated increased recruitment was supplied by Rinderer and Hagstad (1984). They found, when studying the foragers of field colonies, that increased amounts of CSA resulted in an increase in the proportion of nectar foragers, a decrease in the proportion of pollen foragers and a decrease in the proportion of foragers simultaneously collecting nectar and pollen.

The usefulness of these two contrasting foraging strategies varies through the nectar flow season (Rinderer, 1982b). Honey storage strongly stimulated by empty comb is enhanced during the main nectar secretion period. However, strong stimulation by empty comb impairs honey storage during the poorer quality nectar secretion periods of autumn (FIGURE 4). In such conditions, lesser stimulated bees which forage as individual gleaning bees are more successful. These two foraging strategies, or at least the honey storage pattern

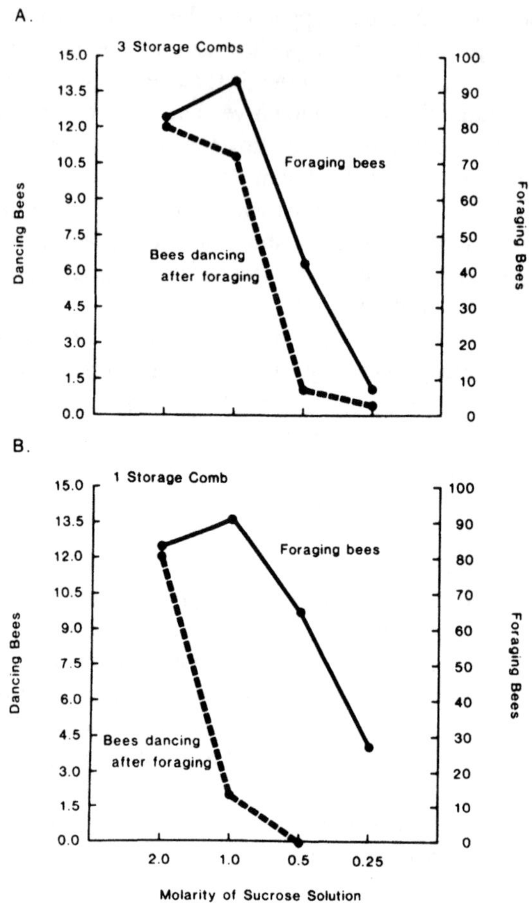

FIGURE 3. Numbers of bees which foraged from observation hives having (A) three storage combs or (B) one storage comb, and numbers of bees which danced after foraging on four concentrations of sucrose solution presented at a feeding station in a sequence of descending concentration. (From data of Rinderer, 1982b. Copyright in public domain)

they produce, can derive from conditions other than differential stimulation by empty comb. A curious, previously unpublished result of the experiment shown in FIGURE 4 was a change in the order of colonies when ranked according to the average calories collected by individual foragers (FIGURE 5). These values were estimated for each of six colonies in each of two apiaries throughout the season (levels of CSA were maintained experimentally throughout the season). Initially, in May and June, the ranking of colonies was identical from week to

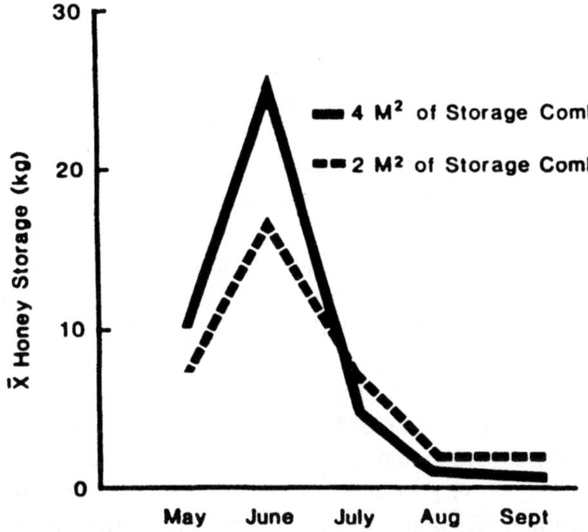

FIGURE 4. Average monthly weight of honey (kg) stored by six colonies with 4 m^2 of storage comb and six colonies with 2 m^2 of storage comb. The colonies were equally divided between two apiaries. (From Rinderer, 1982b. Copyright in public domain.)

CALORIES	JUNE 1	JUNE 2	JUNE 3	JUNE 4	JULY 1	JULY 2	JULY 3	JULY 4	AUGUST 1	AUGUST 2	AUGUST 3	AUGUST 4
Most	1	1	1	1	1	1	6	6	6	6	6	6
	2	2	2	2	2	6	1	5	5	5	5	5
	3	3	3	3	6	2	2	1	1	4	4	4
	4	4	4	6	3	3	5	2	4	1	3	3
	5	5	5	5	5	5	4	4	2	3	1	2
Least	6	6	6	4	4	4	3	3	3	2	2	1

Week: 1 2 3 4 (JUNE), 1 2 3 4 (JULY), 1 2 3 4 (AUGUST)

FIGURE 5. The seasonal change in ranking, based on the average calories collected by individual foragers from colonies with 4 m^2 of CSA (colonies 1, 2, and 3) and from colonies with 2 m^2 of CSA (colonies 4, 5, and 6). These trends were consistent for the colonies in both apiaries that provided the honey yield data presented in FIGURE 4.

week. The three colonies in each apiary having more CSA had nectar foragers collecting more calories as a group and their rankings were consistent. The three colonies in each apiary having less CSA had nectar foragers collecting fewer calories as a group and their rankings were consistent. As the season progressed, the rankings changed. The ranking of the colonies initially collecting the fewest calories progressively moved higher. At the end of the season, the initial rankings were completely reversed. This phenomenon occurred in identical fashion among the colonies in both apiaries. This remarkable consistency, as well as the inversion of ranks as the season changed, suggested that some stable (probably genetic) component of the individual colonies determined the magnitude of the behavioral response to CSA.

An additional experiment also suggested the involvement of genetic factors. Monthly hive weight records in an apiary of from four to eight colonies given uniform management were recorded by Oertel *et al.* (1980). An analysis of these records permitted an evaluation of the possible influences on honey storage patterns of genetic or very local, stable differences in environment, such as exposure to sunlight (Rinderer, 1982b). From the records of each year, we identified those colonies that stored the most and the least amount of honey during the month in which the apiary stored the most honey. Data on honey storage by this pair of colonies in the first and last months with a nectar flow were then identified. The colonies that stored the least honey during the major nectar flow consistently stored the most during both the early and the later minor flows (FIGURE 6). These patterns of honey storage support the hypothesis that more intense nectar harvesting and selection of primarily highly rewarding nectar sources result in more honey storage during major nectar flows, while less intense harvesting and selection of lower quality nectar sources, i.e. gleaning, result in greater storage during minor flows.

These data were not collected in a way that permitted an unambiguous identification of the source of intercolony variation. Potential genetic sources were confounded with possible, although less likely, environmental sources. Nonetheless, the genetical hypothesis is intriguing, since it would explain why different stocks of bees give different comparative honey yields in different areas. Genetical differences may cause bees to be more or less responsive to stimulation from empty comb and thereby increase the tendency of bees to be either group foragers or individual gleaning foragers. Such genetical variation would permit both artificial and natural processes to select bees best suited to the usual nectar availability of their environment.

THE ANNUAL CYCLE OF REGULATION OF EUROPEAN HONEY BEE FORAGING

Predictable, seasonal trends in nectar production commonly occur in temperate climates (Crane, 1975; Oertel *et al.*, 1980). In many areas, early

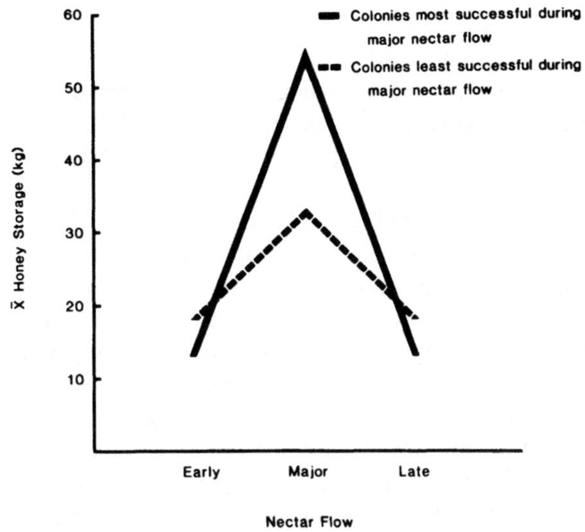

FIGURE 6. Average seasonal weight of honey (kg) stored by colonies with the most and least honey during the major nectar flow during 25 years. (From Rinderer, 1982b. Copyright in public domain.)

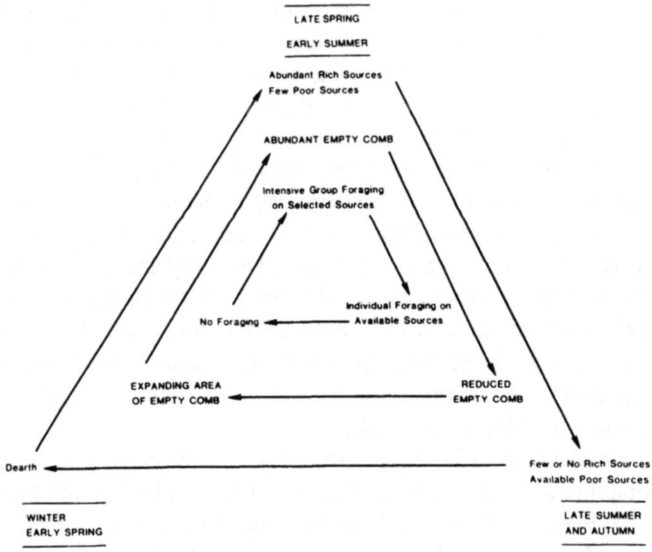

FIGURE 7. A model of the seasonal regulation through the influence of empty comb of European honey bee nectar harvesting and honey production in a temperate climate.

spring nectar sources tend to be both few and poor. This situation changes rapidly, and by late spring to early summer, nectar sources tend to be both varied and rich. A decline in variety and abundance follows, until by late summer, only a few poor quality nectar sources usually exist (FIGURE 7).

These seasonal changes in the quality and quantity of nectar availability are paralleled by the amount of empty storage comb in a feral honey bee nest (FIGURE 7). The winter use of stored honey followed by the intensified brood rearing of early spring and the consequent food consumption by greater numbers of bees result in a rapid reduction in the amount of stored honey. This causes a concomitant increase in empty comb. During the major nectar flow, the empty comb is utilized for honey storage. Consequently, empty comb becomes increasingly less available as the season progresses.

The foraging of honey bees also fluctuates in ways that are most appropriate to the predictable nectar flow conditions of the season (FIGURE 7). The collection of experiments using hoarding cages, observation hives, and field colonies demonstrate that the varying amounts of volatiles emanating from varying amounts of empty comb regulate the fluctuating characteristics of foraging. The chief feature that demonstrates this is that comb stimulation which is usual to a season causes bees to behave as though they were experiencing that season regardless of the season in which the experiment was conducted. In conditions of greater stimulation by comb volatiles, bees tend to be more selective of nectar sources and through the use of increased levels of dance communication exploit quality sources intensively. In conditions of less stimulation by comb volatiles, bees tend to be less selective of nectar sources and less likely to perform recruitment dances for all but the highest quality sources. These conditions hold true, regardless of nectar flow conditions.

The value of comb volatiles as a regulator of foraging resides in its common influence on all the bees in a colony engaged in nectar collection and storage. Scout bees have differing experiences in the field. Thus, the overall nectar flow conditions cannot serve to guide recruitment to only the best nectar sources. Scouts with incomplete experience would dance for the best sources they found as intensively as other scouts would dance for the best available sources. In the same way, the receptivity of house bees to accept nectar loads is insufficient for overall regulation. Bees receiving nectar in the hive also have a limited experience with incoming nectar loads.

Certainly, there are adjustments in foraging and dancing depending upon nectar availability and variety, house-bee receptivity and a host of other extrinsic factors (von Frisch, 1967). Stimulation by empty comb establishes the thresholds for foraging and dancing and probably also house bee receptivity to incoming nectar loads and thereby provides a common reference for the fine-scale adjustment of foraging and communication which is best suited to the colony's needs.

Usually, comb regulation and secondary adjustment of foraging and communication lead to nectar harvesting behavior which is best suited to varying seasonal conditions. In unusual conditions, levels of comb stimulation might be superficially considered inappropriate to nectar flow conditions. A colony with storage combs filled with honey part-way through an unusually strong and prolonged nectar flow might be thought to have a lower nectar harvesting intensity than appropriate. This situation would only serve to increase the number and the probability of the survival of swarms issued by the colony. The parent colony would not suffer a disadvantage, since colonies establish themselves in selected cavity volumes (Seeley and Morse, 1976; Seeley, 1977) that are presumably large enough to hold food reserves capable of supporting the survival of the colony through the season of dearth. Adversity (unusual weather, disease, or predation) may result in a colony having large amounts of empty comb (and low food reserves) at a time before or after the main nectar flow period. Such a colony would be in danger of starvation, and a highly intensive nectar harvesting (and exploitation of only high quality nectar sources) is not inappropriate. Maximal foraging on poor quality nectar sources may not provide sufficient food reserves for survival. However, finding a good quality nectar source (perhaps the honey reserves of a nearby colony), and exploiting it intensely would have a greater chance of ensuring the colony's survival than foraging on poor quality nectar sources.

Comparative Nectar Harvesting of European and Africanized Bees

The experimental pathway to understanding the comparative foraging and honey production of Africanized and European bees rested in comparing representatives of the two groups of bees across varying conditions of forage availability and foraging stimulation. A preliminary experiment compared hoarding by the two groups with standard amounts of comb and comb impregnated with 2-heptanone (Rinderer *et al.*, 1982), since 2-heptanone increased the hoarding rates of European bees (Rinderer, 1982c). Africanized bees hoarded less sucrose regardless of the treatment. The increase in hoarding rate induced by 2-heptanone in Africanized bees was about half the increase in hoarding rate induced in European bees.

These results did not support the conclusion that Africanized bees are superior honey producers in all cases (Kerr, 1967, Kerr *et al.*, 1972; Gonçalves, 1975) since for European bees high hoarding rates correlate well with honey production (Kulincevic and Rothenbuhler, 1973; Kulincevic *et al.*, 1974; Rinderer and Baxter, 1979; Rothenbuhler *et al.*, 1979).

A second, more detailed hoarding experiment (Rinderer *et al.*, 1986) compared Africanized and European honey bee hoarding when the amounts of empty comb were varied. The experiment was modeled after one which showed that for European bees, additional empty comb increased both hoarding intensity and hoarding efficiency (Rinderer, 1983). Bees in the hoarding cages had access to three identical gravity feeders containing either 20% (wt/wt) sucrose in water solution, 50% sucrose solution or water only. This experimental design, giving bees simultaneous access to different sucrose solutions, permitted estimates of hoarding efficiency. These estimates of reward per unit of work were calculated as total grams of sugar hoarded divided by total grams of solution hoarded, and could range from 0.2 (only 20% solution was hoarded) to 0.5 (only 50% solution was hoarded). Portions of efficiency estimates derived from the hoarding of the 20% solution probably inflate the overall efficiency estimate somewhat, since bees tend to take smaller loads of lower quality food (von Frisch, 1967). This method assumed equal-sized loads. Thus any differences in efficiency were likely to be more extreme than those calculated by this method.

In the comparison of Africanized and European bees, both types of bees increased both their hoarding intensities and their hoarding efficiencies. However, European bees hoarded more with greater efficiency; again suggesting that it might not be valid to extrapolate the observations of superior honey production by Africanized bees to all conditions.

Field experiments that took advantage of various nectar flow conditions indicated that the two types of honey bees respond differently to different nectar flow conditions. In an experiment comparing responses to daily fluctuations in nectar flow conditions (Rinderer *et al.*, 1984), the relative energy content of the nectar loads collected by the two geographical types varied. An important source of this variation was in the volume of nectar collected. Also, European bees generally were more successful in securing a nectar load but their percentage of successful foragers tended to be either high or low. Africanized bees returned more often to their nest without a nectar load but had intermediate as well as high and low percentages of successful foragers. Overall, European colonies had greater number of forages throughout the study.

These differences suggest underlying differences in the use of recruitment and group foraging by the two bee types. Presumably, bees that use increased levels of communication and recruitment and thereby improve their foraging success would be similar to the European bees in this study. Bees with strong tendencies toward group foraging (Johnson and Hubbell, 1975) and that are highly dependent upon communication and recruitment should have either very high or low forager success rates. High rates would occur when there is recruitment and low rates would occur when scouts find only nectar sources lacking sufficient value to stimulate recruitment. Bees showing more reliance on individual foraging would likely have a lower rate of success (mid-range) in conditions favoring group foragers, especially if these sources were scattered and difficult to find. Overall, the result of this experiment suggested that Africanized bees are adapted to nectar resource conditions that are, in most cases, best exploited by gleaning foragers which do not rely strongly on communication. The contrast in the results suggested that European bees are better adapted to conditions that are best exploited by group foraging which is stimulated by dance communication.

An additional experiment confirmed these conclusions (Rinderer *et al.*, 1985). Comparisons of Africanized and European honey bees were made during two periods of nectar availability. The first period provided relatively intense nectar availability. The dominant nectar secreting plants were widely scattered araguaney trees (*Tabebuia sp.*) Secretion by these scattered trees, primarily at night, provided a rich but patchy source of nectar in the first hours of the morning. In the second nectar flow period, nectar availability was relatively weak. The araguaney trees had nearly finished flowering and they were replaced as a dominant nectar source by fence-row plantings of mataraton trees (*Gliricidia sepium*). These legumes produced nectar primarily during the day until mid-to-late afternoon. Hence, in the second nectar flow period there were many more flowers for a longer period each day, with each flower producing much less nectar. Field colonies were used to study honey yields, nectar-load characteristics, and flight activity including the times of flight initiation and cessation. Colonies in observation hives were used to study dance communication and recruitment of foragers.

The principal theme of the results was that bee-type interacted with nectar flow. This was illustrated by the honey production records (FIGURE 8A). In period 1, European bees clearly produced more honey. In period 2, Africanized bees produced numerically more honey. However, the large variance associated with the honey production of the Africanized bees prevented a determination of whether the Africanized bees produced more or the same amount of honey as European bees. Certainly, they did not produce less. These general trends of interaction appeared in nectar load characteristics of volume, concentration and energy content, and in the daily flight patterns of foraging bees.

The patterns of dance recruitment were especially instructive (FIGURE 8B). These too interacted with nectar flow periods. Similar patterns to those of honey production occurred for the numbers of nectar foragers initiating dancing in a five minute period, the numbers of recruit bees following individual dancing bees, and the duration of individual dances. The product of the first two of these values provided an estimate of the rate of recruitment. During the first nectar flow, European bees recruited much more intensively than did Africanized bees. During the second nectar flow, both types showed reduced recruitment; the European bee reduction in recruitment rate was quite strong and recruitment rates of the two bee types were similar. These results strongly support the hypothesis that the fundamental nectar foraging difference between Africanized and European bees is that Africanized bees are adapted to conditions where an individual gleaning type of foraging is more successful whereas European bees are adapted to conditions where intensive foraging reliant on dance communication is more successful.

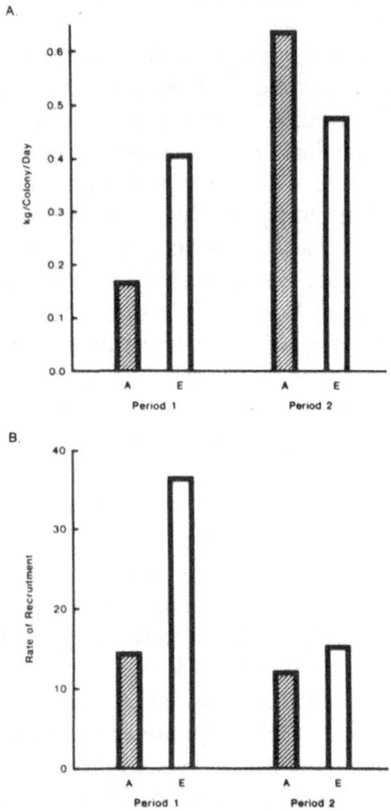

FIGURE 8. (a) Average honey production of ten European (E) and ten Africanized (A) full-sized honey bee colonies during two different nectar flow periods. (b) Average rate of recruitment (numbers of recruited bees/5 minutes) for five Africanized (A) and five European (E) honey bee observation hive colonies during the same two nectar flow periods. (Data from Rinderer *et al.*, 1985. Copyright in public domain.)

Additionally, on each of six days (three for each nectar period) both Africanized and European bees from every colony in the experiment began flying in large numbers before sunrise when illumination measured at colony entrances was 1 lx. The last bees returned to both Africanized and European colonies shortly after sunset when measured illumination was ca. 5 lx. The strict uniformity of flight initiation and cessation of Africanized and European bees in both nectar flow periods strongly suggests that there are no fundamentally important differences in these aspects of foraging behavior. Reported differences (Kerr *et al.*, 1972; Fletcher, 1978) apparently are not sufficiently ubiquitous to

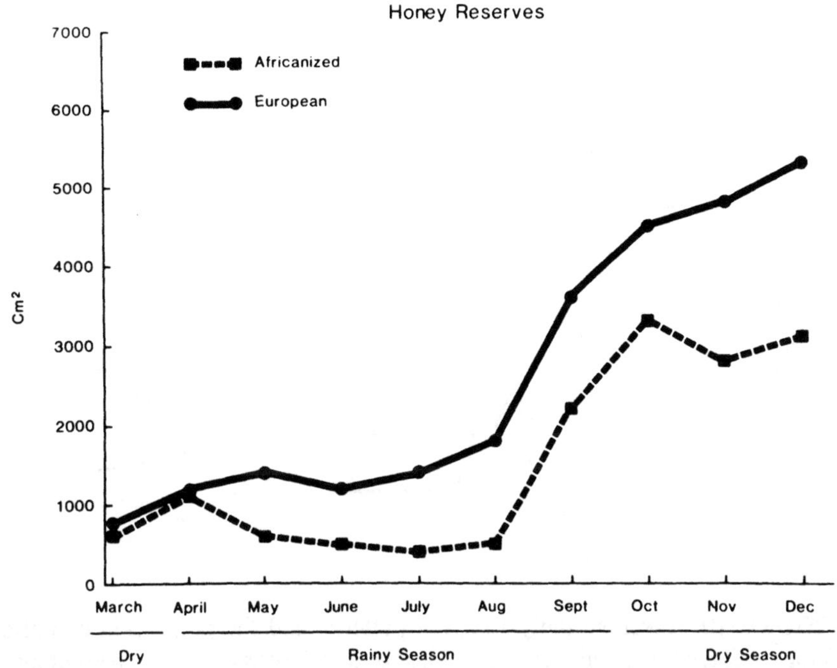

FIGURE 9. The average seasonal accumulation of honey resources by 40 colonies each of Africanized and European honey bees divided equally into two apiaries. (From Pesante, 1985, with permission. Copyright, 1985, by D. Pesante.)

incorporate them into a general model of comparative nectar foraging and honey production.

The results of these shorter term studies were confirmed and extended in a study lasting ten months (Pesante, 1985). He also concluded that "Africanized colonies had lower colony and honey weight gains than European colonies under favorable nectar flow conditions and higher colony honey weight gains than European colonies under poor nectar flow conditions." Across the duration of the study, European bees collected and stored comparatively increasingly greater amounts of honey reserves (FIGURE 9) (Pesante *et al.*, 1987).

Danka *et al.* (1986) studied in detail the apparently greater numbers of foragers from European colonies (Rinderer *et al.*, 1984, 1985). They confirmed the observation and found that the difference was consistent for different sizes of colonies and different nectar flow periods.

Pesante (1985) and then Danka *et al.* (1987) studied the comparative diet selection of Africanized and European bees. Although European colonies had greater overall numbers of foragers, Africanized colonies had greater numbers and higher percentages of pollen foragers throughout the day (FIGURE 10).

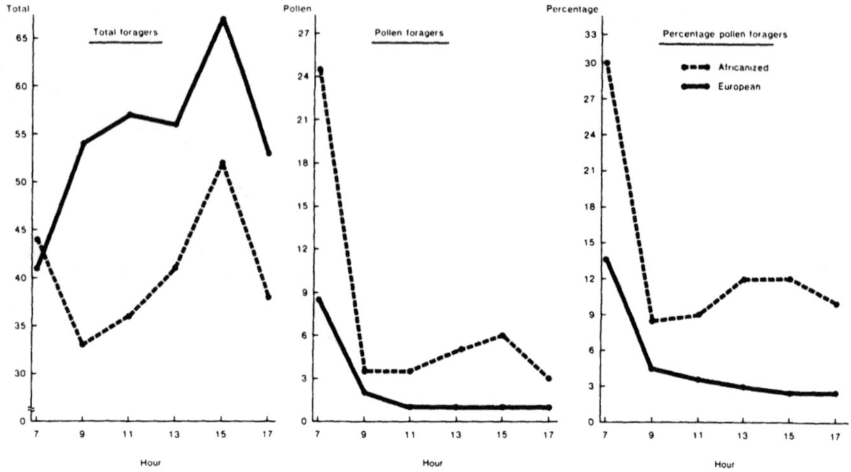

FIGURE 10. Average daily foraging patterns and the proportion of pollen foragers of twelve colonies each of Africanized and European honey bee colonies. (From Danka *et al*., 1987. Copyright in public domain.)

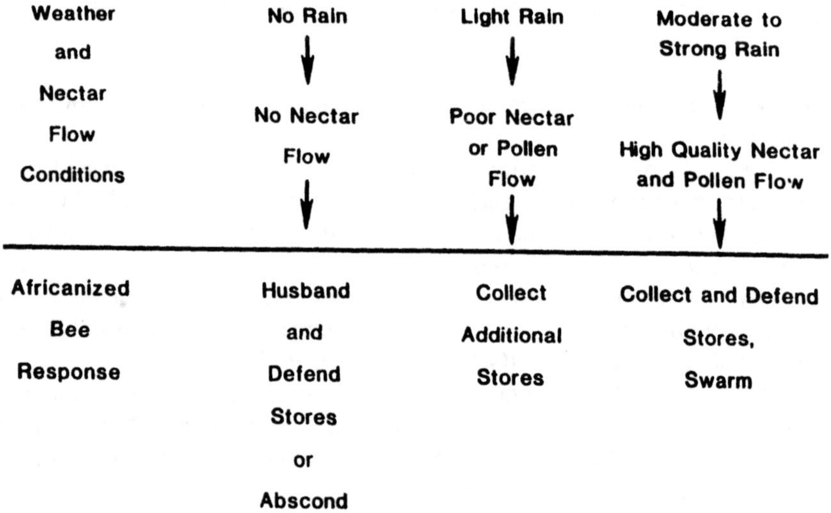

FIGURE 11. A model of the resource dependent nectar harvesting and honey production of Africanized bees. Compare with FIGURE 7.

Africanized pollen foraging resulted in larger stores of pollen in Africanized bee nests in the same experiments in which European nests had larger honey and nectar caches. Additionally, Pesante (1985) found that Africanized bees maintained larger brood nests in conjunction with maintaining larger pollen reserves and smaller honey reserves.

AFRICANIZED HONEY BEE FORAGING

The differences between Africanized and European honey bee foraging are rooted in the ecological differences in their home range. The chief ecological determinant of foraging in the home range of the parental subspecies ofAfricanized bees, *A. m. scutellata*, is rainfall. Trewartha (1981) describes a large number of interacting climatic factors which result in "rainfall [which is] modest in amount" and "highly unreliable as well." This unreliability includes amounts, locations and seasonality. Since flowering and nectar flows are dependent upon these unreliable rains, they are also unreliable. "Annual" nectar flows may not occur in specific localities and sometimes in large regions. Where nectar flows do occur they may be quite weak because only light rains have fallen. When they occur, the "annual" onset of nectar flows may vary by as much as two to three months; as does the onset of the "annual" rains.

These three conditions contributing to extreme nectar flow variability have probably been principle selective forces in the evolution of bees in the area. In any event, the behavior of Africanized bees suggests this since their nectar foraging and other aspects of their natural history are well suited to such conditions. Comparative studies (*loc. cit.*) indicate that these bees will tend to forage as gleaners, that is, as individual foragers on whatever nectar is available. They lack the selectivity which would cause them not to forage on a poor flow regardless of stimulation from empty combs. In Eastern and Southern African conditions, foraging on a subsistence flow may insure survival as the onset of a subsequent nectar flow is at least possible.

Brood nest expansion and subsequent swarming by Africanized bees are triggered by nectar and pollen flows (Pesante, 1985), a clear response to flows starting at unpredictable times. Generally, Africanized bees lack significant brood nest expansion prior to the occurrence of nectar and pollen resources. In contrast, European bees begin their annual population growth as early as late December or January, probably using photoperiod as a cue (Kefuss, 1978).

When quality floral resources exist, Africanized bees place more nearly equal energies into the storage of nutritional resources and swarming. Their more intensive pollen foraging (Pesante, 1985; Danka *et al.*, 1987) leads to a reduced potential for honey storage but an increased potential for swarming. Apparently, Africanized bees will continue to swarm at frequent intervals as long as forage is available to support brood rearing (Otis, 1977).

Thus, our model of Africanized honey bee foraging stresses a far greater unreliability of nectar and pollen availability (FIGURE 11). Africanized bees are essentially opportunistic. Given the opportunity of a subsistence flow, they will forage, store the resources they collect, and increase their chances of survival. Later, they will either reproduce on a better flow should it occur, or, given sufficient nutritional stress, abscond. By absconding, they may find an area with a nectar flow because it received some of the scattered unpredictable rain.

The linkage of absconding and colony defense (FIGURE 11) to resource availability helps complete an understanding of Africanized bee foraging. Absconding is a clear response to nectar and pollen flows triggered by rains which are unreliable in immediate locations. Defensive behavior in African bees and their Africanized progeny has probably been shaped by unpredictable and often poor resource availability as much or more than any other selection pressure. Certainly, Africa has a variety of organisms (the most efficient of which are undoubtably humans) which plunder and destroy honey bee nests and colonies. Such predation is essential to the adaptive value of defensive behavior. However, the natural history of bees in Europe also includes predators (humans again being chief among them). Thus, differences in predation are probably not the origin of the strong difference in levels of defensive behavior (Collins *et al.*, 1982) among African, Africanized and European bees. The more likely candidate is resource unpredictability. With predictable floral resources, the response which best contributes to fitness following minor levels of nest plundering may be to recover the losses through intensive foraging rather than to lose potential foragers through defense. Where this is less of a possibility, defensive behavior would presumably be intensified.

Some experimental evidence for this interpretation was provided by Collins and Rinderer (1985). They found that volatiles from empty comb functioned as primers for defensive behavior for both Africanized and European bees. Colonies with more CSA responded faster to moving targets and stung more often than colonies with less CSA. This experimental demonstration of a linkage between foraging and defense shows that colonies which are living in more marginal nutritional conditions are more defensive.

CONCLUSION

The contrast drawn between Africanized bees cast as individual foragers adapted to often poor and always unpredictable resources, compared to European bees viewed as group foragers adapted to often rich and generally predictable resources, accommodates the wide variety of sometimes superficially contradictory data. It is important to emphasize that this contrast is one of degree and not kind. African bees and their Africanized progeny do have some capacity for group foraging mediated through dance recruitment. Also, their foraging is regulated, albeit to a lesser degree, by the presence of empty comb in their nests. These characteristics are to be expected. Although keynoted by their unpredictability, the rains of southern and eastern Africa do have a seasonality. The dry season will only have very unusual rain; the wet season will have unreliable rains. European bees have some capacity for reacting to nectar flows best exploited by individual foraging and, to a lesser degree, will be stimulated by resource conditions to forage intensely and even to swarm when confronted by contradictory photoperiod cues. Forage resources in temperate zone areas, although they have greater predictability, also vary from year to year (Oertel *et al.*, 1980). Both geographical types of bees show the capacity, common to all western honey bees, of shifting their foraging from intense selective harvesting to gleaning. However, the general tendency of Africanized bees is to be more successful on the side of the foraging continuum where gleaning is more adaptive. European bees tend to be more successful on side of the continuum where intense selective harvesting is more adaptive.

Because of the interactive nature of bees with their environment, the choice of bee stock by an apiculturalist who has the opportunity to use either Africanized or European bees will vary with local conditions, at least if the economic disadvantages of Africanized bee defensive behavior are discounted. Although Africanized bees are now known to not be superior honey producers due to an extended period of daily foraging, there are conditions where their tendency to be superior gleaning bees will result in better honey yields. Alternatively, in better production areas where intensive nectar harvesting is possible, European bees are the clear commercial choice.

Acknowledgement

In cooperation with Louisiana Agricultural Experiment Station.

LITERATURE CITED

Alpatov, W. W. 1976. Rokovaya oshibka v opredelenii parody pchel [A fatal error in determining the race of a honey bee]. *Priroda* 5:72-73 (in Russian)

Cobey, S., Locke, S. 1986. The Africanized bee: A tour of Central America. *Am. Bee J.* 126:434-440.

Collins, A. M., Rinderer, T. E. 1985. Effects of empty comb on defensive behavior of honey bees. *J. Chem. Ecol.* 11:333-338.

Collins, A. M., Rinderer, T. E., Harbo, J. R., Bolten, A. B. 1982. Colony defense by Africanized and European honey bees. *Science* 218:72-74.

Crane, E. 1975. The flowers honey comes from. In *Honey: A Comprehensive Survey*, ed. E. Crane, pp. 3-76. London: Heinemann.

Danka, R. G., Hellmich, R. L. II, Rinderer, T. E., Collins, A. M. 1987. Diet-selection ecology of tropically and temperately adapted honey bees. *Anim. Behav.* 35:1858-1863.

Danka, R. G., Rinderer, T. E., Hellmich, R. L. II, Collins, A. M. 1986. Foraging population sizes of Africanized and European honey bees (*Apis mellifera* L.) colonies. *Apidologie* 17:193-202.

Fletcher, D. J. C. 1978. The African honey bee, *Apis mellifera adansonii*, in Africa. *Ann. Rev. Entomol.* 23:151-171

Gonçalves, L. S. 1975. Do the Africanized bees of Brazil only sting? *Am. Bee J.* 115:8-10, 24.

Johnson, L. K., Hubbell, S. P. 1975. Contrasting foraging strategies and co-existence of two bee species on a single resource. *Ecology* 56:1398-1406.

Kefuss, J. A. 1978. Influence of photoperiod on the behaviour and brood rearing activities of honey bees in a flight room. *J. Apic. Res.* 17:137-151.

Kerr, W. E. 1957. Introdução de abelhas africanas no Brasil. *Brasil Apicola.* 3:211-213.

Kerr, W. E. 1967. The history of the introduction of African bees to Brazil. *S. Afr. Bee J.* 39:3-5.

Kerr, W. E., Gonçalves, L. S., Blotta, L. F., Maciel, H. B. 1972. Biología comparada entre as abelhas italianas (*Apis mellifera ligustica*), africana (*A. m. adansonii*) e suas híbridas. *1º Congr. Bras. Apic. (Florianópolis, 1970)*, ed. H. Wiese, pp. 151-185. Santa Catarina, Brazil.

Kerr, W. E., Portugal-Araújo, V. de 1958. Raças de abelhas de Africa. *Garcia de Orta* 6:53-59.

Kulincevic, J. M., Rothenbuhler, W. C. 1973. Laboratory and field measurements for hoarding behaviour in the honeybee (*Apis mellifera* L.). *J. Apic. Res.* 12:179-182.

Kulincevic, J. M., Rothenbuhler, W. C., Stairs, G. R. 1973. The effect of presence of a queen upon outbreak of a hair-less black syndrome in the honey bee. *J. Invert. Pathol.* 21:241-247.

Kulincevic, J. M., Thompson, J. C., Rothenbuhler, W. C. 1974. Relationship between laboratory tests of hoarding behavior and weight gained by honey bee colonies in the field. *Am. Bee J.* 114:93-94.

Labougle, J. M., Zozaya, J. A. 1986. La apicultura en Mexico. *Ciência y Desarrollo* 69:17-36.

Oertel, E., Rinderer, T. E., Harville, B. G. 1980. Changing trends in Southeastern Louisiana honey production through 25 years. *Am. Bee J.* 102:763-765.

Otis, G. W. 1977. The swarming biology and population dynamics of the Africanized honey bee. *J. Apic. Soc. Trinidad and Tobago* 77:313-325.

Pesante, D. 1985. *Africanized and European honey bee (Apis mellifera* L.) *pollen and nectar foraging behavior and honey production in the neotropics.* PhD Dissertation, Louisiana State Univ., Baton Rouge.

Pesante, D., Rinderer, T. E., Collins, A. M. 1987. Differential nectar foraging by Africanized and European honeybees in the neotropics. *J. Apic. Res.* 26:210-216.

Rinderer, T. E. 1981. Volatiles from empty comb increase hoarding by the honey bee. *Anim. Behav.* 29:1275-1276.

Rinderer, T. E. 1982a. Maximal stimulation by comb of honey bee (*Apis mellifera*) hoarding behavior. *Ann. Entomol. Soc. Am.* 75:311-312.

Rinderer, T. E. 1982b. Regulated nectar harvesting by the honeybee. *J. Apic. Res.* 21:74-87.

Rinderer, T. E. 1982c. Sociochemical alteration of honey bee hoarding behavior. *J. Chem. Ecol.* 8:867-871.

Rinderer, T. E. 1983. Regulation of honeybee hoarding efficiency. *Apidologie* 14:87-92.

Rinderer, T. E. 1985. Africanized honeybees in Venezuela: Honey production and foraging behaviour. In *Apiculture in Tropical Climates, Vol. 3,* ed. M. Adee, pp. 112-116. Gerrards Cross, England: Int. Bee Res. Assoc.

Rinderer, T. E., Baxter, J. R. 1978. Effect of empty comb on hoarding behavior and honey production of the honey bee. *J. Econ. Entomol.* 71:757-759.

Rinderer, T. E., Baxter, J. R. 1979. Honey bee hoarding behaviour: effect of previous stimulation by empty comb. *Anim. Behav.* 27:426-428.

Rinderer, T. E., Baxter, J. R. 1980. Hoarding behavior of the honey bee: Effects of empty comb, comb color, and genotype. *Environ. Entomol.* 9:104-105.

Rinderer, T. E., Bolten, A. B., Collins, A. M., Harbo, J. R., 1984. Nectar foraging characteristics of Africanized and European honeybees in the neotropics. *J. Apic. Res.* 23:70-79.

Rinderer, T. E., Bolten, A. B., Harbo, J. R., Collins, A. M. 1982. Hoarding behavior: European and Africanized honey bees (Hymenoptera: Apidae). *J. Econ. Entomol.* 75:714-715.

Rinderer, T. E., Collins, A. M., Hellmich, R. L. II, Danka, R. G. 1986. Regulation of the hoarding efficiency of Africanized and European honey bees. *Apidologie* 17:227-232.

Rinderer, T. E., Collins, A. M., Tucker, K. W. 1985. Honey production and underlying nectar harvesting activities of Africanized and European honeybees. *J. Apic. Res.* 23:161-167.

Rinderer, T. E. Hagstad, W. A. 1984. The effect of empty comb on the proportion of foraging honeybees collecting nectar. *J. Apic. Res.* 23:80-81.

Rothenbuhler, W. C., Kulincevic, J. M., Thompson, V. C. 1979. Successful selection of honey bees for fast and slow hoarding of sugar syrup in the laboratory by the honeybee. *J. Apic. Res.* 18:272-278.

Ruttner, F. 1986. Geographical variability and classification. In *Bee Genetics and Breeding,* ed. T. E. Rinderer, pp. 23-56. Orlando, Fla: Academic Press.

Seeley, T. D. 1977. Measurement of nest cavity volume by the honey bee, *Apis mellifera. Behav. Ecol. Sociobiol.* 2:201-227.

Seeley, T. D., Morse, R. A. 1976. The nest of the honey bee (*Apis mellifera* L.) *Insectes Soc.* 23:495-512.

Smith, F. G. 1953. Beekeeping in the tropics. *Bee World* 34:233-245.

Swezey, S. L. 1986. Africanized honey bees arrive in Nicaragua. *Am. Bee J.* 126:283-287.

Trewartha, G. T. 1981. *The Earth's Problem Climates.* Madison: University of Wisconsin Press, 2nd ed.

von Frisch, K. 1967. *The Dance Language and Orientation of Bees.* Cambridge, Mass: The Belknap Press of Harvard University Press.

Wiese, H. 1977. Apiculture with Africanized bees in Brazil. *Am. Bee J.* 117:166-170.

13

ASPECTS OF AFRICANIZED HONEY BEE ECOLOGY IN TROPICAL AMERICA

David W. Roubik[1]

Continuous ecological data on Africanized honey bees in the neotropics are currently being gathered and evaluated in at least a few sites. This approach seems necessary because there has seldom been continuous study of what happens to Africanized honey bee populations as they invade new areas. It is still too early to appreciate what difference it will make that this bee is now a resident of approximately 20 million km^2 of diverse habitat between northern Argentina and Mexico. For apiculture, there has been a profound impact. But this is potentially of very small magnitude compared to other interactions between the bees and native plants and animals. The honey bee's spread and apparently permanent colonization of tropical America may be likened to a vast experiment. In some ways it is comparable to the abrupt removal of much tropical forest during the past few decades, since it too, potentially, affects many thousands of species. What might be the outcome of this experiment?

Honey bees of the species *Apis mellifera* are without doubt the most intensively studied insects, but their ecology is poorly known. Under natural conditions, there is a paucity of information about their major predators, parasites, competitors, resources and allies. Their evolutionary ecology, or the selective basis for traits they have inherited, is in many respects obscure. Since the honey bee is primarily a tropical insect, as are all social bees that maintain long-lived colonies (Michener, 1974), field research on Africanized *A. mellifera* can be considered an introduction, and a late one at that, to honey bee natural history. This type of information is fundamental to an understanding of how honey bee populations fit in with (or fail to fit in with) those of other

[1]Dr. Roubik is at the Smithsonian Tropical Research Institute, Balboa, Panama; his mailing address is: APO Miami, FL, 34002-0011, USA.

organisms. Along with an insight into the workings of social organization and colony activity (Seeley, 1985), such natural history provides a basis for learning to live with a colonizing insect--one that is very different from both native Hymenoptera and introduced European honey bees.

Beginning with the most general types of ecological interactions between Africanized *A. mellifera* and its environment in South and Central America, I hope to develop a broad picture of the ecology of this honey bee in the neotropics. To date, most research on the impact of Africanized honey bees in natural habitats has involved their competitive interactions with native bees. This will be the primary focus for exploring one result of their introduction to the Americas. Nonetheless, in order to make a statement about honey bee influence on various dynamics of tropical plants and pollinators, as well as other organisms, I have directed considerable research toward characterizing such populations as they normally exist (Roubik, 1989). Much of the research on community ecology that is emerging contains new information and perspectives, which seem essential to an evaluation of the impact of Africanized honey bees. Many advantages accrue from structuring research around both short- and long-term consequences of the Africanized honey bee invasion. Baseline data must be developed, along with suitable techniques for analyzing them. I have participated in this phase of the research since 1976, and it has taught me that nothing can be said regarding the changes due to Africanized honey bee arrival until patterns that already existed have been adequately characterized. Thus far, it seems that an equally long period may be necessary to determine whether there will be lasting effects due to the colonizing honey bees. The arrival of Africanized honey bees in the neotropics has proffered a research tool and another reason for studying tropical nature before it is altered beyond recognition.

NATURAL HISTORY

Where do Africanized honey bees live?

In terms of elevational limits, nests of the Africanized honey bee reach 2400 m in the northern Andes and up to 3000 m at La Paz, Bolivia, where conditions are relatively dry; they reach about 1200 m in the rainier, low mountains of central America. The lowlands, from sea level to about 500 m elevation, seem to support more colonies of the bees than do higher elevations. Dry forest or other relatively dry habitats seem to support more colonies than rain forest and wet lowlands. Similar climatic associations of the African honey bee in southern Africa are summarized by Taylor (1977). The present information is based upon personal observations and is thus a collection of impressions, rather than hard data. It may be satisfactory for some future hypothesis tests.

Slightly more quantitative studies have been made of nest density. In equatorial Africa, R. Darchen (pers. comm.) estimates that approximately ten colonies of *Apis mellifera* can be found in three km^2 of forest. A figure of six colonies per km^2 has also been given as a general estimate for South and Central America (Roubik, 1983a; Taylor, 1985; Roubik, 1987). Quantitative studies were made by Kerr and coworkers (Kerr, 1984) in the drier forests of the southwestern Amazon. They found an average of one colony of Africanized *A. mellifera* in 9,300 m^2. In the same study, a colony of each of five stingless bee (Meliponinae) species was found in 7,530 to 21,717 m^2. If these figures are multiplied to yield the number of colonies expected in each km^2, 46-133 bee colonies of each species might be found, 108 of them Africanized honey bees. The lower figure corresponds closely to estimates of stingless bee colony density in two neotropical forests calculated by Hubbell and Johnson (1977) and Johnson and Hubbell (1984). Unfortunately, these studies were made within areas far smaller than one km^2. And therefore, unless the distribution of colonies of a given species is uniform throughout the square kilometer in question, the number of colonies cannot be estimated simply by multiplying. To judge from two careful studies of spatial relationships among stingless bee colonies nesting in forest (Hubbell and Johnson, 1977; Fowler, 1979), the nest dispersion of a species is often random, and aggregations may conceivably be common. Thus we do not know how many nests of the Africanized honey bee, or native bees, exist per unit area of habitat. Many nests are cryptic or too high above the ground to be detected by all but the most labor-intensive survey techniques, such as the cutting of trees (Roubik, 1983b). Finally, the above studies of nesting colonies revealed a rather small proportion (about 15-50%) of the local species, which may suggest a scattered, aggregated nest dispersion of most species.

The permanence of nesting Africanized bee colonies in forest seems quite low, but from observations in Panamá, a nesting advantage seems to be enjoyed by colonies that build exposed (not enclosed in cavities) nests or establish them in cavities high in trees. Perhaps these colonies thus avoid excessively humid conditions and predator attack during the rainy season, when many colonies abscond (Boreham and Roubik, 1987). The initiation of the rainy season coincides with massive swarm movement on (and to?) the wet Atlantic side of the Panama Canal. It also is the only time of the year when megaswarms, several times larger than normal reproductive swarms, and very small swarms are seen in appreciable numbers. These are very likely to be absconding colonies, that can combine to form enormous swarms containing multiple queens (Silberrad, 1976; Anderson *et al.*, 1983). Despite this type of seasonal swarm movement, each of the exposed nests I have monitored or been shown in Panamá (two built under arboreal termite nests and one under a large tree branch) were occupied by colonies for at least 14-24 months (E. Adams, pers. comm). These nests were 14-22 m above ground level. Two additional colonies in tree cavities, but situated over 20 m above the forest floor, continued in their nests for over a year. In contrast, two other nests found in tree cavities in the same forests, but within three meters of the ground, were first occupied and then abandoned within four months. Both of these cavities were at other times occupied by a colony of polybiine wasps, *Stelopolybia* sp. (similar to French Guiana, Roubik, 1979a). One cavity was occupied before the honey bees by the stingless bee *Scaptotrigona barrocoloradensis*, and after the honey bees by another subgenus of *Trigona, Tetragona perangulata*, then later by *Stelopolybia* and still later by *Tetragonisca angustula*. Thus, other forest bees and wasps that use some of the nest sites of the Africanized honey bee are unstable or temporary in nest occupancy, and it is not surprising that similar behavior exists in Africanized honey bees. The cause of nest abandonment or colony mortality in all of these cases was uncertain, but the above scant observations on Africanized honey bees suggest that seasonal changes in rainfall or humidity directly cause absconding.

However, Africanized honey bees utilize nest cavities that are not used by most other social bees in the neotropics, and these include tree cavities with large holes leading to them. A small number of native stingless bee species use such cavities (Roubik, 1983b). These nest sites are abundant in forest plots that have recently been cleared and burned. Hollow fallen trees are readily occupied by swarms, and several colonies can be found within a few hectares in these disturbed habitats in lowland Panamá. With subsequent burnings these nest sites disappear, but small plots of 5-10 ha in their first year routinely harbor a few Africanized honey bee colonies in Panamá.

Nesting within protected sites used by birds, bats, kinkajous and opossums is common for the Africanized honey bee. However, the impact of Africanized bees on the availability of these sites to native forest animals is not known. It has been noted that apparently unoccupied tree cavities of this type are common (Hubbell and Johnson, 1977; Johnson and Hubbell, 1984). Competition for them is probably slight, and the Africanized honey bee would be unlikely to upset the breeding success of, for example, parakeets or trogons that nest in excavations within arboreal termite nests, or breeding activity by tree-hole-nesting birds. However, the presence of such cavities, which are not occupied through the year by native animals, enhances the chance of survival and reproduction by Africanized bees in forested areas.

In more extensively altered habitat, even one containing large tracts of forest and vegetation, Africanized honey bee nests seem to be concentrated in manmade structures (Boreham and Roubik, 1987). More than 40% of 281 established nests in the Panama Canal area were found in diverse manmade cavities; including buildings, the figure was about 75% of all nests. Nest sites high above the ground, notably in water towers and church belfries, were not uncommon. Natural sites in the same area included hollow tree trunks, fallen palm fronds, piles of rocks, and crevices in the ground. Only 2% of total nests were found in animal burrows (made by armadillos) and in arboreal termite nests within hollows made by nesting parakeets. Less than ten exposed nests were noted in this 50 km^2 area. They were not only found on large tree branches, as within natural forest, but also occurred on thick, matted vegetation such as vines, bushes and tall grass (Boreham and Roubik, 1987). The combination of exposed nest habits, and opportunistic use of nearly any shelter afforded by an artificial substrate, contribute to making the Africanized honey bee highly adapted as a rapid colonizer of new territory. If there is adequate food, then the urban and agricultural areas of the American tropics are very suitable habitats for this bee.

The dynamics of colony density and population regulation in invading Africanized honey bees are poorly known, especially in light of population changes during and after initial colonization (Boreham and Roubik, 1987; Roubik and Boreham, 1990). In any tropical forest habitat with large trees, the effort required to locate all of the honey bee nests would probably not be commensurate with information thus obtained. The number of bee colonies in an area is not stable, and it must vary widely among different habitats. Again, as a personal impression based only on field experience rather than upon planned field studies, it seems that a few colonies per km^2 is normal for lowland forest, and that a half dozen to a few dozen colonies in this area could be normal for disturbed second growth forest and relatively dry habitat.

What do Africanized honey bees eat?

Honey bees, along with most other bees, are entirely dependent on nectar, homopteran bug secretions and pollen for their food, and on other plant materials for their building supplies. It should be pointed out that no study has been made of the requirements of tropical *A. mellifera* for resin, which is probably an essential germicide (Ghisalberti, 1979). The remainder of this section deals with studies of the pollen and nectar diet of the Africanized honey bee in Panamá (Roubik and Buchmann, 1984, Roubik *et al.*, 1984, 1986). This work, still in progress, has analyzed nectar quality (sugar concentration) and proportional use of pollen types. The pollen study utilized collections from "pollen traps" placed on bee hives. Their utility has recently been tested in studies of honey bees in the tropics (Roubik *et al.*, 1984; Villanueva, 1984). Since the honey bee can forage high above the ground and colony members exploit an area up to 300 km^2 (see below), collection of forage samples at the nest is far more practical than the observation of bees at flowers. The data obtained in this way reflect sampling biases and idiosyncrasies of bee colonies (Schmidt and Buchmann, 1986) but not those of biologists.

The present studies were undertaken after preliminary work in French Guiana during 1976 and 1977 revealed how difficult it was to describe the relative importance of diet items to bee fitness (Roubik, 1979b, 1982a). Lists of plant species known to be visited by honey bees and other bees have been compiled for limited geographic regions in the neotropics (e.g. Ducke, in Roubik, 1979a; Ordtex, 1952; Giorgini and Gusman, 1972; Absy and Kerr, 1977; Heithaus, 1979a,b,c; summary by Crane *et al.*, 1984). When such lists are extensive, or represent very limited observations that are nonetheless repeated over large areas or for several years, they allow recognition of higher plant taxa that are important to bees. At the level of species, however, they show nothing about the importance of a plant in bee reproduction. There are several reasons for this. First, the energy value of nectar and the nutritional value of pollen vary markedly among plant taxa (Stanley and Linskens, 1974; Baker and Baker, 1983; Gottsberger *et al.*, 1984). Nectar sugar concentration in flowers of a particular species often changes through the day, and floral nectar ranges from roughly 10-80% sugar. Pollen protein content varies from about 7-65% dry weight. Second, the food items used by bees must change to some extent from year to year, as should their proportion in the diet. For instance, many neotropical forest trees flower every second, third, or fourth year. Finally, that a bee uses ten flower species in a particular time and place does not mean that each contributes equally to survival and reproduction. One species might account for 90% of all pollen protein fed to larvae, and another species might account for 90% of the nectar energy imbibed by a foraging adult. Evidently, lists of plant species sometimes have little to say about bee biology.

Nectar use by Africanized honey bees follows patterns seen in European honey bees and also stingless bees, particularly of the genus *Melipona* (Roubik and Buchmann, 1984; Roubik *et al.*, 1986). Nectar sugar concentration in the loads of returning foragers ranges at least from 12-63% sugar. The maximum rate of energy harvest per forager occurs when sugar concentration is near 60% (Roubik and Buchmann, 1984), but colonies do not predominantly utilize nectar of this particular sugar concentration. In studies performed on a few thousand returning foragers in five forested lowland areas, both in the wet and dry seasons, colonies most often harvested nectar having a total percentage of sugar in the mid 20s to mid 40s (Vergara, 1983; Roubik *et al.*, 1986; and unpubl. data). For comparison, ripened honey in nests of *A. mellifera* is generally around 80% sugar. The choice of nectar of a particular concentration is influenced by the distance between the bee nest and the resource, and also the ease with which the nectar is imbibed and extracted from the flower. Bees readily imbibe dilute nectar if it is close to their home base, and nectar of higher sugar concentration can induce them to fly further afield as they forage. This relationship is not only complicated by diverse factors such as competition and flower morphology, it is strongly influenced by the distribution of foraging bees, which as shown below, do not relate uniformly to distance from the nest.

Studies of a wide variety of highly social bees in the tropics have shown that colony selection of floral nectar varies, and this variability is the result of physiological or social characteristics of the foragers and various features of their food sources and their nests. What the foragers harvest during the period of maximum colony foraging is not necessarily the richest resource. Bees take what is available, possibly because of other social competitors (Roubik and Buchmann, 1984). However, the net rate of energy gain by the colony may not be as important as the efficiency with which nectar (or other resources) are harvested and stored (Seeley, 1986). In this manner, some social bee colonies may be conservative both in foraging expenditure (Roubik, 1982b) and may be inclined to forage primarily on the richest available nectar (Roubik *et al.*, 1986). In the tropics, the concentration of sugar in nectar brought to the bee nest steadily increases after foraging is initiated in early morning; it may peak in midmorning or later in the day (Roubik and Buchmann, 1984; Inoue *et al.*, 1985). This phenomenon is probably due primarily to water evaporation, yet some flowers maintain stable nectar quality while others rapidly increase sugar content, even in shaded, humid environments (Roubik and Buchmann, 1984). The highest average nectar sugar concentration that I have recorded during a one hour interval for a foraging Africanized honey bee colony was 42%, and this occurred at 1400 hours in lowland Panamá.

The plant species that appeared to provide considerable nectar were sometimes also major pollen sources in central Panamá (TABLE 1). Fresh nectar samples from three apiaries were analyzed by J. E. Moreno (in Roubik *et al.*, 1984). These were taken on the Pacific side of the isthmus during March, June and September of 1983, corresponding to mid-dry, early-wet and late-wet seasons. Pollen in the nectar samples showed that the plants that provided nectar included Rutaceae, Bombacaceae, Compositae, Rubiaceae, Anacardiaceae, Meliaceae, Burseraceae, Mimosoideae, Myrtaceae, Euphorbiaceae and Boraginaceae. Some of the genera involved were *Vernonia, Baltimora, Clibadium, Cavanillesia, Pseudobombax, Ceiba, Zanthoxylum, Citrus, Spondias, Cedrela, Cordia, Eugenia, Psidium, Genipa, Uncaria, Macrocnemum, Bursera, Enterolobium* and *Croton*. Most of these are indicated as good honey sources in Central America (Espina and Ordtex, 1983), but an attempt to rank them by importance would be quite futile. Indicators of nectar sources are sometimes provided by pollen that falls into nectar at the flowers or that falls from the bodies of bees into nectar in the nest (Gottsberger *et al.*, 1984; Sawyer, 1988). The relative amount of pollen in nectar or honey, at best, can show whether a nectar source is widely or scarcely used. There appeared to be no adequate way of showing relative degree of nectar utilization, at least in our studies, by counting the grains of pollen of different species found in nectar or honey. Pollen from nectarless flowers is often abundant in honey stored in tropical bee nests (Absy *et al.*, 1980). Further, pollen abundance is certainly not similar in flowers as diverse as bat-pollinated *Pseudobombax*, which has several hundred stamens on each flower, and small bee-pollinated flowers such as *Baltimora, Bursera* and *Spondias*. The species found to be nectar sources in the preliminary work in Panamá are of some interest. However, the opinions of experienced beekeepers seem to serve just as well, if not better, in forming an idea of which nectar sources are of lasting importance to *Apis*, even though these observations are often inapplicable in a natural forest environment (see Crane *et al.*, 1984 for discussion of second-growth vegetation providing nectar).

TABLE 1. Major pollen sources of colonizing Africanized honey bees in central lowland Panamá. Plant genera - season and year of observation

Site & Characteristics	Late wet-Mid-Dry	Late dry-Mid wet
Pacific secondary forest	(1983) *Pseudobombax Oenocarpus, Cavanillesia Erythrina*	*Inga, Spondias, Urera, Bursera, Eugenia, Leucaena, Melampodium*
	(1984) *Pseudobombax Oenocarpus, Erythrina Serjania*	*Oenocarpus, Spondias, Bursera, Leucaena, Mimosa, Paspalum*
Pacific secondary forest and agricultural land	(1983) *Cedrela, Spondias Enterolobium, Zanthoxyllum*	*Zea, Croton, Clibadium, Paspalum, Bactris, Geonoma, Mimosa, Spondias,*
	(1984) *Bombacopsis, Melampodium, Pterocarpus, Casearia Spondias*	*Bursera, Spondias, Zea, Panicum, Melampodium*
Pacific agricultural land and forest patches	(1984) *Mikania, Spondias*	*Spondias, Bursera Chrysophila, Panicum, Mimora, Eclipta, Heliconia*
Pacific agricultural land and second growth	(1983) *Pseudobombax Spondias*	*Spondias, Croton, Zea*
	(1984) *Pseudobombax Spondias*	*Spondias, Croton, Zea* Palmae, *Mimosa*
Atlantic primary forest	(1983) *Pithecellobium Cordia, Oenocarpus, Spondias*	*Oenocarpus, Spondias Chaemaedorea*
	(1984) *Pseudobombax Cedrela, Spondias*	*Bursera, Spondias,* Caryophyllaceae, *Socratea, Elaeis, Oenocarpus*
Atlantic secondary	(1983) *Erythrina, Coussapoa*	*Spondias, Genipa, Croton, Zea*
Forest and agricultural land	*Cordia, Spondias,* (1984) *Veronia, Gauzuma, Spondias*	*Baltimora, Panicum Melampodium*
Atlantic secondary forest and agricultural land	(1983) *Cavanillesia, Cordia, Spondias*	*Spondias, Citrus, Geonoma, Zea, Clibadium*
	(1984) *Davila, Cordia, Clibadium, Spondias*	*Spondias, Bursera, Citrus, Zea, Panicum, Heliconia, Protium*

*Agricultural areas consisted of small farms

In contrast to the largely qualitative studies of nectar, pollen used by Africanized honey bee colonies has been monitored quantitatively both extensively and intensively at 25 sites in central Panamá (Roubik *et al.*, 1984; Roubik, 1988; Roubik and Moreno, 1990). The study employed modified Ontario Agricultural College pollen collectors (Smith and Adie, 1963) placed on hive bases. These were maintained continuously on hives of three colonies at each site. Pollen was removed from collectors every 10-14 days from November 1982 until July 1985. Pollen collectors of this design sequester approximately 60% of the incoming pollen loads from the legs of returning foragers (Levin and Loper, 1984; Schmidt and Buchmann, 1986). Since each of the pollen pellets is almost entirely of a single plant species, a random sample of pollen pellets can show the proportion of the colony diet constituted by different plants. These periodic diet profiles were constructed through our study by combining all pollen pellets harvested from the honey bee colonies at a particular site for each of the sampling dates. The pollen of 30 pellets was identified to species, using a reference collection of the pollen of Barro Colorado Island (Croat, 1978), developed and maintained in my laboratory in Panamá (Roubik and Moreno, in press). More than 30 pellets were examined if total pollen types were found to increase in the last five of 30 pellets.

Some preliminary results on pollen utilization by Africanized *A. mellifera* are given in TABLE 1. These data are from analyses of pollen collected by colonies at seven sites, four on the Pacific side of Panamá and three in the Atlantic lowlands. The combined species listed as major pollen sources constituted roughly 30-70% of all pollen harvested by the colony during the interval indicated. Pollen data reported here include up to 20 months and a total of about 800 pollen samples. It should be noted that the method of examining pollen pellets preferentially considers pollen that was harvested in relatively large amounts instead of the total combined pollen sampled (which was usually 20-40 g dry weight per day). The goal of the research was to identify major sources of pollen, so this methodology was appropriate.

Another question could have been asked: What proportion of the total flora is visited for food by Africanized honey bees? A partial answer is provided by intensive study of pollen harvested by seven Africanized honey bee colonies at one of the sites used in the pollen monitoring project (Roubik *et al.*, 1986). Seven or eight pollen species were found among 30 pellets sampled before and after the intensive study. During the study a total of 33 species were collected. Eight of the 33 were not seen during April-June during the 20-month apibotanical study. Therefore, a conservative estimate of the absolute total numbers of pollen types collected by honey bee foragers would be about twice the number of species recorded by identifying pollen pellets. There were 71 species among the pellet samples taken during 20 months at this site. Multiplying the total number of recorded pollen species by two for each of the three sites for which 20 months of data have now been analyzed, the Africanized honey bee colonies harvested at least 142-204 pollen species. The diversity of flowering plants in these habitats number no more than 800-1000 species known to occur in these forests (Croat, 1978). Comparative data are available, for European honey bees, in the lowland secondary deciduous caducifolious-spiny forest and agricultural area of Veracruz, Mexico (Villanueva, 1984). During one year in this habitat, about 185 plant species were utilized for pollen or nectar. These studies give the impression that honey bees are using about 25% of the flora, yet intensively use many fewer species (Roubik, 1988; 1989).

Striking generalizations can be made about significant diet items of the Africanized honey bee due to information compiled by Croat (1978) on the central Panamanian flora. Colony reproduction, or swarming, was to a great extent due to pollen harvest from dioecious and monoecious species. About half of the 36 genera in TABLE 1 are functionally male, dioecious or monoecious (Croat, 1978), and this is certainly a higher proportion than might be expected for combined plant species in neotropical flora (Bawa, 1980). Honey bees seem to prefer these flowers when they are available. Flowers of such plants often open and present pollen during the night, thus the largest amount of pollen is available as soon as the bees can begin to forage. In disturbed habitats, the principal honey bee forage during the wet season included weedy herbaceous plants, and a number of grasses including a widespread wind-pollinated crop, *Zea mays*. Where there was at least some forest, whether secondary or primary, almost all the pollen harvested by honey bee colonies came from trees. Only three vines and lianas, *Serjania, Davilla* and *Mikania* were frequently used by the Africanized bees. This means that shrubs, treelets, herbs and vines were generally unimportant sources of protein for honey bee colonies. Furthermore, their major pollen sources were often plants pollinated by bats, birds or beetles (notably Bombacaceae and Palmae). In addition, there was a very marked preference for "dense" flowers or stamens, that were grouped in large inflorescences (Roubik *et al.*, 1986). Since 16 of the genera in TABLE 1 are not listed among "apicultural flowers" in tropical America (Espina and Ordtex, 1983), pollen trapping studies can certainly contribute to knowledge of the honey bee's flower visitations.

In central Panamá, the period of major flowering lasts from the mid-dry until the early-wet season (Croat, 1978). Flowers of forest trees predominate during this time; herbaceous and second-growth tree and some shrub species are the major sources of food during the remainder of the year. The data shown in TABLE 1 may be divided into basic types of habitats: secondary growth and agricultural areas, and tropical lowland forest. There was one particularly noticeable difference between the lowland Atlantic forest and all of the other habitats. Pollen came most often from palm trees at all seasons in this forest whereas in nearly all other areas, the palms were represented most heavily in the honey bee diet during the wet season only. It appears that the most important pollen source in this entire area, from near Colón on the Atlantic coast to Capira near the Pacific coast, was *Spondias* (Anacardiaceae). This small tree flowers at the end of the dry season and beginning of the wet season. During the early dry season, one important source of bee forage consistently came from trees of the Bombacaceae, *Pseudobombax* in particular. In the same family, *Cavanillesia* was a significant pollen source at two sites in 1983; it did not flower in 1984.

What eats Africanized honey bees?

Causes of colony mortality are poorly known, and there are no hard data of any kind on colony (queen) longevity in natural nesting conditions and habitat (but see Malagodi *et al.*, 1986). One reason for this lack of information is that colonies of Africanized honey bees abscond, or emigrate, following attack by predators (Winston *et al.*, 1979). Preliminary observations of these authors agree with a broad survey by Caron (1978): Africanized honey bee colonies are harassed by ants and edentates (armadillos and "anteaters"). In tropical lowland forest, I have noticed the most frequent attack of Africanized honey bee colonies by robber stingless bees of the genus *Lestrimelitta* (Roubik, 1981), army ants *Eciton* (Roubik *et al.*, 1986) and a large mustelid, *Eira barbara* (Posey and Camarago, 1985; Roubik and Boreham, 1990). The attacks of *Eciton* were seen especially during the beginning of the wet season in Panamá (see Franks and Bossert, 1985). The attacks by *Lestrimellitta limao* in Panamá, on *Apis* and the stingless bees (the usual hosts), were also concentrated in the wet season. The same is true of two local pyralid wax moths, the larvae of which eat stored pollen and comb wax. Absconding caused by such moths has been noted in Africa (Fletcher, 1978).

A great deal of work remains to be done on the larger natural enemies of honey bee colonies. Honey bee colonies are without doubt one of the richest resources that a consumer of carbohydrate or insect larvae could come upon in the neotropics. The amount of stored honey and brood is several times that of all but the largest nests of native honey-making bees (Roubik, 1979a, 1983b). The honey bee nests are sometimes not well protected, and it is not known if their stinging behavior is sufficient to drive away their neotropical predators. Moreover, the nest of the honey bee, in stark contrast to the nests of stingless bees, has a strong odor of honey and beeswax. At night, both stingless bee and honey bee colonies are audible owing to their ventilation of the nest by fanning with the wings; thus another cue exists for nest location by natural enemies. Predators should rapidly learn to recognize and exploit this new resource. It is well known that European honey bees fare poorly in lowland forests (e.g., Delgado and del Amo, 1984). Here, arrival of Africanized honey bees might augment the numbers of certain types of colony predators, as it surely increases their resource base.

Populations of disease organisms might also benefit from presence of a feral honey bee population. Their dispersal between apiaries is likely to be augmented if honey bee colonies are distributed throughout the countryside, rather than in small isolated pockets such as apiaries. Dispersal of some microbes and mites could, in theory, be facilitated by foraging activities. This might come about at flowers, water sources, or in apiary hives or feral nests; nest robbing by Africanized honey bees, and also movement of drones between colonies (at least in apiaries) seem to occur frequently (Michener, 1975; Rinderer *et al.*, 1985).

To assess whether the tracheal mite, *Acarapis woodi*, came with Africanized honey bees to Panamá from Colombia, samples from the first 90 Africanized honey bee colonies found in Panamá were dissected (Roubik and Reyes, 1984). These few thousand dissections did not produce a single *A. woodi*, although *A. externus* was found. We had suggested that worker bees might not survive long enough for mites to complete their life cycle, as the average life span of a worker bee is at times less than the generation time of the mite, although the lifespan of workers is not constant through the year (Bailey, 1981; Winston *et al.*, 1983). However, we offered no data to support or refute this suggestion, and the longest recorded survival periods of worker bees permit completion of the mite life cycle. Whether or not this particular mite adversely affects apiary honey bees or the honey bee population at large, it is an interaction that deserves close scrutiny. Its practical value could be as a model for prediction of the population dynamics of other disease organisms associated with colonization by honey bees.

BIOLOGY OF POPULATIONS

Africanized honey bee dynamics

The enormous range expansion of the Africanized honey bee is permitted by its dispersal, survival, reproduction and competitive ability, but the relative importance of such characteristics has not been studied. What is known of honey bee ecology suggests there will be an impact by the Africanized honey bee. There are over 300 native species of highly social bees (Meliponinae) where the Africanized honey bee now lives. There may be ten times this many solitary bee species, and there are thousands of wasps, butterflies, birds and other animals that feed or nest in some of the same places, or that share some of the same predators and other natural enemies. Many animals feed on the fruit and seeds resulting from pollination by bees. As nearly as might be estimated, there are a 10^{12} (one trillion) Africanized honey bees living in the neotropics, an accurate figure if the average density of their colonies is six per km^2. These colonies would consume two billion kilos of pollen and several times this much nectar on an annual basis (Roubik, 1987). Nonetheless, two long-term studies of their populations, one by the Amazonian Kayapo Indians (18 years), who have an advanced indigenous knowledge and awareness of bees (Posey and Camargo, 1985), and an intensive seven year survey in Panamá (Boreham and Roubik, 1987, Roubik and Boreham 1990) show that populations of the bees decline after an initial rapid buildup. A third study in French Guiana indicated slow buildup of the Africanized honey bee population with no decline after eight years (Roubik, 1987). Two principal processes that could produce the observed population changes are: 1) learning by predators and their resulting population buildup to eventually reduce the invading bee population, and 2) decreasing swarm production by Africanized honey bee colonies due to increased competition for food with their own kind. Predators and food both limit bee populations, as both colony mortality and colony reproduction rates change over time, but more specific questions need to be asked that fit different environmental contexts and particular sets of interactions.

Other phenomena are superimposed on the broadly complementary processes of predation and competition. Forest is being replaced with secondary growth, and agricultural areas are becoming more widespread. Removal of the trees and natural forests at once reduces the biomass of bees that live almost exclusively in trees or within forests. These are in large part the native highly social bees, the meliponines. Flowering vegetation that grows where forest once existed, even cultivated corn, is available to Africanized honey bees but not to the indigenous bees that are ubiquitous in normal forest because these bees have long since disappeared from the altered habitat. Moreover, the Africanized honey bee colonies are mobile and have large foraging areas. In mosaic habitats, where there are patches of forest interspersed with other types of vegetation, the advantages of the Africanized honey bee over native bees may be very great indeed. Is this the case with tropical forest?

Competition between bees

As a "foraging machine" the Africanized honey bee has a larger colony and hence more foragers, as well as a more sophisticated communication system, than any of its neotropical bee competitors. It is not an aggressive forager despite occasional jostling at large flowers and is regularly attacked by the few dozen aggressive stingless bees, in the groups *Trigona* and *Tetragona*. Its reaction to harassment by these territorial foragers is generally to continue visiting the same inflorescence (Roubik, 1978, 1980a,b, 1981 and unpubl. data). The meliponine foragers employ a tactic that does not work with this 'new' type of bee. Not only the general types of floral resources, but the intensity with which they are used, are similar between Africanized honey bees and meliponine bees (Roubik, 1979b, Roubik *et al.*, 1986). Flower species that serve as their resources are largely those used by the other local bees of similar size or smaller (Roubik, 1979b, 1982a). In a 13-day study, I found that 47 pollen species were utilized by the honey bees and by stingless bees of all sizes; 14 of these pollen species were only used by honey bees (Roubik *et al.*, 1986). However, the species used most heavily were very similar for the four *Melipona*, eight *Trigona* and *Apis*. Furthermore, the flowers of these species are borne on large inflorescences and are spatially compact; thus they present a considerable quantity of food in a small area. We do not know how specialized any of these bees are in their floral and foraging preferences; this topic is actively under investigation. However, the few detailed studies that have been made on pollen diets of tropical forest bees, both honey bees and stingless bees, show that most are highly dependent on palms (Roubik and Moreno, 1990). Sugar concentrations of nectar collected overlapped broadly, and their timing of maximum foraging was similar. Some nectar preferences were evident, however. Some *Melipona* preferred more viscous nectar than did the Africanized honey bees, and some *Trigonini* appeared to specialize on more watery nectar than that used by the honey bees.

An experiment initiated in 1979 tested a possible effect of the general similarity in food choice and disparity in foraging ability between Africanized and meliponine bees in central Panamá. During a two week period in 1983, i.e., about one year after the first Africanized bees arrived in central Panamá but while they were scarcely present in the forest, 17 colonies of *A. mellifera* and 12 species of native meliponine bees were monitored through the morning for colony foraging activity in the presence, and in the absence, of 20 foraging honey bee colonies (Roubik *et al.*, 1986). In the same region and in other forested areas, studies were made to find the spatial distribution of foraging Africanized honey bees in relation to their colonies (Vergara, 1983, after Visscher and Seeley, 1982). Studies in nearby Barro Colorado Island (BCI) were used to gauge the distribution of foraging stingless bee in another forest site (Roubik and Aluja, 1983). Together, these projects showed that despite the apparent ability to escape competition with the honey bee by shifting foraging areas to flowers not dominated by them (a tactic which may typify competition among highly social bees), 23% of pollen or nectar collection rates of the various stingless bee colonies registered measurable losses due to the introduced honey bee colonies. Interpretation of these results led to several predictions, as outlined below, that can be tested by continuing long-term studies of bee populations in Panamá.

At the time of the experiments there were virtually no local colonies of honey bees. Thus a complication for interpreting the results was measurement of their significance for the normal situation of honey bee colonies scattered within a forest. FIGURE 1 is a diagram of the distribution of foragers from two colonies of honey bees and three colonies of large stingless bees. The flight ranges are known from the studies mentioned above, but the distributions are theoretical in that they represent the areas in which bees of a colony would forage over an indefinite time on a flat surface. As shown in studies with the honey bee *A. cerana*, in steep terrain bees tend to forage further from the nest in the direction of an uphill incline, since the return trip to the nest is less costly in terms of expended flight energy (Dhaliwal and Sharma, 1974).

The distribution of foragers in the shaded areas of FIGURE 1 represent the variables plotted in FIGURE 2. It shows that most foraging occurs at intermediate distances from the nest. In fact, as many honey bee foragers fly within 0.6 km of the nest as between 6.3 and 6.9 km! This analysis suggests that most competition between foragers with nests very close to each other does not occur in the immediate vicinity; for *Apis* it is in the area 2-4 km from the nest. Within distance bands of uniform width, the foraging area increases as the distance from the nest increases (FIGURES 2 and 3). If a topographical dispersion of all colony foraging is envisioned in three dimensions, it has the shape of a bell. The area from which most foragers return is about 0.5-1.5 times the standard deviation of their foraging ranges. The probability of foraging at increasing distances decreases logarithmically (Roubik and Aluja, 1983; see Seeley, 1985, p. 92), while foraging area increases arithmetically with distance away from the home base (FIGURES 2 and 3). Thus there must be some intermediate distance band at which most foraging occurs. The area under normal curve in a "z" distribution was multiplied by the proportion of total foraging area within a distance band to derive forager number shown in FIGURE 2, taking three standard deviations of flight range to represent maximum foraging distance.

Competition between bees of colonies or solitary nests can be conceptualized as the overlap of foraging distributions (FIGURE 3). As shown in the figure, general regions of overlap correspond to the intensity of competition. Maximum competition occurs when nests are barely separated. Precise volumes could be computed for such overlaps, which would correspond to the total competition pressure over time, but the general idea is enough for the present discussion. Competition declines with increasing distances between bee nests. For *A. mellifera* a maximum normal foraging range of 10 km (Visscher and Seeley, 1982; Vergara, 1983) means these colonies compete with stingless bees from nests separated by over 12 km (Roubik and Aluja, 1983). Over half of all *Apis* foraging flights might occur within 2 km of the nest (Seeley, 1985), so that they might compete strongly with stingless bees if not more than 4 km from their nests.

Using this criterion, I calculated the competition pressure generated by the 20 colonies of Africanized honey bees in the above experiment and found it to be approximately equal to that of one honey bee colony per km^2 (Roubik *et al.*, 1986). The amount by which the honey bees diminished the harvest rate of nectar or pollen for stingless bee colonies was calculated, and it was shown that these colonies would lose 25% of their total harvest, if the type and intensity of competition recorded during seven days in the early wet season could be applied to a whole year. Whether or not this holds true, the way in which competition affected the stingless bee colonies was unexpected. It diminished or removed their peak harvest periods, which were very rare, but otherwise caused little

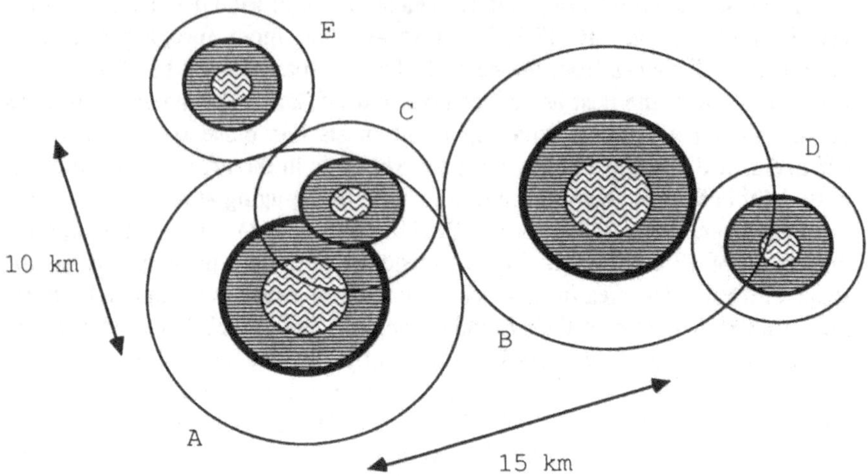

FIGURE 1. Foraging competition bands surrounding five social bee colonies. At this scale, A and B represent colonies of Africanized *Apis mellifera* and colonies C, D, E are of *Melipona*. Darkly shaded bands are the areas in which most competition and foraging takes place (see text).

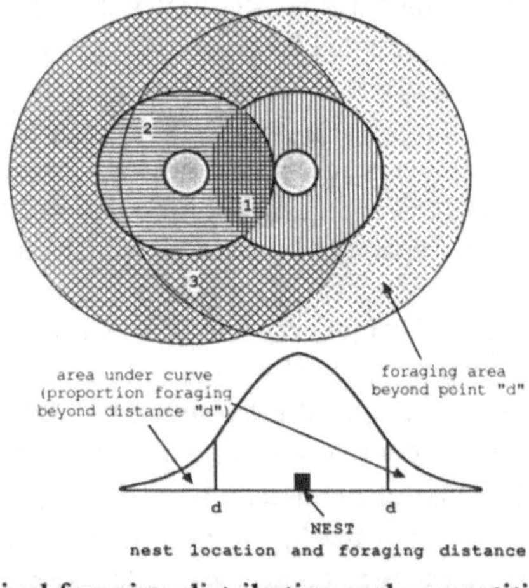

FIGURE 2. Theoretical foraging distribution and competition intensity between two central-place foragers. Area "1" is the area of most intense competition, competition in area "2" is less intense, and area "3" the least competition occurs. The bell-shaped distribution is a 2-dimensional representation of forager distribution within total foraging range.

damage. Peaks averaged 4% of total foraging time, yet included an average of 28% of all resources harvested. In short, the highly social bees obtain a sizeable portion of their food during short bursts of foraging activity, regardless of species, size of the workers or colony size and recruitment system. The only species that did not reveal a pronounced foraging peak was the only aggressive forager included in the study, *Trigona ferricauda* (Roubik, 1982c).

Another aspect of this study was that even the colonies lacking statistical declines in foraging activity during the presence of honey bees displayed pronounced foraging peaks only during their absence. Avoidance of most competition when honey bees foraged might account for this. There is good reason to believe that unaggressive bees, particularly the highly social species, have foraging patterns that allow them to escape competition at flowers. Bees are highly mobile and have large foraging areas. The highly social bees seem to utilize floral resources that can be very rewarding, and they use these heavily in the absence of competitors. It would not be surprising if they recognized the chemical cues of these resource species, responding with a burst of colony foraging when a bee returns to the nest after a successful foraging bout at such inflorescences. The proposed avoidance of competition, and general lack of competition in the experiment are most easily explained by the short duration of the study. During the seven days when honey bees and stingless bees foraged, the bees could have concentrated their foraging activity anywhere within the resource landscape encompassed by their foraging ranges (FIGURES 1 and 3). Previous short-term studies showed some net competition at flowers but negligible influence on colony resource harvest and brood production (Roubik, 1978, 1983a). Time and space are critical in testing the influence of competition, thus insignificant overlap in these foraging dimensions can occur despite resource sharing. Bees using the same flower species and nesting near each other may easily fail to overlap in the particular patches they encounter and exploit, at least over a relatively short time.

Bee population dynamics, pollination ecology and future studies

Long-term studies could show whether new competition generated by invading honey bees brings about permanent changes in native bee foraging success and populations. Two long-term studies have been carried out to identify population trends in some 100 bee species in Panamá (Wolda and Roubik, 1986; Roubik and Ackerman, 1987). The data from both studies were analyzed after seven continuous years of data collection and both continue now. Neither study showed an abrupt drop in native bee numbers during the first few years after the Africanized honey bee arrived, but careful analysis of individual species or sets of species has not been undertaken. What has been shown for the native bee populations is encouraging: they are remarkably stable in yearly abundance, at least as stable as temperate bee populations (Owen, 1978, in Wolda and Roubik,

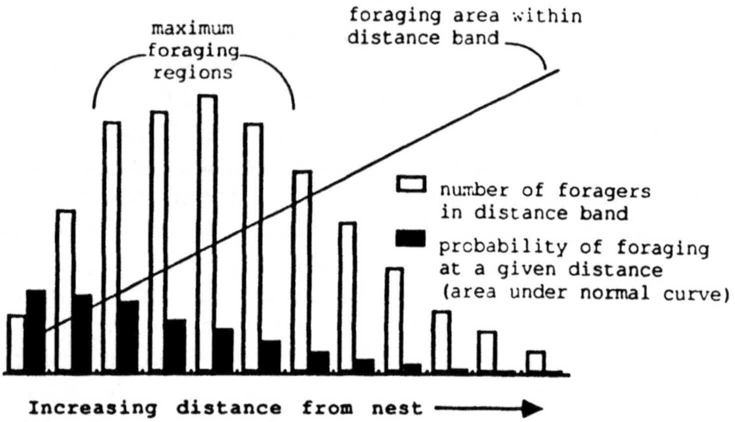

FIGURE 3. Relationship between probability of foraging at increasing distance from the nest, total foraging area within a distance band, and the total foraging activity in different distance bands (see text). The maximum foraging region is an idealized symmetrical distribution of foragers around the nest is between 0.75 and 1.75 standard deviations of foraging distance (for *A. mellifera* 1 s.d. foraging range is about equal to 2.5 km).

1986). This means that changes due to the honey bees should be tractable, because the yearly variation in bee abundance is not so large that any impact that *Apis* might have would be difficult to detect. Furthermore, if the findings had been that bee populations normally fluctuate widely from year to year, one might suspect that the addition of honey bees to such a system would not be any more significant than the effect of processes already in operation.

Detecting the impact of honey bees on plants seems likely to be far more complicated. My impression from long-term studies of fruit and seed production in selected plant species is that these fluctuate so much from year to year that the effect of having a different or additional pollinator would be hidden or very difficult to demonstrate by descriptive data (e.g., Foster, 1982). The male component of plant fitness, involving the dispersal of successfully germinating pollen to conspecifics, may not even be shown by monitoring fruit and seed production. The opportunities for innovative field research are virtually unlimited. For example: 1) What is the reproductive success of dioecious and monoecious plants visited by honey bees? 2) What effects do honey bees have on the amount of outbreeding among tropical trees? 3) How effective are Africanized honey bees as pollinators of crops, both compared to native bees and compared to European honey bees? 4) How do honey bees indirectly affect the reproductive success of plants by competing with generalist foragers?

It seems that Africanized honey bees have the potential to dominate an entire canopy of a flowering tree by recruiting thousands of foragers from their colonies. Several dozen bee species and other animals can normally be found visiting a flowering tree at any one time (Frankie *et al.*, 1983). Some of them may tend to stay in one tree canopy, while others may tend to move between trees. If honey bees have effectively taken over the food source, even if they show no aggressive territorial foraging, then the pollinators providing outcrossing could suffer population decrease, cease to visit the trees where many honey bees forage, and thus move less frequently between plants. The result could be devastating for a dioecious plant species. If it were a tree species, its reproductive population might not change for several decades, yet eventually a change would surely occur.

On the other hand, there is some indication that honey bees could provide superior outcrossing services. Their nectar load is larger than that of most *Melipona*, with which they share many flower preferences (Roubik and Buchmann, 1984; Roubik *et al.*, 1986). A strong possibility is that this would lead to more flower visits per foraging trip, with at least the chance of carrying pollen of more individual flowers. Another possibility exists for greater outcrossing, although there is no empirical information for Africanized honey bees or native neotropical bees in normal nesting conditions. Viable pollen can sometimes be transmitted between foragers in the nest (DeGrandi-Hoffman *et al.*, 1986) which then might disperse it to stigmas on their next foraging trips. As was mentioned above, honey bee colonies forage over much larger areas than do colonies of stingless bees, and their colony populations are generally 10-100 times larger. Even if such intra-colony pollen transfer if minimal, its effect over time could conceivably alter the genetic makeup of local plant populations.

The influence of honey bees on individual species of native tropical plants and animals will have one of three possible outcomes. The native species will suffer, or benefit, or remain more or less unaffected. Accumulated case studies would help to show which of these predominates. In the process of documenting variation in reproduction and mortality in natural plant and animal populations, many new concepts and processes might be uncovered by the spread of the Africanized honey bee.

Acknowledgements

I thank my field collaborators in Panamá for helping in demanding studies, particularly Jim Ackerman, Martin Aluja, Mel Boreham, Steve Buchmann, Byron Chaniotis, Robert Dressler, Vidal Gonzales, Enrique Moreno, Francisco Reyes, Carlos Vergara, Doug Yanega, Dieter Wittmann, Henk Wolda, and the Smithsonian Tropical Research Institute for responding to this biological phenomenon with foresight. I wish especially to thank the Panama Canal Commission and Dr. M. M. Boreham for allowing me to refer to some unpublished data. Peter Becker was instrumental in helping me to formulate the concept in FIGURE 3. The Scholarly Studies program and the Walcott Botanical Endowment of Smithsonian Institution have supported much of this work, and support for honey bee pollen studies was granted from the Panamanian Institute of Agricultural Research (IDIAP).

LITERATURE CITED

Absy, M. L., Bezerra, E. B., Kerr, W. E. 1980. Plantas nectaríferas utilizadas por duas espécies de *Melipona* da Amazonia. *Acta Amazon.* 10:271-281.

Absy, M. L., Kerr, W. E. 1977. Algumas plantas visitadas para obtenção de pólen por operárias de *Melipona seminigra merillae* em Manaus. *Acta Amazon.* 7:309-315.

Anderson, R. H., Buys, B., Johannsmeier, M. F. 1983. Beekeeping in South Africa. *Dept. Agric. Bull.* 394, Pretoria.

Bailey, L. 1981. *Honey bee pathology*. London: Academic Press.

Baker, J. H., Baker, I. 1983. Floral nectar sugar constituents in relation to pollinator type. In *Handbook of Experimental Pollination Biology*, ed. C. E. Jones, R. J. Little, pp. 117-141. New York: van Nostrand Reinhold.

Bawa, K. S. 1980. Evolution of dioecy in flowering plants. *Ann. Rev. Ecol. Syst.* 11:15-39.

Boreham, M. M., Roubik, D. W. 1987. Population change and control of Africanized honey bees (Hymenoptera: Apidae) in the Panama Canal area. *Bull. Entomol. Soc. Am.* 33:34-39.

Caron, D. M. 1978. Marsupials and mammals. In *Honey Bee Pests, Predators and Diseases,* ed. R. A. Morse, pp. 228-256. Ithaca, NY: Cornell Univ. Press.

Crane, E., Walker, P., Day, R. 1984. *Directory of Important World Honey Sources.* London: Int. Bee Res. Assoc.

Croat, T. C. 1978. *Flora of Barro Colorado Island.* Palo Alto, Calif: Stanford Univ. Press.

DeGrandi-Hoffman, G., Hoopingarner, R., Baker, K. 1986. Influence of honey bee in-hive pollen transfer on cross-pollination and fruit set in apple. *Environ. Entomol.* 15:723-725.

Delgado, M., del Amo, S. 1984. Dianámica de poblaciones de *Apis mellifera* L. en una zona tropical húmeda. *Biotica* 9:351-365.

Dhaliwal, H. S., Sharma, P. L. 1974. Foraging range of the Indian honey bee. *J. Apic. Res.* 13:137-141.

Espina, D., Ordtex, G. S. 1983. *Flora Apicola Tropical.* Cartago, Costa Rica: Editorial Tecnologica de Costa Rica.

Fletcher, D. J. C. 1978. The Africanized honey bee, *Apis mellifera adansonii*, in Africa. *Ann. Rev. Entomol.* 23:151-171.
Foster, R. A. 1982. Famine on Barro Colorado Island. In *The Ecology of a Tropical Forest*, ed. E. G. Leigh, Jr., A. S. Rand, D. M. Windsor, pp. 201-212. Washington, D.C.: Smithsonian Institution Press.
Fowler, H. G. 1979. Responses by a stingless bee to a subtropical environment. *Rev. Biol. Trop.* 27:111-118.
Frankie, G. W., Haber, W. A., Opler, P. A., Bawa, K. S. 1983. Characteristics and organization of the large bee pollination system in the Costa Rican dry forest. In *Handbook of Experimental Pollination Biology*, ed. C. E. Jones, R. J. Little, pp. 411-447. New York: van Nostrand Reinhold.
Franks, N., Bossert, E., 1985. The influence of swarm raiding army ants on the patchiness and diversity of a tropical leaf litter ant community. In *Tropical Rain-Forest: Ecology and Management*, ed. S. L. Sutton, T. C. Whitmore, A. C. Chadwick, pp. 151-163. Oxford: Blackwell.
Ghisalberti, E. L. 1979. Propolis: a review. *Bee World* 60:59-84.
Giorgini, J. F., Gusman, A. B. 1972. A importancia das abelhas na polinização. In *Manual de Apicultura*, ed. J. M. F. Camargo, pp. 155-214. São Paulo: Ceres ltda.
Gottsberger, G., Schrauwen, J., Linskens, H. F. 1984. Amino acids and sugars in nectar, and their putative evolutionary significance. *Plant Syst. Evol.* 145:55-77.
Heithaus, E. R. 1979a. Flower visitation records and resource overlap of bees and wasps in northwest Costa Rica. *Brenesia* 16:9-52.
Heithaus, E. R. 1979b. Flower-feeding specialization in wild bee and wasp communities in seasonal neotropical habitats. *Oecologia* 42:179-194.
Heithaus, E. R. 1979c. Community structure of neotropical flower visiting bees and wasps: diversity and phenology. *Ecology* 60:190-202.
Hubbell, S. P., Johnson, L. K. 1977. Competition and nest spacing in a tropical stingless bee community. *Ecology* 58:949-963.
Inoue, T., Salmah, S., Abbas, I., Yusuf, E. 1985. Foraging behavior of individual workers and foraging dynamics of colonies of three Sumatran stingless bees. *Res. Popul. Ecol.* 27:373-392.
Johnson, L. K., Hubbell, S. P. 1984. Nest tree selectivity and density of stingless bee colonies in a Panamanian forest. In *Tropical Rain-Forest: The Leeds Symposium*, ed. A. C. Chadwick, S. L. Sutton, pp. 147-154. Leeds: Leeds Philos. Lit. Soc.
Kerr, W. E., 1984. História parcial de ciência apicola no Brasil. *5° Bras.Apic. congress and III Cong. Latino-Ibero-American Apic.*, ed. L. S. Gonçalves, A. E. E. Soares, D. De Jong, D. Steiner, M. R. Martinho, N. Message, pp. 47-60.
Levin, M. D., Loper, G. M. 1984. Factors affecting pollen trap efficiency. *Am. Bee. J.* 124:721-723.
Malagodi, M., Kerr, W. E., Soares, A. E. E. 1986. Introducão de abelhas na ilha de Fernando de Noronha. 2. Populacão de *Apis mellifera ligustica. Ciênc. Cult.* 38:1700-1704.
Michener, C. D. 1974. *The Social Behavior of Bees: A Comparative Study*. Cambridge, Mass.: Harvard Univ. Press.
Michener, C. D. 1975. The Brazilian bee problem. *Ann. Rev. Entomol.* 20:399-416.
Ordtex, G. S. 1952. *Flora Apícola de la América Tropical*. Habana, Cuba: Editorial Lex.

Posey, D. A., Camargo, J. M. F. 1985. Additional notes on the classification and knowledge of stingless bees by the Kayapó Indians of Gorotire, Brazil. *Ann. Carnegie Mus.* 54:247-274.

Rinderer, T. E., Hellmich, R. L., III, Danka, R. G., Collins, A. 1985. Male reproductive parasitism: a factor in the Africanization of European honey bee populations. *Science* 228:1119-1121.

Roubik, D. W. 1978. Competitive interactions between neotropical pollinators and Africanized honey bees. *Science* 201:1030-1032

Roubik, D. W. 1979a. Nest and colony characteristics of stingless bees from French Guiana. *J. Kans. Entomol. Soc.* 52:443-470.

Roubik, D. W. 1979b. Africanized honey bees, stingless bees, and the structure of tropical plant-pollinator communities. *Proc. IVth Int. Symp. on Pollination, Md. Agric. Exp. Sta. Spec. Misc. Publ.* 1:403-417.

Roubik, D. W. 1980a. New species of *Trigona* and cleptobiotic *Lestrimelitta* from French Guiana. *Rev. Biol. Trop.* 28:263-269.

Roubik, D. W. 1980b. Foraging behavior of competing Africanized honey bees and stingless bees. *Ecology* 61:836-845.

Roubik, D. W. 1981. Comparative foraging behavior of *Apis mellifera* and *Trigona corvina* on *Baltimora recta*. *Rev. Biol. Trop.* 29:177-183.

Roubik, D. W. 1982a. Ecological impact of Africanized honey bees on neotropical pollinators. In *Social Insects in the Tropics,* ed. P. Jaisson, pp. 110-123. Paris: Université Paris-Nord.

Roubik, D. W. 1982b. Seasonality in colony food storage, brood production and adult survivorship: studies of *Melipona* in tropical forest. *J. Kans. Entomol. Soc.* 55:789-800.

Roubik, D. W. 1982c. The ecological impact of nectar-robbing bees and pollinating hummingbirds on a tropical shrub. *Ecology* 63:354-360.

Roubik, D. W. 1983a. Experimental community studies: time-series tests of competition between African and neotropical bees. *Ecology* 64:971-978.

Roubik, D. W. 1983b. Nest and colony characteristics of stingless bees from Panamá. *J. Kans. Entomol. Soc.* 56:327-355.

Roubik, D. W. 1987. Long-term consequences of the Africanized honey bee invasion: implications for the United States. In *American Farm Bureau Symp. on the Africanized honey bees,* ed. D. E. Rawlins. pp. 46-54. Park Ridge, Illinois.

Roubik, D. W. 1988. An overview of Africanized honey bee populations: reproduction, diet, and competition. In *Africanized Honey Bees and Bee Mites,* ed. G. Needham, M. Delfinado-Baker, R. Page, C. Bowman, pp. 45-54. Chichester, England: Ellis Horwood Limited.

Roubik, D. W. 1989. *Ecology and Natural History of Tropical Bees.* New York: Cambridge Univ. Press.

Roubik, D. W., Ackerman, J. D. 1987. Long-term ecology of euglossine orchid-bees in Panamá. *Oecologia* 73:321-333.

Roubik, D. W., Aluja, M. 1983. Flight ranges of *Melipona* and *Trigona* in tropical forest. *J. Kans. Entomol. Soc.* 56:217-222.

Roubik, D. W., Boreham, M. M. 1990. Learning to live with Africanized honey bees. *Interciencia (Caracas)* 15:146-153.

Roubik, D. W., Buchmann, S. L. 1984. Nectar selection by *Melipona* and *Apis mellifera* and nectar intake by bee colonies in a tropical forest. *Oecologia* 61:1-10.

Roubik, D. W., Moreno, J. E. 1990. Social bees and palm trees: what do pollen diets tell us? In *Social Insects and the Environment: Proc. 11th Int. Cong. IUSSI,* ed.

G. K. Veeresh, B. Mallik, C. A. Viraktamath, pp.427-428. Dehli: Oxford & IBH.
Roubik, D. W., Moreno, J. E. in press. Pollen and spores of Barro Colorado Island. Monosgraphs in Systematic Botany No. 36, Missouri Botanical Garden: St. Louis.
Roubik, D. W., Moreno, J. E., Vergara, C., Wittmann, D. 1986. Sporadic food competition with the African honey bees: projected impact on neotropical social bees. *J. Trop. Ecol.* 2:97-111.
Roubik, D. W., Reyes, F. 1984. Africanized honey bees have not brought acarine mite infestations to Panamá. *Am. Bee J.* 124:665-667.
Roubik, D. W., Schmalzel, R. J., Moreno, J.E. 1984. *Estudio apibotanico de Panamá: Cosecha y fuentes de polen y nectar usados por Apis mellifera y sus patrones estaciones y anuales.* Organismo Internacional Regional de Sanidad Agropecuaria, Tech. Bull. 24, San Salvador. 73 pp.
Sawyer, R. 1988. *Honey Identification.* Cardiff, Wales: Cardiff Academic Press.
Schmidt, J. O., Buchmann, S. L. 1986. Floral biology of the saguaro (*Cereus giganteus*) I. Pollen harvest by *Apis mellifera. Oecologia* 69:491-498.
Seeley, T. D. 1985. *Honey bee Ecology: A Study of Adaptation in Social Life.* Princeton, NJ: Princeton University Press.
Seeley, T. D. 1986. Social foraging by honey bees: how colonies allocate foragers among patches of flowers. *Behav. Ecol. Sociobiol.* 19:343-354.
Silberrad, R. E. M. 1976. *Bee-keeping in Zambia.* Bucharest: Apimondia.
Smith, M. V., Adie, A. 1963. A new design in pollen traps. *Can. Bee J.* 74:4,5,8.
Stanley, R. G., Linskens, H. F. 1974. *Pollen: Biology, Biochemistry, Management.* Berlin: Springer-Verlag.
Taylor, O. R. 1977. Past and possible future spread of Africanized honey bees in the Americas. *Bee World* 58:19-30.
Taylor, O. R. 1985. African bees: potential impact in the United States. *Bull. Entomol. Soc. Am.* 31:15-24.
Vergara, C. 1983. Rango de vuelo y cuantificación de los recursos colectados por abejas Africanizadas en un bosque tropical de Panamá. *Data report.* Balboa, Panama: Smithsonian Trop. Res. Inst.
Villanueva, R. 1984. Plantas de importancia apicola en el ejido de Plan del Rio. Veracruz, Mexico, *Biotica* 9:279-340.
Visscher, P.K., Seeley, T. D. 1982. Foraging strategy of honey bee colonies in a temperate deciduous forest. *Ecology* 63: 1970-1801.
Winston, M. L., Otis, G. W., Taylor, O. R. 1979. Absconding behaviour of the Africanized honey bees in South America. *J. Apic Res.* 18:85-94.
Winston, M. L., Taylor, O. R., Otis, G. W. 1983. Some differences between temperate European and tropical African and South American honey bees. *Bee World* 64:12-21.
Wolda, H., Roubik, D. W. 1986. Nocturnal bee abundance and seasonal bee activity in a Panamanian forest. *Ecology* 67:426-433.

14

BEE DISEASES, PARASITES, AND PESTS

H. Shimanuki, D. A. Knox,[1] and David De Jong[2]

Africanized honey bees are subject to the same diseases and pests as European bees. However, differences in the biology, behavior and management of the two types of bees affect the relative importance and the visibility of these diseases and pests. For example, in areas saturated with Africanized bees, European colonies only survive with a great deal of care and rarely become very strong because of the intense competition for resources. For this reason, it is often difficult to determine the relative impact of the bee diseases since it is not possible to maintain Africanized and European colonies of equal strength in the same area.

Both Africanized and European bees appear most able to defend themselves against diseases and pests in habitats close to those of their respective origins. Both types of bees are generally more susceptible to diseases with which they have had little or no contact, until they have passed through an adaptation period.

European bees are more conservative when they rear brood, especially in small colonies. They tend to arrange the brood nest in the form of a sphere to facilitate thermoregulation when temperatures fall. Africanized bees convert most of their resources into colony growth and do not arrange such a compact brood nest. The brood patterns of Africanized bee colonies tend to cover the area of the frame leaving few cells empty. European colonies (even first generation hybrids) reduce their brood rearing much more during the "cold" months of the year than do the Africanized bees. Thus, colonies with European bee characteristics tend to be weaker with lower brood (and resulting adult) populations during the colder months. This fact is reflected in the expression of bee diseases in the colonies.

---

[1]Drs. Shimanuki and Knox are at the USDA-ARS, Bee Research Laboratory, Bldg. 476, BARC-East, Beltsville, Maryland 20705, USA.

[2]Dr. De Jong is at the Depto. Genética, Faculdade de Medicina, de Ribeirão Preto, Univ. de São Paulo, Ribeirão Preto, 14.049, SP, Brasil.

Some diseases and pests cause subtle losses of bees, honey and hive equipment. Other more serious diseases can destroy the entire population or productivity of a colony in a year or two. A beekeeper must be able to distinguish between healthy and diseased colonies and then be able to take the necessary steps to control the diseases. Preoccupation with diseases of Africanized bees arises after other beekeeping problems have been overcome, such as developing management techniques for honey production, swarm prevention, queen breeding and controlling defensive behavior. However, the high reproductive capacity of Africanized bees gives the beekeeper the luxury of being selective about the kinds of colonies that he will keep in his apiaries. Colonies that become weak due to diseases or any other problem are usually eliminated and replaced at little or no cost with a captured swarm or a divided colony. There is little incentive to nurse one or a few debilitated colonies when the mere placement of empty hives quickly brings in replacements. Such management practices tend to select for resistance in the bees and help to eliminate potential disease problems.

The purpose of this chapter is to present a survey of some of the more important diseases and pests that may be encountered when working with Africanized and European bees. Much research on the diseases of Africanized bees has been done in Brazil, so we have relied heavily on examples from that country in this chapter. A list of books, pamphlets and scientific papers containing further information on diseases and pests is provided at the end of the chapter.

BROOD DISEASES

Brood diseases are generally much easier to recognize than adult diseases. However, a close examination of the brood is necessary to correctly identify the disease. Colonies that appear weak or show little flight activity should be suspect and examined closely for disease. Moreover, a strong and populous colony is not necessarily disease-free.

Brood combs from healthy colonies typically exhibit a solid and compact brood pattern. Virtually every cell from the center of the comb outward has eggs, larvae or pupae and few cells are empty. In contrast, brood combs from diseased colonies usually have a scattered brood pattern (pepperbox appearance). This can best be described as a mixture of sealed and unsealed cells throughout the brood combs. Sometimes the pepperbox appearance may result from an inbred or failing queen, the presence of laying workers, chilling, or absconding preparations.

The appearance of brood cell cappings is also useful in diagnosing brood diseases. Cappings of healthy brood are uniform in color and convex (higher in the center than at the margins) while cappings of diseased brood tend to be darker and concave (sunken), and are frequently punctured. The unfinished cappings of healthy brood may appear to have punctures, but since cells are always capped from the outer edges to the center, the holes are always centered and have smooth edges.

Another useful diagnostic feature of diseased brood is the presence of the dried remains of the larva or pupae. These remains are called scales in the case of American foulbrood, European foulbrood and sacbrood diseases. The scales are found on the bottom side of brood cells lying lengthwise and can be seen with the unaided eye.

Hygienic Behavior and Resistance to Brood Diseases

There are no strains of bees completely immune to brood diseases. However, different degrees of resistance have been shown to occur. Africanized bees appear to be better at removing dead and diseased brood from the cells than are European bees (Camargo, 1972; Cosenza and Silva, 1972). Various researchers have compared the tendency of the bees to remove freeze-killed brood. Africanized and Africanized/Caucasian hybrid bees are more efficient than Caucasian bees in removing freeze-killed brood (Cosenza and Silva, 1972). Africanized bees have also been found to be better than Italian bees at removing dead brood (Lengier, 1977; Message, 1979). Therefore, Africanized bees are considered to be more resistant to American and European foulbrood than the European strains.

American Foulbrood Disease

American foulbrood (AFB) is worldwide in distribution and can be found in every continent. The disease is caused by the bacterium, *Bacillus larvae*. Only the spore stage of *Bacillus larvae* is capable of initiating the disease. Worker, drone and queen larvae are susceptible to AFB for up to three days after egg hatch. From the fourth day, the larvae become immune to the disease. American foulbrood may not destroy a colony in the first year. However, if left unchecked, the number of infected individuals increases, ultimately leading to the death of the colony.

Healthy larvae are pearly white in appearance, while diseased individuals show a color change from dull white to brown and finally black with the progression of the disease. When larvae are brown, the symptom of ropiness can be demonstrated. A quick way to identify AFB in the field is to macerate the suspected brood with a matchstick or twig and carefully withdraw it from the cell. If AFB is present the brown remains can usually be drawn out like a thread for two or more centimeters. However, this symptom is not always present in AFB diseased larvae.

Combs with AFB diseased larvae may exhibit the pepperbox appearance, although when the infection is light, this symptom may not be present.

Cappings over diseased brood are darker brown in color and are usually punctured and sunken.

The final stage of the disease is the formation of a scale. If AFB kills the pupa, the resulting scale may have the so-called pupal tongue. This is a threadlike projection that extends away from the scale towards the center of the brood cell.

Bee diseases are subject to legal regulations. Consequently, the options for disease prevention and control may be limited by legal considerations. For this reason, beekeepers should first establish the legality of any proposed chemical treatment before using it.

Burning of hives with AFB disease is required in some areas. In others, beekeepers are allowed to retain the hive bodies, bottom boards, and inner and outer covers, burning only the frames. The parts salvaged are scrubbed with a stiff brush and hot soapy water. Some beekeepers also scorch the inner surface of the hive bodies before reuse.

Another option available to some beekeepers is the use of a lye solution. Here the wooden equipment is completely immersed for 20 minutes in a boiling solution containing one pound (454 g) lye to 10 gallons (38 l) of water. Since lye solutions are caustic, appropriate safety equipment should be worn.

Sodium sulfathiazole (no longer registered for bee use in the U.S.) and Terramycin® (oxytetracycline, HCl) are two materials that have been used for the prevention and control of AFB. Both drugs are available in several formulations and each carries specific instructions. Both drugs may result in residues in surplus honey if not used properly.

European Foulbrood Disease

European foulbrood disease (EFB) is caused by the bacterium, *Streptococcus pluton*. This disease is worldwide in distribution and in some areas is considered to be as serious a problem as AFB. In South America, EFB is more frequently encountered than AFB.

European foulbrood is occasionally found in Africanized bees in Brazil, but it is not very common at present. Generally, in the cooler regions of Brazil it is most serious in early winter (Guimarães and Gonçalves, 1977). It was reported in European bees before the arrival of the African bees (Kerr and Amaral, 1960; Camargo, 1972). In 1973, EFB eliminated many colonies in various regions of São Paulo State. *Streptococcus pluton* was isolated from the infected larvae (Machado and Lemos, 1975; Message, 1979). It is also a problem in Minas Gerais and Santa Catarina (Guimaraes and Gonçalves, 1977; Sande, 1986) and in Rio Grande do Sul (Lengier, 1977).

In severe cases, the population of a colony with EFB may be depleted to a point where productivity is affected. This is because the disease usually occurs just when colonies are developing seasonally maximal populations. However, in most cases, the onset of a surplus honey flow results in the disappearance of the symptoms.

Worker, drone and queen larvae are all susceptible to EFB. Affected larvae usually die at about four days of age while still in the coiled stage in uncapped cells . This condition is in contrast to AFB where the affected individuals die in the later upright stage.

The general appearance of combs with AFB or EFB disease are quite similar. The combs may show a scattered brood pattern and the cappings, when present, may be discolored, punctured and/or sunken.

Larvae that are diseased show color change from pearly white to yellow, and finally black. In the yellow stage, the tracheal system of the larvae becomes conspicuous and has the appearance of silver threads. The larvae usually appear undernourished, twisted and swirled to the cell walls. Sometimes they resemble an irregular mass of "melted" tissue.

The diseased larva may show some ropiness but not to the extent associated with AFB, and the texture of the ropy material is granular. European foulbrood disease is sometimes referred to as sour brood because of its odor. The secondary invaders associated with the disease are responsible for this characteristic.

Scales of EFB are rubbery and much easier to remove than those of AFB. In rare cases, the pupa may show a "false tongue" which is actually the thick, blunt head. This should not be confused with the threadlike pupal tongue associated with AFB scales.

Terramycin has shown to be effective in the prevention and control of EFB. Also, some control can be obtained by requeening diseased colonies. Requeening allows for a break in the brood cycle, thereby permitting the nurse bees time to remove diseased brood. It is also helpful if the new queen is a more prolific egg-layer.

Sacbrood Disease

Sacbrood is the only brood disease known caused by a virus. The disease generally is present in only a small proportion of the larvae and rarely, if ever, destroys a colony. Nevertheless, it is important that beekeepers distinguish it from the more serious foulbrood diseases.

Larvae affected with sacbrood disease change from a pearly white color to gray and finally black. The death of the larvae occurs when it is in the upright state, just prior to pupation. Consequently, affected larvae are usually found in capped cells. The cappings may be sunken, discolored, and also punctured. Rarely does the disease become serious enough to show the pepperbox symptom.

If an affected larvae is carefully removed from its cell, it will appear like a sac filled with water. This symptom gave rise to the name "sacbrood disease." Head development is typically retarded and the head region is usually darker than the rest of the body. In advanced cases, this region may be bent toward the center of the cell. The scales of dead larvae typically are brittle but easy to remove. There is no characteristic odor associated with diseased.

Sacbrood has been reported from several regions of Brazil (Sande, 1986), but it is not clear that it is the same variety known from Europe and North America because the symptoms are different. A larva infected with the virus does not become dark or have a tough consistency in Brazil; rather it is light in color and easy to rupture (De Jong, pers. obs.). Various attempts to verify the virus involved have shown it to be different from the known sacbrood viruses (D. Message, and B. Ball, pers. comm.). There have been numerous reports of widespread colony mortality due to this "new" kind of sacbrood. It has been found to be a problem in Rio de Janeiro and Minas Gerais (Message, 1979), as well as in São Paulo and has been reported from Santa Catarina (Sande, 1986).

There are no chemotherapeutic agents for the control of sacbrood disease, although requeening of affected colonies may sometimes be of help. In most cases, the disease disappears without causing any serious loss in population or honey yields.

Chalkbrood Disease

Chalkbrood (CB) is caused by the fungus, *Ascosphaera apis*. This disease is rarely serious, although some beekeepers report that it can cause a reduction in honey production. Larvae three to four days are infected after egg hatch and become overgrown with mycelium, which dries up to form a solid mummified mass, which is usually white; hence, the name chalkbrood. Sometimes the mummies are mottled with black spots or are completely black due to the presence of fruiting bodies of the fungus. Mummies can often be found at the hive entrances or on the bottom boards of infected colonies and can easily be removed from the cells by tapping the comb against a solid surface.

No chemotherapeutic agent is available for the control of chalkbrood. The best counter measure appears to be the maintenance of populous colonies.

Mite Diseases

There are two parasitic mites that beekeepers should learn to recognize, *Varroa jacobsoni* and *Tropilaelaps clareae*. Both feed on developing brood, but in the adult stage they are also found on the adult bees and can be seen with the unaided eye. Both mites have been reported on *Apis mellifera* and on *A. cerana*. *Tropilaelaps* is also found on *A. dorsata*. Until the 1950s the mites were only reported in Southeast Asia. At present, only the continent of Australia is free of *Varroa*. *Tropilaelaps* is still found only in Southeast Asia.

The adult female *Varroa* mite is oval and flat, about 1.1 mm long and 1.5 mm wide, pale to reddish-brown in color, and can be seen easily with the unaided eye. Male mites are considerably smaller and are pale to light tan (Delfinado-Baker, 1984). The females attach to the adult bee between the abdominal segments or between body regions, which makes them difficult to detect. They can easily feed on hemolymph and the adult bee suffers not only the loss of blood but may also be subjected to microbial invasion, resulting in a reduced life expectancy (De Jong and De Jong, 1983).

Throughout much of South America, *Varroa* infestations are relatively innocuous. Climate seems to be an important factor in the infestation rate; there tends to be fewer *Varroa* in colonies in tropical and subtropical regions than in temperate areas (De Jong *et al.*, 1984; Moretto, 1988). In tropical and subtropical regions of South America, the percentage of female *Varroa* that effectively reproduce and the number of offspring per reproducing female is lower than that found in Europe (Ritter and De Jong, 1984). However, it is also apparent that Africanized bees are more resistant than European bees, as the infestation rates are consistently lower in the Africanized bee colonies (Steiner *et al.*, 1984; Message, 1986; Engels *et al.*, 1986; Medonoza *et al.*, 1987, Moretto, 1988) even when they are kept under the same conditions in the same apiary. The proportion of successfully reproducing mites is also lower in the Africanized colonies (Camazine, 1986; Rosenkranz, 1986; Moretto, 1988). Only about 0.64 adult female descendants are produced per original adult female in the worker brood cells (Message, 1986), much lower than the rate found for European bees. Cell size may be a factor in the low infestation levels, as Africanized brood is more heavily infested in European size brood cells than in Africanized size brood cells, even when kept in the same colony (De Jong *et al.*, 1985).

The development time of the worker brood may also influence *Varroa* reproduction. Fewer mites reach maturity on Africanized brood which develops faster than European brood. However, even though Africanized worker bees have a sealed-cell stage that is slightly shorter (about 20 hours less) than that of European workers, the number of mite offspring per reproducing female mite is only slightly different (Rosenkranz, 1986).

Female *Varroa* enter the brood cells shortly before capping, feed on the larvae, and lay two to six eggs. Adult mites develop in six to ten days and mate in the cells before emerging with the bee. The male mites soon die after mating. The most severe parasitism occurs on the older larvae and pupae, drone brood being preferred to worker brood (Ritter and Ruttner, 1980). In heavy infestations, pupae may not develop into adult bees, and those that do emerge weigh less than healthy bees and may have shortened abdomens, misshapen wings, and deformed legs (De Jong *et al.*, 1982).

No chemical for the control of *Varroa jacobsoni* has been universally accepted, but Amitraz, fluvalinate, cymiazol HCl, bromopropylate, formic acid, malathion (1%), and coumaphos are the most common in use today. Fluvalinate and Amitraz have both shown promise in controlling *Tropilaelaps clareae*

(Burgett and Kitpraser, 1990). Also, a break in the brood cycle appears to offer good control (Woyke, 1985).

ADULT BEE DISEASES

Most adult bee diseases are difficult to diagnose because the gross symptoms are not specific to a particular disease. For instance, inability to fly, unhooked wings and dysentery are general symptoms that could result from many disorders. In many cases a microscopic examination is required to make a proper diagnosis.

Acarine Disease

Acarine disease is caused by the honey bee tracheal mite, *Acarapis woodi*. The mites enter the tracheal system (breathing tubes) of young adult bees and multiply there. They derive nourishment from the host's blood and also interfere with their respiration, ultimately leading to the premature death of the bee. Mature female mites emerge from the tracheae in search of a new host. The disease can be carried to a healthy colony by robbers, drifting bees, or by a beekeeper transferring infected workers or queens.

No one symptom characterizes this disease; an affected bee may have disjointed wings and be unable to fly, or have a distended abdomen, or both. Absence of these symptoms does not necessarily imply freedom from mites. Positive diagnosis requires microscopic examination of the tracheae. Acarine disease may persist in a colony for years and cause little damage, but combined with other diseases and/or poor seasons, it may lead to the death of the colony.

Acarine disease is found in both *Apis mellifera* and *Apis cerana*. The disease has now been reported in every major beekeeping country except Australia and New Zealand.

Acarapis woodi has been described on African bees in Africa (Benoit, 1959). It has also been reported from various regions of Brazil, and occasionally is a cause of bee mortality, principally in the cooler regions of the southern states in late fall, winter and early spring. In these regions, the beekeepers sometimes treat the colonies with methyl salycilate, though the disease appears sporadically, and often many years pass between outbreaks (M. W. van de Sande, pers. comm.). *Acarapis woodi* has been found in Rio Grande do Sul (Nascimento, 1970), Santa Catarina (Wiese and Meyer, 1975; Sande, 1986), São Paulo (Flechtmann *et al.*, 1977; Silva, 1977) and in Goias (Message *et al.*, 1983). Although the latter state has a very hot climate, the bees examined were very heavily infested.

Chlorobenzilate, which is no longer available, has been the most widely used chemical for the control of *Acarapis woodi*. Currently, menthol is being used extensively for its control. Menthol, when used properly, has been shown to be effective, harmless to bees, and does not contaminate honey (Giavarini and

Giordani, 1966; Wilson *et al.*, 1988). Amitraz and bromopropylate have been found to be reasonably effective against adult tracheal mites, but not against immature mites (Eischen *et al.*, 1987).

Nosema Disease

Nosema is the most widespread of all bee diseases. It is caused by the protozoan, *Nosema apis*. The disease cycle is initiated by adult bees ingesting spores which multiply in the epithelial cells of the midgut and then compete with the host for nutrients. Workers, drones and queens are equally susceptible, causing a reduction in their life expectancy. This disease is estimated to cause a loss in honey production in excess of 40% and also contributes to winter losses. Queens that are infected may have their egg laying impaired and ultimately be superseded. Positive diagnosis for the disease can be made by microscopic examination of the ventriculus. In the case of queens, a coprological examination is necessary.

Nosema is found in Africanized bees in some of the cooler regions of Brazil, especially during the winter months. It occurs in mountainous areas of Minas Gerais from May to September (Guimarães and Gonçalves, 1977). High spore counts have been found in May and June in the coastal region of Santa Catarina state (Wiese, 1975). Heavy bee mortality is occasionally reported from Rio Grande do Sul (Nascimento, 1970).

Control of the disease is possible by feeding fumagillin in sugar syrup to the bees. However, since nosema is spread to healthy bees via contaminated combs, the combs should also be decontaminated by heat treatment (at 49°C for 24 hours).

Amoeba Disease

Amoeba disease is caused by the protozoan, *Malpighamoeba mellificae*. Little is known about the pathogenicity of this organism except that it is found in the Malpighian tubules of adult bees and it presumably impairs the function of the tubules. This disease is most commonly found in worker bees. Generally, it is of little consequence but when combined with nosema the impact is more serious than either disease alone. No gross symptoms characterize this disease and a microscopic examination of the Malpighian tubules for the presence of cysts is necessary for diagnosis.

There is no known control for the amoeba disease. Fortunately, infected colonies appear to recover without the assistance of the beekeeper.

Paralysis Disease

A virus is the inciting agent of paralysis disease, although in some cases, paralysis-like symptoms are caused by toxic chemicals. This disease probably occurs frequently in bee colonies but is seldom diagnosed. It can be found throughout the bee season but is rarely serious and should not cause the death of a colony.

In severe cases, large numbers of bees leave their hives. They tend to run in circles and are unable to fly. Frequently, they appear black, hairless and shiny.

No chemicals are effective against paralysis disease. It is believed that susceptibility is a heritable character, and that requeening of colonies with queens from a different source could cure the problem.

Septicemia Disease

Septicemia is caused by a bacterium *Pseudomonas apiseptica*, and it is rarely encountered. The disease results in the destruction of the connective tissues of the thorax, legs, wings and antennae. Consequently, affected bees fall apart when handled. Dead or dying bees may also have a putrid odor.

There is no known control for this disease. However, it is not considered serious and no preventive measures are necessary.

PESTS AND PREDATORS

Africanized bees are much better at protecting themselves from the most dangerous of the predators, man. Most beekeepers in Brazil will not keep gentler colonies, even if the bees were productive because they want the bees to be able to protect themselves. Strong nervous colonies set up on rickety hive stands are a good deterrent to commonplace thievery. Unfortunately, the thieves tend to adapt to their prey and eventually learn ways to get the honey without being stun (much). Nevertheless, since the stinging bees at least limit the number of potential hive robbers, there is still a strong preference for bees than can protect themselves (De Jong, 1984).

A few wild animal species are able to attack and destroy Africanized bee colonies, including a mustellid called "ariranha" (*Pteronura brasiliensis*) and some large armadillo species such as "tatu-canastra" (*Priodontes giganteus*). There are no bears in South America so we do not know yet how they will fare with these bees.

Wax Moths

The greater wax moth (GWM), *Galleria mellonella* is the most serious pest of the honeycombs, but comb damage is also caused by the lesser wax moth,

Achroia grisella and the Mediterranean flour moth, *Ephestia kuehniella*. These moths are an especially serious problem in tropical and subtropical climates where the warm temperatures favor their rapid development.

Female GWM lay their eggs in clusters, usually in cracks between the wooden parts of the hive. After the eggs hatch, the larvae feed on the wax combs, obtaining nutrients from honey, cast off pupal skins, pollen and other impurities found in the beeswax, but not the wax itself. Consequently, darker combs are more likely to be infested than light combs or foundation.

Combs stored in dark, warm and poorly ventilated rooms are most likely to be destroyed by GWM. Active colonies that lose part of their adult population due to diseases or toxic substances are also likely to be invaded.

Africanized bees appear to be quite good at controlling greater wax moths, but lesser wax moths are often found in the brood combs and commonly damage a small percentage of the brood.

The most effective control method is to maintain populous colonies, as the GWM is less likely to occupy combs that are covered by bees. Chemicals such as paradichlorobenzene, ethylene dibromide and phostoxin should be used with great caution, and should be used to treat only stored combs, not combs containing honey for food use. Heat treatment of combs may also be used for combs without pollen or honey.

Bee-louse

Braula coeca is not really a louse, but a wingless fly that feeds on honey. No detrimental effect on adult bees has been attributed to the bee-louse; however, larvae can damage the appearance of comb honey. Adult lice are found on adult worker bees and queens. There are no known control methods for the bee-louse and none are considered necessary.

It is important to note that *Braula coeca* resembles *Varroa jacobsoni* in size and color. However, *Braula*, being an insect, has six legs, whereas *Varroa*, being an arachnid, has eight legs.

Braula has existed in southern Brazil, especially in Rio Grande do Sul, during the years that Italina and German black bees were the main races encountered. Since the Africanized bees were introduced into these areas, these interesting honey bee colony inhabitants are no longer found (De Jong, unpubl. obs.). However, in Colombia one still commonly finds *Braula schmitzi*, even on Africanized bees (C. Mantilla Cortes, pers. comm.).

LITERATURE CITED

Benoit, P. L. G. 1959. The occurrence of the Acarine mite *Acarapis woodi* in the honey bee in the Belgian Congo. *Bee World* 40:156.

Burgett, D. M., Kitpreasert, C. 1990. Evaluation of Apistan as a control for *Tropilailaps clareae* (Acari: Laelapidae), an Asian honey bee brood parasite. *Am. Bee J.* 130:51-53.

Camargo, J. M. F. 1972. Patologia Apicola. In *Manual de Apicultura*, ed. J. M. F. Camargo, pp. 215-240. São Paulo: Editora Agronomica Ceres, Ltda.

Camazine, S. 1986. Differential reproduction of the mite, *Varroa jacobsoni* (Mesostigmata: Varroidae) on Africanized and European honey bees (Hymenoptera: Apidae). *Ann. Entomol. Soc. Am.* 79:801-803

Cosenza, G. W., Silva, T. 1972. Comparação entre a capacidade de limpeza de favos da abelha africana, da abelha caucasiana e de suas híbridas. *Ciência e Cultura* 24:1153-1158.

De Jong, D. 1984. Africanized bees now preferred by Brazilian beekeepers. *Am. Bee J.* 124:116-118.

De Jong, D., Andrea Roma, D. de, Gonçalves, L. S.. 1982. A comparative analysis of shaking solutions for the detection of *Varroa jacobsoni* on adult honey bees. *Apidologie* 13:297-306.

De Jong, D., De Jong P. H. 1983. Longevity of Africanized honey bees (Hymenoptera: Apidae) infested by *Varroa jacobsoni* (Parasitiformes: Varroidae). *J. Econ. Entomol.* 76:766-768.

De Jong, D. Gonçalves, L. S., Morse, R. A. 1984. Dependence on climate of the virulence of *Varroa jacobsoni*. *Bee World* 65:117-121.

De Jong, D. Message, D., Issa, M. 1985. The influence of cell size on infestation rates by the mite *Varroa jacobsoni*. *Abstract of the XXX Apimondia Int. Apic. Cong, Nagoya, Japan*. p.35.

Delfinado-Baker, M. 1984. The nymphal stages and male of *Varroa jacobsoni* Oudemans - a parasite of honey bees. *Int. J. Acarol.* 10:75-80.

Eischen, F. A., Pettis, J. S., Dietz, A. 1987. A rapid method of evaluating compounds for the control of *Acarapis woodi* (Rennie). *Am. Bee J.* 127:99-101.

Engels, W., Gonçalves, L. S., Steiner, J., Buriolla, A. H., Issa, M. R. C. 1986. *Varroa*-befall von *carnica*-völkern in tropenklima. *Apidologie* 17:203-216.

Flechtmann, C. H. W., Amaral, E., dos Santos, F. D. 1977. Ocorrência da acariose no estado de São Paulo. *Anais do IV Cong. Bras. Apic., (Curitiba, 1976).* ed. L. S. Gonçalves, pp. 197-198, Paraná, Brazil.

Giavarini, I., Giordani, G. 1966. Study of acarine disease of honey bee. *Final Technical Report*. Natl. Inst. Apic., Bologna, Italy.

Guimarães, N. P., Gonçalves, A. H. C. 1977. Ocorrência e tratamentos de cria putrida européia (EFB) e Nosemose no municipio de Baipendi (MG). *Anais do IV Cong. Bras. Apic., (Curitiba, 1976).* ed. L. S. Gonçalves, pp. 205-206, Paraná, Brazil.

Kerr, W. E., Amaral, E. 1960. *Apicultura Científica e Práctica.* Secretaria de Estado dos Negócios da Agricultura do Estado de São Paulo. Directoria de Publicadade Agrícola. 148 pp.

Lengier, S. 1977. Comparação da abelha italiana com a abelha africanizada quanto a resistência á doença da cria. *Anais do IV Cong. Apic., (Curitiba, 1976).* ed. L. S. Gonçalves, pp. 199-203, Paraná, Brazil.

Machado, J. O., Lemos, M. V. F. 1975. *Streptococcus pluton*, seu isolamento e combate nos apiários de Igarapava, Jardinópolis e Ribeirão Preto - Estado de São Paulo. *Anais do IV Cong. Bras. Apic., (Curitiba, 1976).* ed. L. S. Gonçalves, pp. 199-203, Paraná, Brazil.

Mendoza, M. R. Q., Delgado, S. G., Villasboa, H. R. D. 1987. *Investigacion comparativa de la incidencia del acaro Varroa jacobsoni Oudemans sobre distintas razas y lineas de abejas Apis mellifera L. en el Paraguay.* Depto. Apicultura, Univ. Nac. Asunción, Paraguay.

Message, D. 1979. *Efeito de condições ambientais no comportamento higiênico em abelhas africanizadas, Apis mellifera*. Master thesis. Depto. Genetica, Fac. de Medicina, Univ. São Paulo, Ribeirão Preto, Brazil.
Message, D. 1986. *Aspectos reproductivos do ácaro Varroa jacobsoni e seus efeitos em colônias de abelhas africanizadas*. (Reproduction by the mite *Varroa jacobsoni* and the effects of infestation on Africanized honey bees). PhD Dissertation. Depto. Genetica, Fac. de Medicina, Univ. São Paulo, Ribierão Preto, Brazil.
Message, D., Bezerra, M. A. F., Pfrimer, R., Gonçalves, L. S. 1983. Ocorrência de acariose em abelhas *Apis mellifera* na refiao de Goiâna. *Ciência e Cultura (suppl)* p. 711.
Moretto, G. 1988. *Efeito de diferentes regiões climáticas Brasileiras e de tipos raciais de abhelhas Apis mellifera na dinâmica de populações do ácaro Varroa jacobsoni*. Masters thesis. Depto. Genetica, Fac. de Medicina, Univ. São Paulo, Ribeirão Preto, Brazil.
Nascimento, C. B. 1970. Pesquisa de endo e ecto-parasitos de *Apis mellifera* L. In *1° Cong. Bras. Apic. (Florianópolis, 1970)*, ed. H. Wiese, pp. 199-207. Santa Catarina, Brazil.
Ritter, W., De Jong, D. 1984. Reproduction of *Varroa jacobsoni* O. in Europe, the Middle East and tropical South America. *Z. Angewandte Entomol.* 98:55-57.
Ritter, W., Ruttner, F.. 1970. Diagnoseverfahren (Varroa). *Allg. dtsch. Imerzig.* 5:134-138.
Rosenkranz, P. 1986. Factors affecting *Varroa* reproduction in colonies of *Apis mellifera*: a comparison of European and Africanized honey bees. *Arbeitsgemeinschaft des Instituts fur Bienenforschung*, V. Abstracts of the Workshop in Feldafin/Starnberg, West Germany.
Sande, M. W. van de 1986. Diseases of Africanized bees in Santa Catarina, Brazil. *Anais do 6° Cong. Bras. Apic, (Florianópolis 1984)*. In press.
Silva, R. M. B. da 1977. Ocorrência de *Acarapis woodi* Rennie em Pindamonhangaba, Estado de São Paulo. *Anais do IV Cong. Bras. Apic., (Curitiba, 1976)*. ed. L. S. Gonçalves, pp. 207-209, Paraná, Brazil.
Steiner, J., Issa, M. R. C., Buriolla, A. H., Engels, E., Gonçalves, L. S., Engels, W. 1984. Varroatose na Alemanha e no Brasil. *Apic. no Brasil* 1:35-37.
Wiese, H. 1975. Primeira curva de esporulaçao de *Nosema apis* Zander no estado de Santa Catarina, Brasil. *Anais do III Cong. Bras. Apic., (Piracicaba, SP)*. pp. 211-214.
Wiese, H., Meyer, C. R. 1975. Contribuição para o reconhecimento da acariose em *Apis mellifera* L. *Anais do III Cong. Bras. Apic., (Piracicaba, SP)*. pp. 207-210.
Wilson, W. T., Moffett, J. O., Cox, R. L., Maki, D. L., Richardson, H., Rivera, R. 1988. Menthol treatment for *Acarapis woodi* control in *Apis mellifera* and the resulting residues in honey. In *Africanized Honey Bees and Bee Mites*. ed. G. R. Needham, R. E. Page, M. Delfinado-Baker, C. E. Bowman, pp. 535-540. Chichester, England: Ellis Horwood Limited.
Woyke, J. 1985. *Tropilaelaps clareae*, a serious pest of *Apis mellifera* in the tropics, but not dangerous for apiculture in temperate zones. *Am.. Bee J.* 125:497-499.

Further Reading

Bailey, L. 1963. *Infectious diseases of the honey bee*. London: Land Books.
Hansen, H. 1980. *Honey Bee Brood Diseases*, ed. R. A. Morse, English edition. Ithaca, NY: Wicwas Press.

Morse, R. A. 1990. *Honey Bee Pests, Predators, and Diseases.* Ithaca, NY: Cornell Univ. Press. 2nd ed.

Gochnauer, T. A., Furgala, B., Shimanuki, H. 1957. Diseases and enemies of the honey bee. In *The Hive and the Honey Bee,* ed. Dadant & Sons. Hamilton, Ill: Dadant & Sons.

Shimanuki, H. 1980. Diseases and pests of honey bees. In *Beekeeping in the United States.* USDA Agriculture Handbook No. 335, Rev. Ed.

PART FOUR

Defensive Behavior

15

DEFENSIVE BEHAVIOR

Michael D. Breed[1]

"The colonies always appear to be alerted, ever ready to defend the hive and on occasion the whole colony goes beserk and stings every living thing in sight." (Smith, 1958)

In this chapter I discuss the dynamics of the defensive response, which includes guarding, attack, and pursuit, each of which is often mediated by alarm pheromones. Defensive behavior, and associated stinging incidents, are the most attention getting aspect of Africanized bee behavior. Defense has certainly been the most important component of human-Africanized bee interactions and has the greatest impact on apiculture. The exact nature of human-Africanized bee interactions in North America cannot be predicted, but the generation of an understanding of the defensive response seems prudent.

Quantitative studies of defensive responses have shown that Africanized bees have greatly enhanced responses to movement, vibration, and to alarm pheromone, as compared to North American stocks (Collins *et al.*, 1982). The genetic aspects of these differences are discussed in Chapters 16 and 17 in this volume. As might be expected with behavioral differences between races, genetic experiments clearly show an underlying genotypic component to the observed racial differences (Collins *et al.*, 1982, 1988; Stort, 1975a,b,c).

Less is known about the organization of the defensive response. I view two aspects as central to understanding how defensive behavior functions. First, defensive behavior is part of the overall division of labor in the colony. As such, specialists in defense should be present, and these specialists should fit within the pattern of temporal polyethism (Lindauer, 1952; Seeley, 1982; Winston and Punnett, 1982; Kolmes, 1985) of colonies. Second, defense is a sequence of events ranging from alert to mass attack with multiple stinging. We know a lot about the magnitude of responses to stimuli at colony entrances

---

[1]Professor Breed is in the Department of Environmental, Population and Organismic Biology, The University of Colorado, Boulder, Colorado 80309-0334, USA.

(Butler and Free, 1952; Collins and Kubasek, 1985), less about guard bees (Moore *et al.*, 1987, Breed *et al.*, 1988), and very little about soldier bees (Breed *et al.*, in press). Integrated studies of defensive response sequences, which will be critical to understanding the role of colony defense in colony division of labor, are not yet available. Once all of these factors are fully understood, interracial comparisons will allow determination of the genetic components all aspects of the defensive response.

The high level of variability within honey bee races in defensive response (see Fletcher, Chapter 4, for a discussion of variability within *Apis mellifera scutellata*) and the variability exhibited by individual colonies depending on weather, season, and foraging conditions, add a further dimension of complexity to studies of honey bee defensive behavior. Not all Africanized honey bee colonies are highly defensive, nor are all colonies of North American bees mildly tempered. The genetic components of this variation are explored in the following two Chapters. The components of the defensive phenotype that vary within any given genotype are not well understood and will provide an important basis for further work.

ORGANIZATION OF THE DEFENSIVE RESPONSE

This section deals primarily with information concerning the defensive response of European races of bees, for which more detailed information is available. This is an amplification of the model for defensive behavior developed by Collins *et al.* (1980) and discussed in Chapter 16. The first line of colony defense are the guard bees (Moore *et al.*, 1987). The reactions of guard bees to conspecifics is modified by the ability to discriminate nestmates from non-nestmates (Butler and Free, 1952; Ribbands, 1954; Breed *et al.*, 1985). Defensive responses to major disturbances usually involve large numbers of non-guard defenders, or soldiers, which have only recently been characterized (Breed *et al.*, in press). Disturbance of the colony may change the "alertness" of bees at the colony entrance; consequently defensive responses may vary over time (Butler and Free, 1952). Defensive responses also vary with seasonal changes, particularly with food abundance (Ribbands, 1954). Below I deal with each of these factors in detail. Emphasis in this discussion is given to features that vary between colonies or between racial stocks.

Guard bees

The defense of a honey bee colony's entrance is usually initiated by guard bees (Moore *et al.*, 1987). Guard bees patrol the entrance of the colony and exclude intruders. They may also alert colony members to the presence of intruders although this link has not been conclusively demonstrated. Butler and Free (1952) characterized the posture of guarding bees: "These [alert] guard bees assume a typical attitude, frequently standing with their forelegs off the ground, with their antennae held forwards, and their mandibles and wings closed." This posture is easily recognized by observers and can serve as an indicator of the level of guarding activity in a colony. Alert guards accost entering bees and antennate them; bees discriminated as non-nestmates are ejected from the colony entrance (Breed, 1983; Breed *et al.*, 1988). Guards at the entrance also may have some significance in defending against vertebrate predators, although the level of disturbance caused by such predators is usually sufficient to stimulate a large number of additional bees to come out of the colony.

Butler and Free (1952) noted that guards attacking robbing bees were alerted by the unique pattern of flight exhibited by robbing bees. They related this to the findings of Lecomte (1951), who was the first to explore in a systematic way the effects of movement on flight activity and aggressiveness of bees. Butler and Free (1952) did similar experiments with moving lures in front of colonies. They argued that if the movement simulated that of the peculiar flight of robbers that odor then played no role in discrimination of these non-nestmates; that movement alone was adequate for discrimination. In general, disturbed bees are attracted by moving objects and dark colors; this allows them to orient to predators (Lecomte, 1952).

Moore *et al.* (1987) characterized guards of colonies of North American bees of European origin; they initiate guarding at a slightly younger than foragers (mean age=15.1 days), guard for very short periods of time (mean time=1.4 days), and are perhaps rarer than might have been expected (mean number present=72). Only a small percentage of any age cohort exhibits guarding behavior (mean=10.1). Thus guarding is a specialized task within the overall scheme of temporal polyethism in honey bees.

Occasionally guards may briefly fly away from the colony entrance and inspect, or even sting, vertebrates in the region of the colony (Moore *et al.*, 1987). These flying guards are rare in European colonies, but may be more common in Africanized colonies (unpubl. data). These guards are a non-random subset of the guard population; they have been guards for several days and represent about 1% of the guarding population. These guards respond to human observers in the vicinity of the colony by flying close to the face and other exposed areas of skin. Occasionally they may sting human observers (Moore *et al.*, 1987).

There are several unanswered questions about the place guarding has in honey bee temporal polyethism. After guarding workers become foragers. Many of those workers which did not guard (the bulk of the age cohort) also become foragers. The role that the non-guards of guarding age play is not known. It is also not known what the behavioral precedents to guarding are: are bees that become guards identifiably unique in other behavioral patterns?

Another important, but little understood, question is how guards are distributed in the colony. Casual observation indicates that in addition to being present at the main entrance of the colony they also guard ventilation holes that are often placed in the inner cover of commercial hives. To test whether guards are also present within the nest Breed and Morhart (unpubl. data) introduced living and dead bees into the brood nest area of colonies. Dead bees were rapidly removed by undertakers but the living bees remained for over an hour within the nest with little evidence of removal attempts. This result is somewhat at variance with observations of Page (unpubl. data) that indicate that under some circumstances even very young bees are expelled when introduced into the brood nest. Further investigation will be required to determine whether guards are active only at entrances, in an envelope around the nest (as found in open nesting species of *Apis*) or are dispersed throughout the nest.

Stinging by guards releases alarm pheromone, which may attract soldiers (see below), but the relationship, if any, between guard and soldier activity has not been documented. It is possible that guards have a key role in alerting other bees in the colony to major threats, but it is also possible that the activities of the guards are completely independent of the soldiers. Experimental exploration of this point will clarify much of how defensive responses are organized.

A recent cross-fostering study (Breed and Rogers, unpubl. data) supports the hypothesis that there is a genetic basis for the expression of guarding behavior. Bees fostered from highly defensive colonies into low defense colonies are more likely to guard than bees from low defense colonies fostered into high defense colonies. In addition, Robinson and Page (1988) and Breed *et al.* (in press) found allozymic evidence for genetic differentiation of guards from other task groups in colonies. Thus there is good evidence for a genetic basis for guarding behavior.

Soldiers

As noted above, guards are clearly not the only bees involved in colony defense. Non-guard defenders, which are referred to as "soldiers," are the least well understood component of the defensive response. In the search for critical differences between European and "Africanized" bees this group is probably of pre-eminent importance. The massive defensive response sometimes observed in "Africanized" bees is presumably due to enhanced responsiveness of non-guard defenders and/or larger numbers of such defenders, but there are no data available on even this basic question.

Recently I have, with others, (Breed *et al.*, in press) initiated studies on this group of bees. Our studies have, so far, been limited to bees of European origin in North America. We assayed soldiers by removing all the frames from colonies and shaking the bees from the frames onto the ground. Sweep samples were then taken of the bees flying around the investigators. Although we defined bees that flew in response to such a disturbance as soldiers, our assays probably included some foragers that returned to the colony during the approximately ten minute time period of disturbance and collection.

By marking age-cohorts of newly emerged bees we determined that bees flying in response to disturbance were from a broad range of age categories. Few bees were under ten days of age, indicating that "house bees" do not normally serve as non-guard defenders. The response of bees older than ten days was not reflective of any age-specialization, and bees of foraging (the usual terminal duty for bees) age were well represented in sample. Wing wear indicated that the non-guard defenders were a population statistically discriminable from both foragers and guards. In some colonies repeated disturbance on subsequent days appeared to stimulate responses of increased strength, but the assay was not adequate to quantify such differences.

Soldiers are recruited to the site of the intruder either by the visual stimuli (Gary, 1975) or by alarm pheromones. At the colony entrance, guards may release alarm pheromone from the sting or mandibular glands; other bees then orient to the location of the disturbance. Away from the entrance, as in a major disturbance by a mammalian predator, the motion and color of the predator play key roles (Gary, 1975). Alarm pheromone released in stinging the predator also serves to orient attacking bees and stimulate further stinging (Boch *et al.*, 1962).

Honey bee alarm pheromones are produced by the sting apparatus and are released when the sting is autotomized (Boch *et al.*, 1962). Smaller amounts of pheromone may be released when the sting is extruded by an alarmed bee. Alarm pheromones of secondary importance are produced by the mandibular glands and released when a bee bites (Shearer and Boch, 1965). A major disturbance, such as removal of the cover on a hive, may stimulate bees to fly even if the alert has not been mediated by guard bees and alarm pheromone.

Bees that respond to alarm pheromone or disturbance do not all respond in the same way. The most extreme response to a large intruder is landing on it and stinging. Bees may also engage in what I term intimidatory behavior, in which they fly at the intruder but do not land. For human observers the most annoying component of this pattern is when the bee hovers a few inches from the face, appearing to threaten to sting sensitive areas such as the eyes or nose. This probably effectively deters most mammalian or avian predators. Large numbers of bees may simply fly in the region of the intruder, without actually coming near. These bees scare inexperienced human observers.

Thus there is a gradient of response along a continuous scale from flight to oriented flight to stinging. In European bees the bulk of bees responding to a disturbance are usually in the first category (just flying); in "Africanized" bees the response is shifted toward the third category (stinging). Unfortunately there are no quantitative data available on the distribution of bees along this gradient in either racial group. Gary (1975) estimates that only one half of one percent of the bees in a colony will actually sting; quantitative comparisons of propensity to sting between European and Africanized bees would be very interesting. Within a colony the distribution is affected by the severity of the disturbance.

(Beekeepers have noted that the "temperament" of a colony may vary from day to day; beekeeping lore relates this to climatic conditions and food availability.)

In European bees even major disturbance of a colony results in a limited proportion of the colony population flying and taking part in the defense of the colony. An informal estimate is that from a colony of 50,000 workers roughly 5,000 (10% of the colony population) are involved (as fliers, intimidators, or stingers) in a maximal defensive response. The other bees remain on the comb or, if shaken from the comb, cluster. "Africanized" bees, on the other hand, may have a much higher proportion of the colony population mobilized in a defensive response.

In addition to characterizing the soldiers, Breed *et al.* (in press), determined that soldiers can be allozymically distinguished from other task groups in the colony. This is consistent with a similar result for guards (see above), but guards and soldiers are genetically distinguishable from each other.

INTENSITY OF DEFENSIVE RESPONSES

Breed *et al.* (1989) showed that the persistence of guard bees at the colony entrance is correlated with the number of bees responding to alarm pheromone in a standard assay of colony defensiveness (Collins and Kubasek, 1982). The average number of days a bee guards varies considerably from colony to colony. In some colonies few if any guards are observed guarding for more than a single day. In other colonies bees guard up to four or five days and the mean number of days guarding is higher. Guards that persist are more likely to fly from the entrance at vertebrate intruders in the area of the colony (see above). Responsiveness to alarm pheromone could be interpreted as a measure of the intensity of response of soldiers; if this is the case then intensity of guarding and soldiering are related.

The correlation of different components of the defensive response could be used to support the hypothesis that there is an underlying unified variable that determines the defensiveness of a colony. In casual observations I have seen that individual bees from highly defensive colonies are much more likely to attack unfamiliar bees even when they are removed from the colony environment as pupae and are maintained in a highly artificial laboratory environment. Clearly one of the next steps with this type of work is to obtain measurements of many different variables concerned with colony defense so that the appropriate multivariate analysis can be performed to accurately identify interrelated variables.

A detailed discussion of the heritability of the intensity of defensive responses is given by Collins and Rinderer in the following Chapter (16).

ASSAYS FOR DEFENSIVE BEHAVIOR

Collins and Kubasek (1982) reviewed tests for colony defensiveness. Most assays of colony defensiveness use probability of flight (or some closely related measure, such as the number of bees in the air around the colony) or probability of stinging a target as measures of "aggressiveness." These measures are usually taken in the presence of a quantifiable stimulus, such as a blow of measured intensity to the side of a hive, a swinging target of known size, or a combination of disturbance and target. These effectively separate "Africanized" and European stocks, but are limited in their utility in helping us to understand the functional bases for differences between races. In some cases these measures have been used in conjunction with alarm pheromones (see above); hence responsiveness to the pheromones becomes an additional variable.

A large number of other useful variables could be quantified that would give additional information (TABLE 1). Interracial comparisons of variables such as the number of guard bees, the number flying guards, and the persistence of individual guards would give insight into how European and "Africanized" bees differ in the structure of their first line of defense. Interracial comparison of the allocation of bees to flying, intimidating, and stinging during a defensive response would give interesting insight into how the allocation of defensive behavior might have evolved. Such analyses would give greater meaning to genetic studies as the focus would be on specific individual phenotypes, rather than colony level phenotypes.

In further work it will be important to determine the correlations among these individual behavioral patterns. If some of the various components of the defensive response are independent, then more sophisticated analyses of the genetic bases for defensiveness will be possible. Correlational studies among a small subset of these variables (Collins and Kubasek, 1982) indicate that features that could be independent, such as individual responsiveness to alarm pheromone and number of stings in a target, are highly correlated.

INTERSPECIFIC COMPARISONS

Seeley *et al.* (1982) provided a quantitative study of defensive responses of three other species of honey bees. *Apis florea* and *Apis dorsata* construct their nests in exposed locations, usually on branches. *A. florea* workers are small and have a relatively ineffective sting; they abandon their nest when attacked by predators. *A. florea* nests may be difficult to find, as they seem to prefer nest vegetation for their nesting site. *A dorsata* workers are large, aggressive, and inflict painful stings. *A. dorsata* colonies are usually aggregated in high trees, making them highly visible but relatively inaccessible to non-flying predators. *Apis cerana* is more like *A. mellifera* in its biology, with nests in cavities

TABLE 1. Examples of quantifiable factors related to the honey bee defensive response.

number of guard bees
mean length of time a bee guards
number of guard bees that fly while guarding
flight radius of flying guards
responsiveness of guards to alarm pheromone (measured as time latency, effect of pheromonal dilution, or individual aggressiveness)
number of bees responding to single recruiting guard
number of bees responding by flying to a stimulus object
number of bees stinging the object
proportion of colony population responding to a stimulus
distribution of colony population among stingers, intimidators, and fliers

protected by a relatively small entrance, but *A. cerana* workers are not aggressive. Colonies of the two open nesting species are protected by curtains of bees that cover the comb.

Seeley *et al.* (1982) postulate that the defensive requirements of open nesting species are higher, because of the large number of bees required to form the curtain. Relatively high investment in defense would require adjustment of the division of labor schedule of the colony (Seeley, 1982). However, they may have underestimated the defensive investment in soldiers in cavity nesting species; it will be interesting to determine actual relative investments in defense in cavity and open nesting honey bees.

DISCUSSION

Defense is the most easily noticed difference among racial stocks of the honey bee. Many other differences exist (Chapters 16 and 17) which have significant impact on management and the apiarist, but defense is the behavior that poses significant public health and public relations for the apiarist. Thus it is surprising how little is known about the ethology of defense in any honey bee race.

Defense in the bees of European origin used in North America is organized around a progression of defenses (flying guards, entrance guards, and non-guard defenders) that are called into play as the disturbance to the colony increases. It is not immediately apparent, from the accounts that are available, if the defensive response of the Africanized bees in South and Central America is organized in the same way, but with accelerated responses, or in a completely different way.

Understanding this central issue is key to work on the the impacts of hybridization between the races and the possible success of genetic saturation programs.

Another poorly understood phenomenon is the lability of defensive responses under changing environmental conditions. Both European and "Africanized" colonies exhibit marked variation in defensiveness over time. As the "Africanized" bee encounters new environments in its colonization of the Americas a vital question is how much, if at all, will the defensive response change. Work such Spivak's (Chapter 7) is an important contribution in this direction.

The genetics of the defensive response has undergone intense scrutiny and is the subject of the following two chapters.

LITERATURE CITED

Boch, R., Shearer, D. A., Stone, B. C. 1962. Identification of iso-amyl acetate as an active component in the sting pheromone of the honey bee. *Nature* 195:1018-1020.

Breed, M. D. 1983. Nestmate recognition in honey bees. *Anim. Behav.* 31:86-91.

Breed, M. D., Butler, L., Stiller, T. M. 1985. Learning and genetic influences in the discrimination of colony members by social Hymenoptera. *Proc. Natl. Acad. Sci. (USA)* 82:3058-3061.

Breed, M. D., Robinson, G. E., Page, R. E. 1990. Division of labor during honey bee colony defense. *Behav. Ecol. Sociobiol.* in press.

Breed, M. D., Rogers, K. B., Hunley, J. A., Moore, A. J. 1989. A correlation between guard behavior and defensive response in the honey bees, *Apis mellifera. Anim. Behav.* 37:515-516.

Breed, M. D., Stiller, T. M., Moor, M. J. 1988. The ontogeny of kin discrimination cues in the honey bee, *Apis mellifera. Behav. Genet.* 18:439-448.

Butler, C. G. , Free, J. B. 1952. The behaviour of worker honey bees at the hive entrance. *Behaviour* 4:262-292.

Collins, A. M., Kubasek, K. J. 1982. Field test of honey bee (Hymenoptera: Apidae) colony defensive behavior. *Ann. Entomol. Soc. Am.* 75:383-387.

Collins, A. M., Rinderer, T. E., Harbo, J. R., Bolten, A. B. 1982. Colony defense by Africanized and European honey bees. *Science* 218:72-74.

Collins, A. M., Rinderer, T. E., Tucker, K. W. 1988. Colony defence of two honey bee types and their hybrids. I. Naturally mated queens. *J. Apic. Res.* 27:137-140.

Collins, A. M., Rinderer, T. E., Tucker, K. W., Sylvester, H. A., Lackett, J. J. 1980. A model of honey bee defensive behaviour. *J. Apic. Res.* 19:224-231.

Gary, N. E. 1975. Activities and behavior of honey bees. In *The Hive and the Honey Bee*, ed. Dadant & Sons, : Hamilton, Il.: Dadant & Sons.

Kolmes, S. A. 1985. An information-theory analysis of task specialization among worker honey bees performing hive duties. *Anim. Behav.* 33:181-187.

Lecomte, J. 1951. Heterogeneite dans le comportement agressif des ouvrieres d'*Apis mellifera. C. R. Acad. Sci. (Paris)* 234:890-891.

Lecomte, J. 1952. Récherches sur le comportement agressif des ouvrières d'*Apis mellifera. Behaviour* 4:60-66.

Lindauer, M. 1952. Ein beitrag zur frage der Arbeitsteilung im Bienenstaat. *Z. vergl. Physiol.* 34:299-345.

Moore, A. J., Breed, M. D., Moor, M. J. 1987. The guard honey bee: Ontogeny and behavioral variability of workers performing a specialized task. *Anim. Behav.* 35:1159-1167.

Ribbands, C. R. 1954. The defence of the honey bee community. *Proc. Royal Soc. (B)* 142:514-524.

Robinson, G. E., Page, R. E. 1988. Genetic determination of guarding and undertaking in honey bee colonies. *Nature* 333:356-358.

Seeley, T. D. 1982. Adaptive significance of the age polyethism schedule in honey bee colonies. *Behav. Ecol. Sociobiol.* 11:287-293.

Seeley, T. D., Seeley, R. H., Akratanakul, P. 1982. Colony defense strategies of the honey bees in Thailand. *Ecol. Monographs* 52:43-63.

Shearer, D. A.,Boch, R. 1965. 2-Heptanone in the mandibular gland secretion of the honey bee. *Nature* 206:530.

Stort, A. C. 1975a. Genetic study of the aggressiveness of two subspecies of *Apis mellifera* in Brazil. II. Time at which the first sting reached the leather ball. *J. Apic. Res.* 14:171-175.

Stort, A. C. 1975b. Genetic study of the aggressiveness of two subspecies of *Apis mellifera* in Brazil. IV. Number of stings in the gloves of the observer. *Behav. Genet.* 5:269-274.

Stort, A. C. 1975c. Genetic study of the aggressiveness of two subspecies of *Apis mellifera* in Brazil. V. Number of stings in the leather ball. *J. Kans. Entomol. Soc.* 48:381-387.

Smith, F. G. 1958. Beekeeping observations in Tanganyika 1949-1957. *Bee World* 42:29-36.

Winston, J. L., Punnett, E. N. 1982. Factors determining temporal division of labor in honey bees. *Can. J. Zool.* 60:2947-2952.

16

GENETICS OF DEFENSIVE BEHAVIOR I

Anita M. Collins[1] and Thomas E. Rinderer[2]

Of all of the distinctive characteristics of the Africanized bee, the one that has received the most attention, especially in the popular press, is its defensive behavior. Reports in Africa with *A. m. adansonii* (now *scutellata*) included descriptions of the high levels of colony defense and the modifications this required in management techniques (Smith, 1958; Papadopoulo, 1971; Guy, 1972). The scientists who imported the original queens from Africa into Brazil took precautions to prevent introduction of this undesirable trait into the local population (Portugal-Araújo, 1971).

From the first, the proposed solutions to the problems of the Africanized bees have been genetic ones. Kerr and his associates intended to produce a gentle, high-honey-producing hybrid of *A.m. scutellata* with the other subspecies already in Brazil (*A. m. ligustica* and *A. m. mellifera*) (Nogueira-Neto, 1964). Even after the escape of the African swarms, Kerr suggested "the solution" was still to develop "a new race of bees" (1968). Thousands of Italian queens were produced and distributed to beekeepers in the area of Africanization to induce Italianization of managed colonies (Kerr, 1967, Michener, 1975). The report of the National Research Council Committee on the African Honey Bee recommended the development of a selected desirable stock to be used in a barrier zone to control the northward migration of the bee and perhaps elsewhere to alleviate problems already in existence (Anonymous, 1972).

There were obvious indications that defensiveness might be modified genetically. Defensiveness was often referred to in discussions of subspecies differences (Ruttner, 1975). Also, some beekeepers had exercised artificial

[1]Dr. Collins is with the USDA--ARS, Honey Bee Research Laboratory, 2413 E. Highway 83, Weslaco, TX, 78596, USA.

[2]Dr. Rinderer is with the USDA--ARS, Honey Bee Breeding, Genetics and Physiology Research Laboratory, 1157 Ben Hur Rd., Baton Rouge, Louisiana, 70820, USA.

selection on their managed stocks to produce more gentle bees. However, little was known about the basic genetics of this trait.

The situation in South America was favorable for genetic investigations of defensive behavior. Populations that showed clear phenotypic differences, varying from gentle to defensive, could be compared in the same location. There was also a hybridized population, the result of ongoing mating between the two general parental types, African and European, as well as the hybrids produced. The extreme variability expressed in this honey bee population, referred to as Africanized, may have been partly due to environment, for the same colonies often reacted differently at different times and places (Anonymous, 1972; Brandeburgo *et al.*, 1982; Villa, 1988). But the consistent association of African ancestry and extreme defensiveness implied that a majority of the between-colony differences were due to differences in genotype, and that the variability of the hybrid population could be tapped as a possible resource for a solution to the Africanized bee problem through genetic selection.

THE CHARACTER

Description

The term colony defense encompasses a complex sequence of possible actions performed by individual bees or, more usually, a group of bees. A theoretical model discussed in detail by Collins *et al.* (1980), presents this behavior as a response sequence by an individual worker bee (of whatever type) to stimuli in her environment. However, there are several modes of communication between workers that provide for coordination of many individuals into an effective social unit of defense. The most notable of these is alarm pheromone.

When an individual bee perceives an appropriate and sufficient stimulus, she becomes alerted. Her response may include a typical alert posture (Ghent and Gary, 1962; Maschwitz, 1964) with body raised, wings extended, mandibles open and antennae waving. The sting may also be extended. If appropriate stimulation continues, the bee begins random movement which serves to eventually bring her into proximity with the source of the disturbance. With continued stimulation from the intruder, the bee can orient to the disturbance, and finally attack. The obvious form of this attack is stinging, but threat behavior by a flying, buzzing bee is also effective. The defending worker may also bite, crawl into ears, nose or mouth, or burrow into fur or clothes.

An alerted worker may run into the colony releasing alarm pheromone from her extended sting. This recruits other bees to defend, and is the primary method of coordinating a group response. This recruiting bee may or may not be involved in later stages of defense. Perception of the original disturbing stimuli by many bees can also produce a group reaction. The searching, orienting and attacking of defending bees may attract more workers to the location. Stings left in an intruder continue to release alarm pheromone, serving as markers for more attacks (Maschwitz, 1964).

Depending upon her physiological condition, a worker may not respond to disturbance of the colony by defensive behavior, but by withdrawal. This could occur at the initial time of alerting, with the bee reverting to a defensive sequence if provocation continues. Alternatively, a defending bee may reach a point at which she flees. In case of long-term disturbances, the ultimate defense of the colony may be to abscond from the nest site.

The bases for differences in defensiveness could lie at any stage of this complex series of activities. More defensive bees might perceive alarm pheromones or other stimuli, such as vibration, at lower levels than less defensive bees. If sensory abilities are the same, more defensive bees might have lower thresholds, that is, they show a behavioral response at lower levels of stimulation. A given stimulus could provoke faster, more vigorous responses or more bees to respond. The production and release of more, or more effective, alarm pheromone would enhance the group response. Greater ability to orient to and follow an intruder, greater propensity to respond by stinging than by threat or withdrawal, and longer maintenance of an aroused state could also be factors in increased levels of defense.

Quantification

Appropriate genetic dissection of defensive behavior should include measurement of many of these aspects of the behavioral sequence. Morphological and physiological assays would be less subject to environmental conditions than behavioral ones, but may be difficult to define, to perform, and to correlate with the end result of a defensive sequence. The more clearly discrete units within the behavioral sequence can be defined, the easier will be the genetic analysis. For a more complete discussion of these principles, see Rinderer and Collins (1986). A number of such assay systems have already been developed.

Study of the production of alarm pheromone began when Boch *et al.* (1962) identified isopentyl acetate (IPA) as an active component of the sting alarm pheromone. Another compound, 2-heptanone (2HPT), produced in the mandibular glands of the worker, was also shown to have an alarm function (Shearer and Boch, 1965). Fifteen more compounds associated with the sting apparatus were identified (Blum *et al.* 1978) and shown to stimulate an alert response from worker bees (Collins and Blum, 1982, 1983).

To measure production of alarm pheromone, workers of foraging age were collected from colony entrances and frozen. Stings, or stings and heads, from samples of ten bees from each colony were removed with forceps and collected over ice in pesticide-grade methylene chloride (with sodium sulfate as a drying agent) in crimp-top vials. Samples of these solvent-extracts were analyzed by gas chromatography for alarm pheromone components.

Response to alarm pheromone components was measured in laboratory tests of caged young workers following the procedure of Collins and Rothenbuhler (1978). Bees emerged from brood cells in an incubator in single-colony groups over periods of 24 hours and were caged in glass-fronted wooden cages (Kulincevic *et al.*, 1973). They were given 24 hours to adjust to the cage before they were tested for the next three days. Observations were made on the initial activity level in the cage prior to testing, the speed of the response to the test chemical (time until a response is first seen), the intensity of the response based on both number of responders and the vigor of the response, and the duration of the response.

Southwick and Moritz (1985) and Moritz *et al.* (1985) measured increases in oxygen consumption following exposure to alarm pheromone. Group size and pheromone concentration up to 2.4 µg/ml influenced the level of the response. Oxygen consumption was highly correlated with colony temperament and is a good candidate for a quantitative measure of this trait.

The structures on the antennae associated with the perception of alarm pheromone are the *sensilla placodea* (Kaissling and Renner, 1968). It was hypothesized that more would be present in Africanized bees, enhancing defense by increasing the response to pheromone. The number of these present on worker bee antennae were estimated using light microscopy (Stort and Barelli, 1981) as well as scanning electron microscopy (Collins, unpubl. data).

A test of colony defense, the sequential response by a colony following disturbance, was developed by Collins and Kubasek (1982) to elicit the distinct phases proposed by the model (Collins *et al.*, 1980) and to quantify each of them. The bees were alerted by the application of synthetic alarm pheromone at the colony entrance and the time until the first defender exited the colony in the search mode was recorded. After 30s of exposure to the pheromone, a second stimulus, vibration, was applied to the colony by means of a marble propelled from a sling. This vibration enhanced the defense response. At 60s, two small (2 in x 2 in) dark suede patches (targets) were waved in front of the colony entrance by a mechanical device. The time until the first bee landed on a target and attempted to sting was measured. After 30s, the targets were removed and the test ended by blowing smoke around the colony entrance. Later, the stings embedded in the targets were counted. A series of photographs of the area around the hive entrance was taken at 30s, 60s and 90s and was used to determine the numbers of bees responding. After experience with Africanized bees, a second series of photographs was added to estimate the numbers of airborne defenders.

A number of other possible characters that could be used to quantify defensive behavior certainly exist. Electroantennagraph studies would measure response to alarm pheromone at a direct physiological level. Biochemical assays for levels of neurotransmitters could discover differences in basic excitability of bee types. Some basic work has been done on perception (Wehner, 1971; Anderson, 1977) and orientation in honey bees (Lindauer, 1963) that could be used for genetic comparisons. There is room for expanded work on the events involved in colony defense. As the behavioral sequence becomes more finely dissected and understood, clearer analysis of the genetic control should be possible. However, considerable genetic investigation has already been accomplished.

THE GENETICS

Population

Perhaps the earliest behavioral-genetic analyses of colony defense were the comparative descriptions of the various subspecies. Ruttner (1975) describes *A. m. mellifera*, one of the European subspecies, as being frequently aggressive and *A. m. carnica* as the most gentle. Many such descriptions are to be found in honey bee literature. Numerous reports of the results of crosses between subspecies also exist (Ruttner, 1968). However, these evaluations were usually subjective observations based on beekeeper experience during normal hive manipulations and were not quantitative.

With the development of appropriate measuring systems, defensive behavior of populations (subspecies, races, stocks, lines or types) of bees can be accurately described and differences can be related to the underlying genetic structures of the population. Initial descriptions include the simple statistics of means, variances and ranges. Eventually, the frequencies of specific genes in a honey-bee population may be used in descriptions. Importantly, any such descriptions, whether simple or complex, apply only to specific populations at the time of the measurement since all biological systems are subject to constant change.

The exact genetic constitution of the spreading Africanized hybridized population is unknown because we have little information on the structure of the previously existing South American population of honey bees and extent of crossing with the African subspecies. In any case, there have been significant changes in the population resulting from its spread and requeening efforts designed to control it. For example, when the spreading population reached Venezuela and Colombia, it encountered an established feral population of European bees, probably hybrids of *A. m. mellifera* and *A. m. ligustica*. Also, the Brazilian government is reported to have distributed *A. m. ligustica* queens to beekeepers in the early 1960s (Michener, 1975). The net result is the appearance of local populations of varying genetic profiles, all of which likely include individuals which are genetically similar to each of the parental subspecies and all possible intermediate combinations (Lobo *et al.*, 1989; W. Sheppard and T.

E. Rinderer, pers. comm.). The honey bees in North America are also a mix, with *A. m. mellifera* and *A. m. ligustica* as the primary racial ancestors.

Mendelian

The classical Mendelian approach to genetic analysis is to cross two distinct phenotypes and to use the assortment of phenotypes in the F_1, F_2 and backcross generations to estimate the number of genes involved in the control of the phenotypic difference and their relationships. There are some difficulties associated with this approach in honey bees, because it is not possible to produce an F_2 generation. Because drones are haploid, they produce identical spermatozoa, so the array of segregation types in a cross of an F_1 queen and drones from an F_1 queen is limited. In addition, any one colony (the unit on which many aspects of colony defense must be measured) represents a collection of subfamilies as the queen usually mates with 7-10 drones (Taber, 1958). Each of these drones could have different genotypes ranging from more African-like to more European-like. If the queen is also of a mixed genotype and producing highly variable eggs by genetic segregation, the result is a colony of genetically diverse individuals.

This diversity complicates genetic analysis. Greater precision may be obtained by using inbred queens and inseminating with a single drone, a technique developed by Rothenbuhler (1960) for such studies. If extreme phenotypes of defensive behavior are used, useful information on the mode of inheritance of this character can be obtained. Nevertheless, different crosses could give different results if there are multiple alleles for the genes, or polygenic systems of control.

The genetic diversity within a honey bee colony complicates the genetic analysis in another way. A colony is normally composed of several "patrilines," each represented by the offspring of the queen and one of the drones with which she mated. Some recent studies have shown that at least one aspect of colony defense, guarding, is a very specialized task performed by relatively few bees (Moore *et al.*, 1987) and that colonies with guard bees that persisted in this behavior tended to be more defensive (Breed *et al.*, 1989). Robinson *et al.* (1989) have determined that guard bees are a non-random representation of the bees in a colony, probably from one or a few patrilines. These results imply that an assay of the defensive behavior of a colony may measure only a portion (one patriline) of the genotypic variability carried by the worker progeny of a queen.

Quantitative

Because defensive behavior is complex and difficult to dissect into basic units reflecting the influence of one major gene, a more viable approach to genetic analysis of this character is quantitative genetics. This approach is directed to populations, not individuals, and to characters that are controlled by many genes that interact in both additive and nonadditive (e.g., through dominance or epistasis) ways. Descriptions of traits are expressed in statistical terms such as variance and covariance.

Heritability, h^2, is a major genetic parameter utilized in this approach. It is defined as the proportion of the variation of a character that is amenable to change by genetic selection. Estimates of h^2 are possible with appropriately designed groups of single-drone, inbred-queen matings with an array of phenotypes (Rinderer, 1977), or with measurements across a population (Oldroyd and Moran, 1983).

Heritability estimates are primarily used to predict the results of selection programs. Once selection has been practiced for several generations, it is possible to reestimate h^2 in a form called realized heritability (Falconer, 1981) to monitor the success of the program.

Frequently, a selection program will focus on several traits of importance. Thus, it is necessary to also know the correlations between the characters involved (i.e. different measurements of defense behavior). With appropriate familial relationship among the measured individuals, both phenotypic correlations and genetic correlations reflecting commonality of controlling genes can be calculated.

THE RESULTS

Population Surveys

Alarm pheromone production was measured in samples of worker honey bees from 150 colonies in Louisiana, USA, and 147 colonies in Monagas, Venezuela (Collins *et al.*, 1989). The former colonies were representative of the European less defensive type and the latter were representative of the Africanized defensive type. Twelve alarm pheromone components were measured by gas chromatography: isopentyl acetate, 2-heptanone, butyl acetate, 2-methyl butanol, hexyl acetate, 1-hexanol, 2-heptyl acetate, 2-heptanol, octyl acetate, 1-octanol, 2-nonyl acetate and 2 nonanol. A second portion of the study done with four-week-old Africanized and European workers reared in the same location measured three additional compounds, 1-acetoxy-2-octene, 1-acetoxy-2-nonene, and benzl acetate.

Isopentyl acetate, the major sting alarm pheromone component, 2-heptanone, the mandibular gland pheromone, and 2-menthyl butanol were found in greater quantities in European workers. 1-acetoxy-2-octene and 1-acetoxy-2-nonene were equivalent in the two types. All the other compounds were found in greater quantities in the Africanized bees. Greater quantities of alarm pheromone could increase defensive behavior by reaching threshold levels faster following release, accumulating in greater quantity around the colony (thereby recruiting more bees to defend) and taking longer to dissipate after the event.

Response to alarm pheromone was also different between the two populations (Collins *et al.*, 1987a). Comparisons of workers of the two bee types using the laboratory assay of Collins and Rothenbuhler (1978) found no significant difference in the speed of the response to alarm pheromone. However, the Africanized workers responded in greater numbers and with more vigor than the European workers and continued to show a response for a longer time. This greater responsiveness may be the result of the Africanized bees normally existing at a higher level of nervous excitation as they were consistently more active prior to stimulation with pheromone.

Antennal receptors for alarm pheromone (sensilla placodea) were found to occur in similar quantities in both bee types when counted using a scanning electron microscope (Collins, unpubl. data). Stort and Barelli (1981) had indicated significantly more placodes on European antennae, but they worked only with the light microscope.

Colony defense, as measured by the procedure of Collins and Kubasek (1982), was considerably different in the two populations of European and Africanized bees tested in Louisiana and Venezuela (same colonies as for pheromone production) (Collins *et al.*, 1982). The Africanized workers responded twice as fast initially to the alarm pheromone as did the Europeans and about twenty times as fast when the visual stimulus (moving suede patches) was presented (Africanized mean = 0.3s, European mean = 9.2s). Many more bees from the Africanized colonies responded to the test stimuli. For bees photographed on the colony, the difference was about two-fold. However, there were large numbers of uncounted Africanized workers in the air harassing the experimenters, an activity that was rarely seen in bees from European colonies. Bees from Africanized colonies stung the suede targets 8.5 times more than the European colonies.

The large amount of variation found between Africanized colonies for measures of defensive behavior reflects the hybridized nature of the Africanized population. FIGURE 1 shows the range of one measure of defensive behavior, number of stings in the suede targets, for Africanized and European bees. There were some colonies in the Africanized population that were no more defensive than the Europeans. However, most of the Africanized colonies fell outside the general European range, and the most extreme Africanized colony produced four times more stings than the most defensive European colony.

Controlled Crosses

Response to alarm pheromone and other alerting stimuli has been investigated in inbred lines of European bees and their F_1 hybrids. Boch and Rothenbuhler (1974) stimulated colonies of a line with greater propensity to

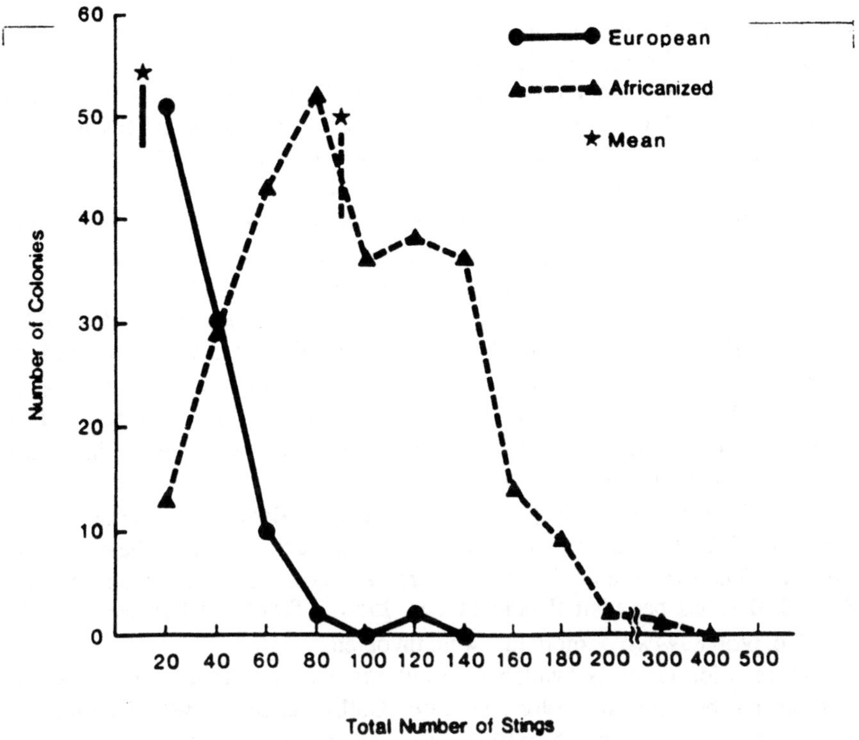

FIGURE 1. Variation in the amount of stinging by European and Africanized honey bees in a test of colony defense. Moving suede targets (2 cm X 2 cm) were presented to the colonies for 30s. (Data from Collins *et al.*, 1982. Copyright in public domain.)

sting (Brown), of one with less propensity to sting (Van Scoy), and their F_1 hybrids, in three different ways: a puff of human breath at the entrance; removal of the hive cover; and IPA at the entrance. The response of the Brown line bees was much greater than that of the Van Scoy bees. The F_1 hybrids (produced by single-drone inseminations) were similar to the Van Scoy parental type for both response to breath and to IPA, responses involving odor perception. For these two responses, the milder response was dominant. The F_1s were intermediate and different from each parental line for response to hive cover removal, a reaction more dependent on visual and tactile senses. This indicated a lack of dominance for this character.

Response to IPA by caged young workers was studied by Collins (1979) using two inbred lines, Brown-Caucasian, initially an outcross from the above Brown line and also defensive, and YD, a gentle line. Single-drone inseminations were used to produce F_1 and backcross colonies. Initial activity level, speed of response and intensity of response showed dominance of the more responsive phenotype. For all the studies with only European bees, estimates were that two or three genetic loci controlled the line differences.

The same laboratory test of response to alarm pheromone (Collins and Rothenbuhler, 1978) was used to assay Africanized and European bees and their F_1 hybrids (single-drone inseminations) reared in Venezuela (Collins *et al.*, unpubl. data). For these crosses, the F_1s tended to be intermediate. This was different than the results just discussed for European lines. Probably different alleles or loci produced the phenotypic variation between Africanized and European lines, than produced the variation between European lines.

Production of alarm pheromone was evaluated by Boch and Rothenbuhler (1974) in the Brown and Van Scoy lines. The F_1s resembled the higher-producing Brown line, and in fact, produced greater amounts of IPA in some colonies. Because of high variability in the F_1s, the conclusions were that IPA production was controlled by many genes and perhaps the genes for increased production were different for the two lines.

The differences in stinging behavior between the Brown and Van Scoy lines were what initiated the genetic investigations into defense using these lines as subjects. The Brown colonies stung the researcher 1.5 times per visit, the Van Scoy 0.01 times per visit (Rothenbuhler, 1964). Since the F_1s were highly variable, probably two or more loci were involved.

More information is available on hybrids between Africanized bees and European subspecies for colony defense (Collins *et al.*, 1988). Colonies of Africanized bees showing extreme African-like behavior, two European subspecies (*A. m. ligustica* and *A. m. caucasica*) and two F_1 hybrids (*A. m. ligustica* x Africanized and *A. m. carnica* x Africanized) were tested with the Collins and Kubasek (1982) procedure. Speed of response to IPA clearly showed dominance of the slow-reacting parents, similar to Boch and Rothenbuhler's (1974) results. The hybrids were also more similar to the European types in number of bees around the entrance early in the test and were intermediate for total stings (FIGURE 2) and number of bees in the air following presentation of the moving targets. However, the two hybrids were different for the other characters. The *A. m. ligustica* cross produced colonies more like European parents, the *A. m. carnica* cross was more like the Africanized type for number of bees in the air early in the test (30s and 60s). The results indicate a polygenic mode of inheritance for most aspects of colony defense, and the presence of some different alleles in the two European subspecies. Kerr (1967) reported hybrids of *A. m. ligustica* and *A. m. scutellata* to be more like the Italian parent in temperament, however.

Instrumental inseminations of inbred queens by single drones from inbred lines to produce a variety of F_1 hybrids between European and Africanized bees provided further support for the hypothesis of polygenic inheritance and multiple allele differences in different lines (Collins *et al.*, unpubl. data). The phenotypes

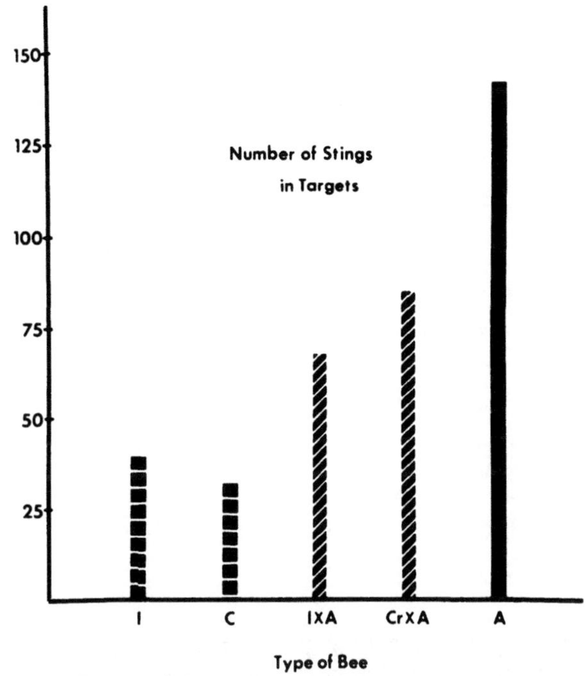

FIGURE 2. Mean number of stings in two moving targets presented to colonies of Africanized, European and hybrid honey bees. I - Italian, C - Caucasian, I x A - Italian x Africanized, Cr x A - Carniolan x Africanized, and A - Africanized. (Data from Collins *et al.*, 1988. Copyright in public domain.)

of the F_1s were similar to one or the other parent or were intermediate to the parental types depending on the specific matings.

Heritability and Correlations

Estimates of heritability have been made for components of defensive behavior (TABLE 1). Collins (1979) used regression of offspring on parent to calculate h^2 from crosses between two inbred European lines and Moritz *et al.* (1987) used the same procedure for matings with non-inbreds. Sib analyses using the sire component of the analysis of variance were used to calculate h^2 from an incomplete diallel test-cross system with European bees (Moritz *et al.*, 1987) and from an array of inbred-queen, single-drone matings with related drones (Rinderer et al., 1983; Collins *et al.*, 1984, 1987b). An alternate approach to calculation of h^2 has been proposed by Oldroyd and Moran (1983) but has never been applied to defensive behavior.

TABLE 1. Heritability (h^2) values for honey bee characters associated with colony defensive behavior.

Character	$h^2 \pm$ standard deviation	Source
Alarm pheromones		Collins *et al.*, 1986
butyl acetate	1.94 ± 0.66	"
isopentyl acetate	0.57 ± 0.55	"
hexyl acetate	1.98 ± 0.66	"
1-hexanol	0.89 ± 0.82	"
2-heptyl acetate	0.67 ± 0.57	"
2-heptanol	0.54 ± 0.54	"
octyl acetate	0.48 ± 0.52	"
2-nonyl acetate	0.96 ± 0.62	"
2-nonanol	1.07 ± 0.64	"
2-heptanone	0.59 ± 0.54	"
Response to alarm pheromones		
initial activity level	0.04 ± 0.01	Collins *et al.*, 1984
time to react	0.83 ± 0.01	"
	0.30 ± 0.01	Rinderer *et al.*, 1983
	0.68	Collins, 1979
Colony defense		
time to respond to targets	0.38 ± 0.19	Collins *et al.*, 1984
number of stings	0.57 ± 0.24	"
number of bees responding at entrance		
pre	0.26 ± 0.25	"
30s	0.12 ± 0.00	"
60s	0.14 ± 0.00	"
90s	0.55 ± 0.02	"

TABLE 2. Phenotypic correlations between selected aspects of defensive behavior, and with several other traits.

trait 1	trait 2	correlation	source	bee population
response to alarm stimuli (IPA, breath, hive opening)	IPA production	NS	Boch/Rothenbuhler 1974	European inbred lines
stinging	hygenic behavior	NS	Rothenbuhler 1964	"
initial activity level	speed of response to alarm pherome in cage	-0.49	Collins et al. 1984	European; Africanized
speed of response to target or pherome	total stings	-0.51 -0.42	" "	" "
speed of response to pheromone by colony	no. bees responding after 30s (pheromone present)	-0.86	"	"
no. bees responding at 90 s (after target presented)	total stings	0.41	"	"
no. bees responding at any time in test	no. bees responding at any other time	.60-.68	"	"

TABLE 2. Continued.

speed of response to pheromone in cage	hoarding	-.70	"	Africanized
speed of response to pheromone by colony	duration of response	-.66	Rinderer et al. 1983	European
"	response to *Nosema apis*	NS	"	"
"	longevity	NS	"	"
IPA production	no. bees responding at 90s	0.27	Collins et al. 1986	European; Africanized
2-heptyl acetate production	"	0.21	"	"
2-nonyl acetate production	"	0.21	"	"
10 sting alarm pheromones	with other alarm pheromes	0.21-.92	"	"
octyl acetate	2 heptanone	0.57	"	"

Because of haplo-diploidy and the social nature of the honey bee, traditional quantitative genetic theory based on diploid systems is not directly applicable to the honey bee. All the authors mentioned above have made attempts to apply this theory with appropriate modification in design. However, these may not be the most accurate estimates possible. Until the theory is more carefully redrafted, these stand as our best estimates. All the values in TABLE 1 except the ones for initial activity level for response to alarm pheromone and for number of bees responding at the entrance prior to the test and at 30s and 60s, indicate sufficient genetic influence on the variation of this trait to encourage attempts to select for reduced levels of defensive behavior.

Another parameter that is important in planning a successful selection program is the correlation between traits under selection. TABLE 2 presents a number of significant correlations that are known for defensive behavior. All the measures of defensive behavior are correlated in such a way that more defensive levels of one are associated with more defensive levels of another. IPA production was not correlated with response to IPA, but was correlated with another aspect of defense, number of bees responding at 90s, which involves release of alarm pheromone during stinging of targets in the last 30s.

The two traits related to disease, hygienic behavior and response to *Nosema apis*, were included in the table because it has been suggested that both defensiveness and disease resistance are aspects of basic vigor in honey bees and should therefore be correlated. These traits were not significantly correlated, however.

In the Africanized population, a fast response to alarm pheromone was correlated with a high level of hoarding, possibly because of the commonality of chemical perception in both traits. This could prove to be a problem if selection were needed for high hoarding and low alarm responses. We infer from the positive correlations of the defensive behavior components that selection for this overall phenotype should be uncomplicated by conflicting correlations. That is, selection for slower responses to moving targets also selects for less stinging, since both are less defensive phenotypes.

Selection

Based on the predictions made from the h^2 estimates for aspects of colony defense, a controlled selection program was initiated using instrumental inseminations of Africanized bees (Collins, 1986). Selection was bidirectional, for more and less defensive bees. Comparisons were made to representatives of the base population for each generation. Defensive behavior was assayed by the responses of full-sized colonies. All the measures were standardized by a Z transformation and combined into an index (I).

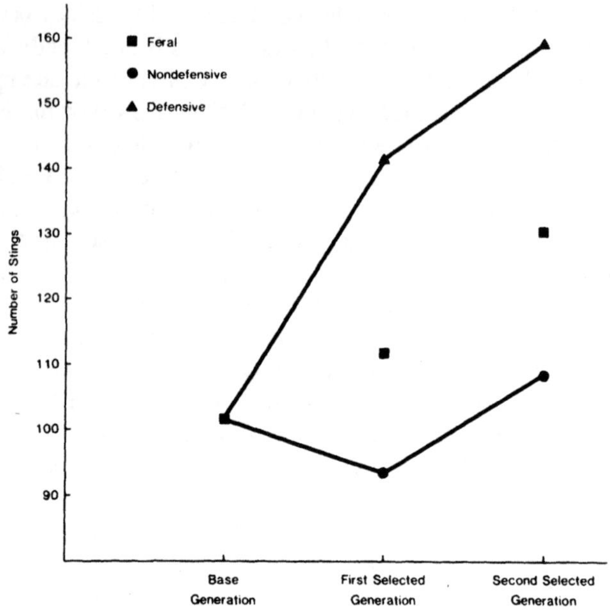

FIGURE 3. Results of a bidirectional selection program for defensiveness in Africanized honey bees. Trait presented is number of stings in two moving targets presented for 30s.

$$I = 1/2\,p + t - \frac{(30s + 60s + 90s)}{3} - s,$$

where p = speed of response to pheromone; t = speed of response to target; 30s, 60s and 90s = number of bees responding at each time; and s = total stings. The ten most defensive and ten least defensive colonies were used as parentals for each generation from a pool of approximately 80 colonies.

FIGURE 3 shows the average number of stings for the base population and the two selected generations. Most of the third generation colonies were lost to insecticide. The selection for a more defensive phenotype was progressing faster than for a more desirable gentle phenotype. The realized h^2 values 0.87 and 0.10 respectively, emphasize these results.

SUMMARY

It is clear that defensive behavior is a complex character for genetic analysis. Even when it has been divided behaviorally into some of its component parts, estimates are for several genes involved in regulation of the variation. This trait must be treated as a polygenic one and dealt with at a population level by quantitative methods (h^2 and correlations). Specific crosses of individual colonies give highly variable results, as do crosses between stocks, lines or types. However, appropriate selection programs should prove successful in reducing extreme colony defensiveness.

Discussions of Africanized bees often imply that these colonies all exhibit the extremely high levels of defense that enhance the reputation of the African, *A. m. scutellata*, subspecies. In fact, the population of bees to which the term Africanized applies is variable in genotype and phenotype. There are some Africanized colonies which are not more defensive than European types, and a greater number that have intermediate levels of defense. Kerr (1968), in an early report from Brazil, indicated that 10% of the assayed colonies were gentle, 60-70% were intermediate and 20-25% were highly defensive. It is the smaller number of colonies exhibiting the extreme of the phenotype that causes most problems for beekeepers and the public.

Beekeepers, especially queen breeders, can cope with this problem in a number of ways. In addition to attempts to maintain partial reproductive isolation for desirable, gentle stocks (Hellmich *et al.*, 1988), a continuing selection program to destroy or requeen the most defensive colonies and rear queens and drones from the least defensive colonies can be readily practiced. Use of stocks that are less defensive, and that produce less defensive hybrids when crossed to Africanized bees should be encouraged.

Further work on the genetics of defensive behavior is desirable. Other assay techniques could be found, especially some that function at basic physiological levels that are less affected by environment than gross behavioral assays. Easily performed assays that could be used by many beekeepers would be valuable. More effective, efficient production of mated queens could be achieved if there was a way to assay queens and drones directly for this trait or for a highly correlated trait. The possibility also exists that single gene mutants might occur which alter or interfere with the normal sequence of colony defense, such that massive stinging did not occur.

Acknowledgement

In cooperation with the Louisiana Agricultural Experiment Station.

LITERATURE CITED

Anderson, A. M. 1977. Shape perception in the honey bee. *Anim. Behav.* 25:67-79.

Anonymous. 1972. *Final report of the Committee on the African honey bee,* Washington, D. C.: Natl. Res. Counc. Natl. Acad. Sci. 95 pp.

Blum, M. S., Fales, H. M., Tucker, K. W., Collins, A. M. 1978. Chemistry of the sting apparatus of the worker honeybee. *J. Apic. Res.* 17:218-221.

Boch, R., Rothenbuhler, W. C. 1974. Defensive behaviour and production of alarm pheromone in honeybees. *J. Apic. Res.* 13:217-221.

Boch, R., Shearer, D. A., Stone, B. C. 1962. Identification of isoamyl acetate as an active component in the sting pheromone of the honey bee. *Nature* 195:1018-1020.

Brandeburgo, M. M., Gonçalves, L. S., Kerr, W. E. 1982. Effects of Brazilian climatic conditions upon the aggressiveness of Africanized colonies of honey bees. In *Social Insects in the Tropics,* ed. P. Jaisson, pp. 255-280. Paris: Université Paris Nord.

Breed, M. D., Rogers, K. B., Hunley, J. A., Moore, A. J. 1989. A correlation between guard behavior and defensive response in the honeybee, *Apis mellifera. Anim. Behav.* 37:515-516.

Collins, A. M. 1979. Genetics of the response of the honeybee to an alarm chemical, isopentyl acetate. *J. Apic. Res.* 18:285-291.

Collins, A. M. 1986. Bidirectional selection for colony defense in Africanized honey bees. *Am. Bee J.* 126:827-828.

Collins, A. M., Blum, M. S. 1982. Bioassay of compounds derived from the honeybee sting. *J. Chem. Ecol.* 8:463-470.

Collins, A. M., Blum, M. S. 1983. Alarm responses caused by newly identified compounds derived from the honeybee sting. *J. Chem. Ecol.* 9:57-65.

Collins, A. M., Brown, M. A., Rinderer, T. E., Harbo, J. R., Tucker, K. W. 1987b. Heritabilities of honey-bee alarm pheromone production. *J. Heredity* 78:29-31.

Collins, A. M., Kubasek, K. J. 1982. Field test of honey bee (Hymenoptera: Apidae) colony defensive behavior. *Ann. Entomol. Soc. Am.* 75:383-387.

Collins, A. M., Rinderer, T. E., Daly, H. V., Harbo, J. R., Pesante, D. G. 1989. Alarm pheromone production by two honeybee *(Apis mellifera)* types. *J. Chem. Ecol.* 15:1747-1756.

Collins, A. M., Rinderer, T. E., Harbo, J. R., Bolten, A. B. 1982. Colony defense by Africanized and European honey bees. *Science* 218:72-74.

Collins, A. M., Rinderer, T. E., Harbo, J. R., Brown, M. A. 1984. Heritabilities and correlations for several characters in the honey bee. *J. Hered.* 75:135-140.

Collins, A. M., Rinderer, T. E., Tucker, K. W. 1988. Colony defence of two honeybee types and their hybrids. I. Naturally mated queens. *J. Apic. Res.* 27:137-140.

Collins, A. M., Rinderer, T. E., Tucker, K. W., Pesante, D. G. 1987a. Response to alarm pheromone by European and Africanized honeybees. *J. Apic. Res.* 24:217-223.

Collins, A. M., Rinderer, T. E., Tucker, K. W., Sylvester, H. A., Lackett, J. J. 1980. A model of honeybee defensive behaviour. *J. Apic. Res.* 19:224-231.

Collins, A. M., Rothenbuhler, W. C. 1978. Laboratory test of the response to an alarm chemical, isopentyl acetate, by *Apis mellifera. Ann. Entomol. Soc. Am.* 71:906-909.

Falconer, D. S. 1981. *Introduction to Quantitative Genetics*. London: Longman Group Limited. 365 pp. 2nd ed.

Ghent, R. L., Gary, N. E. 1962. A chemical alarm releaser in honeybee stings (*Apis mellifera* L.). *Psyche* 69:1-6.

Guy, R.D. 1972. How does the Brazilian bee compare to South African bees? *S. Afr. Bee J.* 44:9.

Hellmich, R. L., Collins, A. M., Danka, R. G., Rinderer, T. E. 1988. Influencing mating of European honey bee queens in areas with Africanized honey bees. *J. Econ. Entomol.* 81:796-799.

Kaissling, K. E., Renner, M. 1968. Antenna Rezeptoren fur queen substance und sterzeldrun bei der Honigbiene. *Z. Vergl. Physiol.* 59:357-361.

Kerr, W. E. 1967. The history of the introduction of African bees to Brazil. *S. Afr. Bee J.* 39:3-5.

Kerr, W. E. 1968. Solução e criar uma raça nova. *O Apicultor* 1:7-10.

Kulincevic, J. M., Rothenbuhler, W. C., Stairs, G. R. 1973. The effect of presence of a queen upon outbreak of a hair-less black syndrome in the honey bee. *J. Invert. Pathol.* 21:241-247.

Lindauer, M. 1963. Uber die Orientierung der Biene in Duftfeld. *Naturwissenshaften* 50:509-514.

Lobo, J. A., Del Lama, M. A., Mestriner, M. A. 1989. Population differentiation and racial admixture in the Africanized honey bee (*Apis mellifera* L.). *Evolution* 43:794-802.

Maschwitz, 1964. Alarm substance and alarm behavior in social Hymenoptera. *Nature* 204:324-327.

Michener, C. D. 1975. The Brazilian bee problem. *Ann. Rev. Entomol.* 20:399-416.

Moore, A. J., Breed, M. D., Moore, M. J. 1987. The guard honey bee: ontogeny and behavioral variability of workers performing a specialized task. *Anim. Behav.* 35:1159-1167.

Moritz, R. F. A., Southwick, E. E., Breh, M. 1985. A metabolic test for the quantitative analysis of alarm behavior of honeybees (*Apis mellifera* L.). *J. Exp. Zool.* 235:1-5.

Moritz, R. F. A., Southwick, E. E., Harbo. J. R. 1987. Genetic analysis of defensive behavior of honeybee colonies (*Apis mellifera* L.) in a field test. *Apidologie* 18:27-42.

Nogueira-Neto, P. 1964. The spread of a fierce African bee in Brazil. *Bee World* 45:119-121.

Oldroyd, B., Moran, C. 1983. Heritability of worker characters in the honeybee (*Apis mellifera*). *Aust. J. Biol. Sci.* 36:323-332.

Papadopoulo, P. 1971. The African Bees in Rhodesia. *Am. Bee J.* 111:47.

Portugal-Araújo, V. de 1971. The central African bee in South America. *Bee World* 52:116-121.

Rinderer, T. E. 1977. Measuring the heritability of characters of honeybees. *J. Apic. Res.* 16:95-98.

Rinderer, T. E., Collins, A. M. 1986. Behavioral genetics. In *Bee Breeding and Genetics*, ed. T. E. Rinderer, pp. 115-176. Orlando, Fla: Academic Press.

Rinderer, T. E., Collins, A. M., Brown, M. A. 1983. Heritabilities and correlations of the honey bee: Response to *Nosema apis*, longevity, and alarm response to isopentyl acetate. *Apidologie* 14:79-85.

Robinson, G. E., Page, R. E., Strambi, C., Strambi, A. 1989. Hormonal and genetic control of behavior integration in honey bee colonies. *Science* 246:109-112.

Rothenbuhler, W. C. 1960. A technique for studying genetics of colony behavior in honey bees. *Am. Bee J.* 100:176-198.

Rothenbuhler, W. C. 1964. Behavior genetics of nest cleaning in honey bees. IV. Responses of F_1 and backcross generations to disease-killed brood. *Am. Zool.* 4:111-123.

Ruttner, F. 1968. Methods of breeding honeybees: intra-racial selection or inter-racial hybrid? *Bee World* 49:66-72.

Ruttner, F. 1975. Races of bees. In *The Hive and the Honey Bee.* pp. 19-38. ed. Dadant & Sons, Hamilton, Il. Dadant & Sons.

Shearer, D. A., Boch, R. 1965. 2-Heptanone in the mandibular secretion of the honey-bee. *Nature* 206:530.

Smith, F. G. 1958. Beekeeping observations in Tanganyika 1949-1957. *Bee World* 39:29-36.

Southwick, E. E., Moritz, R. F. A. 1985. Metabolic response to alarm pheromone in honey bees. *J. Insect Physiol.* 31:389-392.

Stort, A. C., Barelli, N. 1981. Genetic study of olfactory structures in the antennae of two *Apis mellifera* subspecies. *J. Kans. Entomol. Soc.* 54:352-358.

Taber, S. 1958. Concerning the number of times queen bees mate. *J. Econ. Entomol.* 51:786-789.

Villa, J. D. 1988. Defensive behaviour of Africanized and European honeybees at two elevations in Colombia. *J. Apic. Res.* 27:141-145.

Wehner, R. 1971. The generalization of directional visual stimuli in the honey bee. *J. Insect Physiol.* 17:1579-1591.

17

GENETICS OF DEFENSIVE BEHAVIOR II

Antonio Carlos Stort[1] and Lionel Segui Gonçalves[2]

Although many studies have been performed on the behavior of honey bees, few investigations are available on the genetic analysis of behavior. Rothenbuhler (1964) showed that colony resistance to *Bacillus larvae* depends on bee behavior, which is controlled by two pairs of genes: u = uncapper, and r = remover.

Nye and Mackenson (1968), after six generations of disruptive selection for specific alfalfa pollen collection, obtained excellent selection gain for both extremes (for high and low alfalfa pollen collecting lines) and showed that this behavior is controlled by polygenic inheritance with an additive effect.

In experiments on flight activity, Kerr *et al.* (1972) showed that the F_1 hybrids from crosses of Italian and Africanized parents were closer to the Africanized parent, indicating dominance of the Africanized genes over the Italian ones.

Kerr *et al.* (1975), in a study of the genetic component of the learning ability concerning the site of a new source of food, concluded that learning is very important for bees, so that any decrease in this ability affects colony survival.

With respect to defensive behavior, the first true evidence for genetic control was obtained by Rothenbuhler (1964), who investigated two lines resulting from inbreeding, Brown and Van Scoy, which differed in temperament. The author showed that stinging behavior was controlled by more than two gene loci.

It was only after the introduction of African bees (*Apis mellifera adansonii*, later called *A. m. scutellata* by Ruttner, 1975) to Brazil that investigators became concerned about gaining a better understanding of the defensive behavior of bees and of the genetics of this behavior. African bees were introduced to Brazil in 1956. They crossed with European (German and Italian) bees present in the country, giving origin to a hybrid which has been called the "Africanized bee" because of its resemblance to *A. m. adansonii* in several traits. The first of bees and of the genetics of this behavior. African bees were introduced to Brazil in 1956. They crossed with European (German and Italian) bees present in the country, giving origin to a hybrid which has been called the "Africanized bee" because of its resemblance to *A. m. adansonii* in several traits. The first visible effect of this hybridization was the change in worker defensive behavior, with bees becoming more aggressive. This fact had a strong impact on Brazilian apiculture. Despite many speculations at the time of introduction, nobody really knew what was happening in nature. African bees were crossing with European bees already present in Brazil but the tendency towards a change in defensive behavior (increase or decrease) on the part of the descendants in the various generations was unknown. What kind of defensive behavior would the hybrid bees originating from the crosses between European and African bees show after a few years? What type of inheritance could be involved in it? How could the problems of aggressiveness be controlled?

Several investigations were carried out in an attempt to answer these questions, and the main results are summarized in this chapter.

MEASUREMENT OF DEFENSIVE BEHAVIOR

When a genetic study is used to compare the aggressiveness or the defensive behavior of two different populations and to analyze the behavior of the descendants of their crosses it is necessary to use a method that will permit quantification of the differences between them. When the present study was undertaken, everybody knew that African bees were more aggressive than European bees, but what was not known, and needed to be determined, was the extent of this greater aggressiveness.

Several methods have been proposed to measure the aggressiveness shown by bees when they defend the hive, some of them based on the presentation of artificial "enemies" to the hive. Lecomte (1963) carried out extensive work on defensive behavior using artificial models acting as hive enemies. He reached the interesting conclusion that the behavior of hive workers in the presence of artificial enemies is very close to that induced by natural enemies. He showed that motion and the odor of venom are important variables in triggering aggressive behavior. He also reported that the number of attacks further increased as a function of the number of bees present in the hive up to a certain maximum limit, and that meteorological factors affect aggressiveness.

Free (1961), in experiments in which small wool balls covered with muslin were balanced in front of the hives, showed that the bees were more stimulated to sting in the presence of dark colors, odor of venom, animals and human sweat, rapid movements, and rough-textured materials.

Ghent and Gary (1962) analyzed the behavior of workers inside the hive presented with round pieces of suede. They found that the release of alarm substances by the bees as well as the motion and texture of the "enemy" had an important effect on the occurrence of stinging. Sakagami and Akahira (1960) also used artificial enemies, and studied the differences in after-sting behavior between two bee species, *A. mellifera* and *A. cerana.*

Boch and Rothenbuhler (1974) described another method for testing aggressiveness, the "opening test." The hive top is gently removed without using smoke, the upper parts of the combs are exposed for three minutes and bee aggressiveness (number of workers which leave the hive) is scored on a scale varying from 0 (very calm) to 4 (highly irritated). The authors showed that the responses of hybrids from crosses of Van Scoy (gentle) with Brown (aggressive) lines were intermediate, indicating lack of dominance. These investigators also used a respiration test, whereby the observer breathes out air near the entrance to the hive and observes the reaction of the bees. When this test was used, the aggressive reaction of the hybrid workers was similar to that of the Van Scoy parents.

Starting in 1966, several tests were performed using black leather balls 2 cm in diameter and filled with cotton. These balls were balanced separately or as a set in front of the hive to stimulate the bees. Defensive behavior was quantified by counting the number of stings left in these artificial enemies. Several of these tests showed that Africanized bees were more aggressive, and showed a greater range of aggressive behavior than European bees (Stort, 1972).

These tests were later improved when it became obvious that other traits in addition to the number of stings left by the bees could be measured during and after the time of stimulation. Thus, a much more encompassing "aggressiveness test" was developed. The test consisted of balancing a 2 cm black leather ball filled with cotton in front of the entrance or the beehive for 60 seconds to stimulate the bees (FIGURES 1 and 2). The following behavioral traits were measured (Stort, 1971, 1972, 1974):

1A - Time at which the first sting reached the leather ball.
2A - Time taken for the colony to become aggressive.
3A - Number of stings left in the gloves of the observer.
4A - Number of stings left in the leather ball.
5A - Distance that the bees follow the observer.

The Scientific committee that came to Brazil in 1972 under the sponsorship of the National Academy of Sciences performed a series of tests with a device called an aggressometer. This apparatus was placed close to the hive and stimulated the bees by means of a round revolving part located in its center. When the bees flew close to the apparatus they were captured by it and dropped into a lower compartment containing water, where they were later counted to determine the number of attacks against the enemy. The American study group also used square leather or suede pieces which were presented to the front of the hive as enemies and in which the number of stings were later counted.

FIGURE 1. Aggressiveness test with Africanized bees.

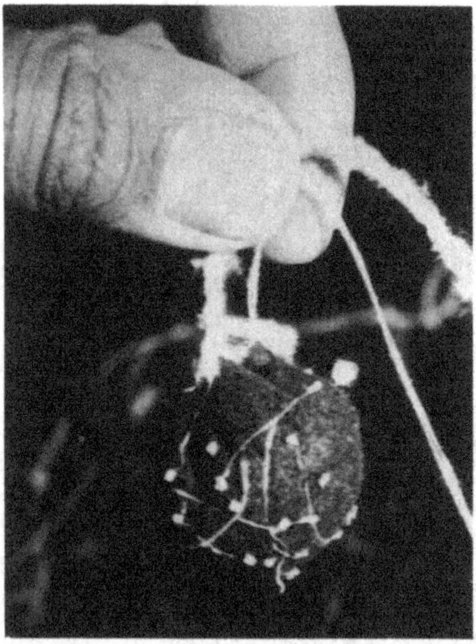

FIGURE 2. Leather ball with stings after the aggressiveness test.

The above tests have been used to determine the defensive behavior of a hive as a whole. Other methods have been used to determine the reaction of workers confined in small boxes (Collins and Rothenbuhler, 1978; Collins, 1979, 1980, 1982; Collins and Blum, 1982, 1983). Data were also obtained in relation to the aggressiveness of Africanized colonies by use of a different methodology (Collins and Kubasek, 1982; Collins *et al.*, 1982; Collins and Rinderer, 1985). The important and interesting data obtained in these studies are reported in Chapter 16.

MATING SCHEME

Genetic analysis of overall hive behavior requires genetic homogeneity of the workers that compose the hive. This homogeneity can be found in inbred populations. An inbred queen which is homozygous for the loci that affect a behavioral trait produces similar ovules for this trait, which, by joining the identical gametes of a drone, produce highly similar workers.

An F_1 generation of workers can thus be produced and the behavior of colony members observed. F_1 queens are later produced and drones are produced from them. These drones are used to fertilize queens originating from the inbred parental lines, resulting in genetically homogeneous backcrossed workers for each cross, but highly variable workers between crosses. It is then possible to identify the number of different types of gametes produced by the F_1 queens and to determine the number of loci involved in different behaviors (Rothenbuhler, 1960, 1964; Cale and Rothenbuhler, 1979).

This inbred mating technique has been used in Brazil to study the defensive behavior of Africanized bees, with a small elaboration of the mating scheme. A queen from hive nº 194 (a gentle colony of the Africanized backcross) was used to produce drones. These drones were mated with queens from the parental Italian colony and produced colonies nºs 197, 236, 219 and 152, which correspond to the test-cross colonies (FIGURE 3). Unfortunately, only a few colonies of this type were obtained, but they were sufficient to test the validity of the genetic hypotheses proposed.

DATA OBTAINED AND GENETIC ANALYSES PERFORMED

The experiments were carried out using nuclei; i.e., hives consisting of three combs, a feeder, and sufficient bees to cover the three combs. Crosses were performed by artificial insemination using a technique based on the studies of Mackensen and Roberts (1948), Laidlaw (1949, 1958), and Woyke *et al.* (1966). Each hive was tested five times, using a new leather ball for each aggressiveness test.

TABLE 1. Mean of logarithm to base 10 (± standard deviation of mean) before leather ball was first stung. Results are given for all colonies used: It = Italian; Af = Africanized; B = backcross

Colony	Queen no.	Time to first sting (log)	Colony	Queen no.	Time to first sting (log)
It	7.1.67	1.099 ± 0.463	Af B	70.3.67	1.045 ± 0.434
Af	63.2.66	0.562 ± 0.387	Af B	232.1.67	0.914 ± 0.370
F_1	120.1.67	0.800 ± 0.453	Af B	210.1.67	0.602 ± 0.367
F_1	148.1.67	0.844 ± 0.378	Af B	231.1.67	0.361 ± 0.494
F_1	112.1.67	0.716 ± 0.858	Af B	162.2.67	0.524 ± 0.242
It B	88.2.67	0.631 ± 0.178	Af B	69.1.67	0.705 ± 0.211
It B	25.2.67	0.431 ± 0.208	Af B	198.1.67	0.276 ± 0.273
It B	47.1.67	0.406 ± 0.284	Af B	43.1.67	0.555 ± 0.522
It B	84.1.67	0.431 ± 0.129	Af B	193.4.67	0.825 ± 0.431
It B	117.1.67	0.240 ± 0.134	Af B	36.3.68	1.372 ± 0.468
It B	91.4.67	0.380 ± 0.177	Af B	196.3.67	0.911 ± 0.477
It B	145.2.68	0.635 ± 0.189	Af B	135.2.68	0.641 ± 0.303
It B	45.1.67	0.276 ± 0.172	Af B	194.1.67	1.105 ± 0.426
It B	8.3.67	0.511 ± 0.150	Af B	211.1.67	1.121 ± 0.292
It B	124.2.67	0.727 ± 0.343	Af B	237.1.67	0 0
It B	67.1.67	1.208 ± 0.593	Af B	173.4.67	0.336 ± 0.078
It B	32.2.67	0.562 ± 0.298	Af B	187.2.67	0.371 ± 0.096
It B	189.1.67	0.786 ± 0.392	Af B	96.1.67	0.521 ± 0.098
It B	184.1.67	0.825 ± 0.591	Af B	199.2.67	0.456 ± 0.150
It B	39.2.67	0.276 ± 0.172	Af B	229.1.67	0.554 ± 0.282
It B	29.4.67	0.571 ± 0.188	ItxAfItAf	219.1.67	0.441 ± 0.078
It B	74.1.67	0.435 ± 0.294	ItxAfItAf	236.1.67	0.716 ± 0.465
It B	94.1.67	1.111 ± 0.400	ItxAfItAf	152.1.67	0.631 ± 0.645
			ItxAfItAf	197.1.67	1.229 ± 0.231

1A - Time at which the first sting reached the leather ball

Two parental colonies were tested for this variable (an Africanized colony and an Italian one), as well as three colonies of F_1 bees, 18 colonies of Italian backcrossed bees, 20 colonies of Africanized backcrosses, and four test cross colonies. The data obtained were transformed logarithmically (TABLE 1) and those for the F_1 and backcrossed colonies were analyzed by means of analysis of variance after the log transformation (TABLE 2).

The differences in behavior among the three F_1 colonies were not significant, indicating genetic homogeneity in the two parental colonies (hive nºs 7 and 63). Since each parental colony produced only one type of gamete, the

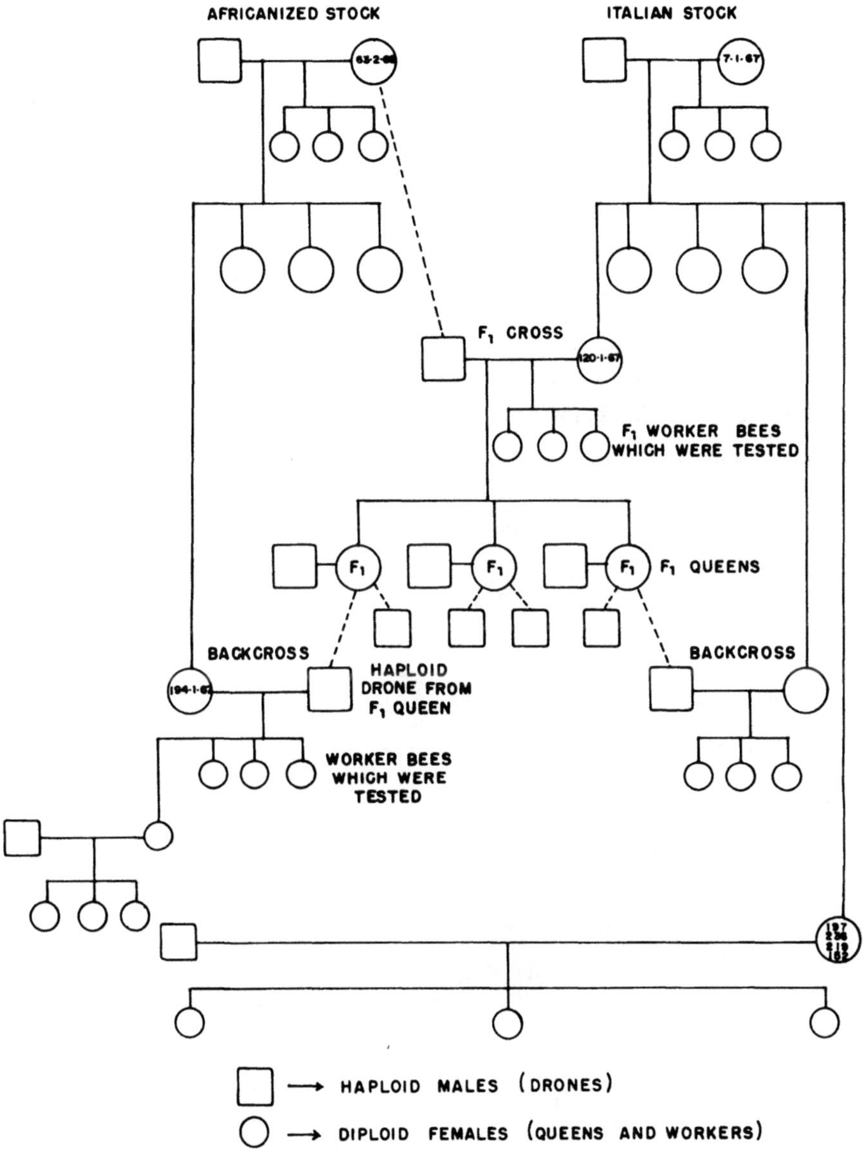

Figure 3. Scheme of matings used.

TABLE 2. Analysis of variance of logarithmic data from the different crosses (Time to first sting).

Colony type	Source of variation	No. degrees freedom	Sums of squares	Mean square	F	P
	between colonies	2	0.04	0.02	0.05	n.s
F_1	within colonies	12	4.34	0.36		
	Total	14	4.38			
	between colonies	19	10.97	0.57	5.15	<0.01
Af backcross	within colonies	80	9.00	0.11		
	Total	99	19.97			
	between colonies	17	6.23	0.36	4.00	<0.01
It backcross	within colonies	72	6.74	0.09		
	Total	89	12.97			

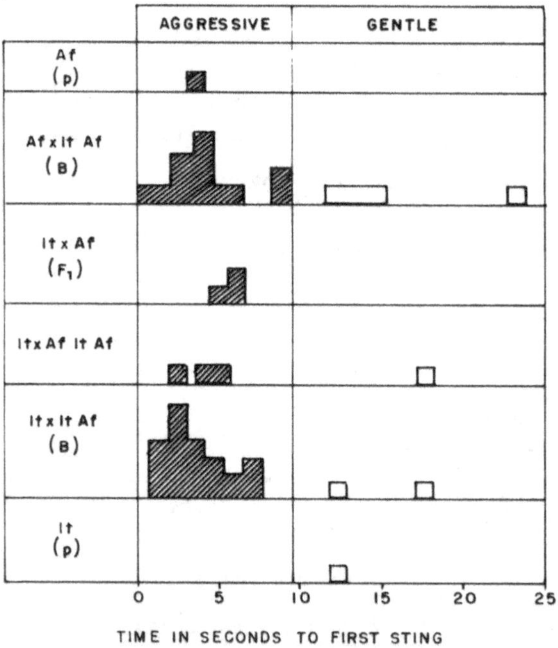

FIGURE 4. Time at which the first sting reached the leather ball (geometric means). P = parental; It = Italian; Af = Africanized; B = backcross.

F_1 colonies had the same genotype. On the other hand, the backcrossed colonies showed significant variability which was probably due to their different genetic constitutions.

The values obtained for the behaviors of F_1 colonies indicate the presence of dominant genes in Africanized bees (FIGURE 4). Sixteen aggressive and four gentle colonies were obtained in the Africanized backcrosses. The criterion used to distinguish an aggressive from a gentle colony was a 10 second interval before the first sting.

The backcrosses to the Italian parents showed a distribution which was strongly shifted towards the side of "short reaction time" (FIGURE 4), indicating that this behavior is not simply a case of gene additivity.

The queens of the test crosses (nº 197-1-67, 236-1-67, 219-1-67 and 152-1-67) were daughters of the Italian inbred colony (FIGURE 3). Crosses of males produced by a daughter of a backcrossed Africanized colony (nº 194) with "gentle queens" would be expected to produce almost solely gentle colonies. However, the obtained segregation, which included a large proportion of aggressive colonies, confirms the complexity of the type inheritance (Stort, 1975a).

Analysis of untransformed data showed 3:1 segregation of aggressive to gentle colonies in both types of backcrosses. On the basis of these data, Stort (1971) suggested the presence of four pairs of genes, Ag_1/Ag_1, Ag_2/Ag_2, ag_3/ag_3, ag_4/ag_4 among the Italian parental bees, and ag_1/ag_1, ag_2/ag_2, Ag_3/Ag_3, and Ag_4/Ag_4 among the Africanized parental bees, with more complex effects than simple additivity among them in the control of this variable.

2A - Time taken for the colony to become aggressive

We found that after the bees had stung the black leather ball many times during the aggressiveness tests, some of them also attacked the protecting veil and gloves of the observer. We considered this to be the time when they are fully aggressive (type 1 behavior), so the time needed to reach this point was also recorded.

Only type 1 behavior was observed in the Italian colonies, whereas the aggressiveness of the Africanized bees increased progressively, so that 10-15 seconds after the type 1 behavior was reached, an explosive behavior occurred (type 2) when hundreds of bees started to fly in all directions attacking and pursuing every person or animal they encountered. The bees tended to be so aggressive that they often collided with obstacles in their path, such as trees, walls and cars.

Only the time needed to reach type 1 behavior was recorded for every colony studied (parental, F_1 and backcrosses). The logarithmic data for the parental, F_1 and backcrossed colonies and test crosses are shown in TABLE 3

TABLE 3. Mean results (± standard deviation) of all colonies used. Data are in logarithmic scale. It = Italian; Af = Africanized; B = backcross

Colony	Queen no.	Time to become aggressive	Colony	Queen no.	Time to become aggressive
It	7.1.67	1.616 ± 0.134	Af B	70.3.67	1.433 ± 0.246
Af	63.2.66	1.338 ± 0.176	Af B	232.1.67	1.260 ± 0.281
F_1	120.1.67	1.466 ± 0.192	Af B	210.1.67	1.213 ± 0.360
F_1	148.1.67	1.180 ± 0.240	Af B	231.1.67	0.666 ± 0.423
F_1	112.1.67	0.992 ± 0.716	Af B	162.2.67	1.398 ± 0.100
It B	88.2.67	1.404 ± 0.104	Af B	69.1.67	1.210 ± 0.209
It B	25.2.67	1.074 ± 0.306	Af B	198.1.67	0.577 ± 0.317
It B	47.1.67	0.952 ± 0.184	Af B	43.1.67	1.172 ± 0.286
It B	84.1.67	1.045 ± 0.349	Af B	193.4.67	1.456 ± 0.221
It B	117.1.67	0.564 ± 0.238	Af B	36.3.68	1.542 ± 0.313
It B	91.4.67	1.216 ± 0.148	Af B	196.3.67	1.414 ± 0.141
It B	145.2.68	1.434 ± 0.197	Af B	135.2.68	1.581 ± 0.094
It B	45.1.67	0.655 ± 0.170	Af B	194.1.67	1.346 ± 0.289
It B	8.3.67	0.961 ± 0.219	Af B	211.1.67	1.382 ± 0.230
It B	124.2.67	1.250 ± 0.122	Af B	237.1.67	0.466± 0.104
It B	67.1.67	1.588 ± 0.246	Af B	173.4.67	0.877 ± 0.246
It B	38.2.67	0.916 ± 0.418	Af B	187.2.67	0.895 ± 0.181
It B	189.1.67	1.377 ± 0.252	Af B	96.1.67	1.321 ± 0.221
It B	184.1.67	1.110 ± 0.460	Af B	199.2.67	1.146 ± 0.333
It B	39.2.67	0.695 ± 0.376	Af B	229.1.67	1.177 ± 0.173
It B	29.4.67	1.302 ± 0.109	ItxAfItAf	219.1.67	1.381 ± 0.137
It B	74.1.67	1.009 ± 0.134	ItxAfItAf	236.1.67	1.044 ± 0.389
It B	94.1.67	1.470 ± 0.189	ItxAfItAf	152.1.67	1.066 ± 0.426
			ItxAfItAf	197.1.67	1.564 ± 0.122

and the results of the analyses of variance are given in TABLE 4. When the data for the trait under study were compared with the trait "time at which the first sting reached the leather ball" a great similarity was observed, with F_1 located on the side of the Africanized parent, and with similar segregations in the colonies representing the two types of backcrosses and in the test cross colonies (FIGURE 5).

Highly significant correlations were obtained between the two behaviors (time at which the first sting reached the artificial enemy, and time taken for the colony to become infuriated) in both types of backcrosses (Stort, 1971, 1977).

These data may suggest that the two traits belong to the same basic behavior or that the genes that regulate them are partially linked (Stort, 1976).

TABLE 4. Analysis of variance of data from the different crosses (data in logarithm). (Time to become aggressive.)

Colony type	Source of variation	No. degrees freedom	Sums of squares	Mean square	F	P
	between colonies	2	0.520	0.260	1.287	n.s
F_1	within colonies	12	2.434	0.202		
	Total	14	2.954			
	between colonies	19	9.700	0.510	7.846	<0.01
Af backcross	within colonies	80	5.206	0.065		
	Total	99	14.906			
	between colonies	17	7.344	0.432	6.545	<0.01
It backcross	within colonies	72	4.784	0.066		
	Total	89	12.128			

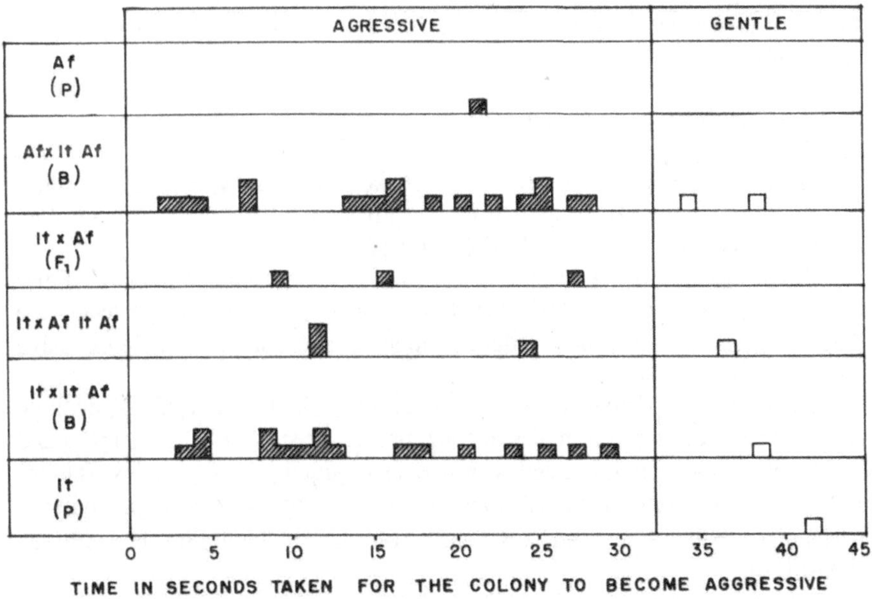

FIGURE 5. Histogram of the data. Distributions were made using geometric means. P = parental; It = Italian; Af = Africanized; B = backcross

3A - Number of stings left in the gloves of the observer

This variable is defined as the number of workers that sting the gloves of the observer when they are located one meter above the entrance to the hive. When the bees become very aggressive they fly in all directions, stinging vigorously and leaving their stings in the objects attacked. The observer who performs the aggressiveness tests wears leather gloves which can retain the stings introduced into them, thus permitting later counts.

During the aggressiveness tests, the gloved right hand of the observer is attacked more because it moves to balance the thread holding the black leather ball.

The F test values obtained in the analyses of variance for the data obtained for Africanized and Italian backcrosses were 23.51 ($p < 0.001$) and 3.06 ($p < 0.01$), respectively, suggesting that the variability is due to the different genetic constitutions of the colonies. TABLE 5 shows the data of the tests applied to the backcrossed colonies. There were 15 gentle colonies and five aggressive colonies in the Africanized backcrosses, corresponding to a 3:1 segregation. In the Italian backcrosses, all colonies were gentle, showing that there was no Mendelian segregation.

The mean number of stings counted in the tests of the parental colonies was 6.20 ± 4.71 for the Africanized colony nº 63, and 0.60 ± 0.89 for the Italian colony nº 7.

Mean values of zero or close to zero were obtained for the F_1 colonies tested, indicating that dominant genes are present in the Italian subspecies. As there was no segregation in the Italian backcrosses and there was a 3:1 gentle to aggressive segregation in the Africanized backcrosses, the hypothesis was raised that the trait "number of stings left in the gloves of the observer" is controlled by two pairs of genes, which will be called F_1f_1 and F_2f_2.

According to this hypothesis, the parental Italian queen had the F_1F_1/F_2F_2 genotype, and the parental Africanized queen was f_1f_1/f_2f_2. The presence of a dominant gene is sufficient to induce gentleness, and only the double recessive (f_1f_1/f_2f_2) is aggressive.

The F_1 queens, daughters of queen nº 120-1-67, had the F_1f_1/F_2f_2 genotype and produced four types of males in equal proportions: F_1/F_2, F_1/f_2, f_1/F_2 and f_1/f_2. When these males mated with queens producing F_1/F_2 ovules (Italian) and f_1/f_2 (Africanized), the result was 3:1 gentle to aggressive segregation in the Africanized backcrosses, and all gentle colonies in the Italian backcrosses (TABLE 6).

The worker progeny of queen nº 194-1-67 was gentle (zero mean), indicating that the male with which this queen had mated was a genetically gentle segregant of the F_1 queen.

The genotype of the female progeny of queen nº 194-1-67 could be F_1f_1/F_2f_2 or F_1F_1/f_2f_2 or f_1f_1/F_2f_2.

TABLE 5. Mean results (± standard deviation) and behaviors of 13 colonies of the Italian backcrosses (*A. m. ligustica* queens mated with drones from F_1 queens) and 20 colonies of the Africanized backcrosses (*A. m. adansonii* queens mated with drones from F_1 queens)

Colonies of Italian backcrosses			Colonies of Africanized backcrosses		
Queen	No. of stings in gloves	Behavior	Queen	No. of stings in gloves	Behavior
47.1.67	2.20 ± 1.92	Gentle	232.1.67	3.00 ± 3.16	Gentle
84.1.67	1.00 ± 1.22	Gentle	210.1.67	0.20 ± 0.44	Gentle
117.1.67	1.40 ± 1.14	Gentle	162.2.67	1.20 ± 1.30	Gentle
91.4.68	0.80 ± 0.83	Gentle	69.1.67	1.00 ± 1.00	Gentle
145.2.68	0.00 ± 0.00	Gentle	43.1.67	0.60 ± 0.89	Gentle
45.1.67	2.60 ± 1.81	Gentle	193.4.67	1.20 ± 2.68	Gentle
8.3.67	2.80 ± 1.64	Gentle	36.3.68	1.00 ± 2.23	Gentle
124.2.67	0.20 ± 0.44	Gentle	196.3.67	0.00 ± 0.00	Gentle
67.1.67	0.00 ± 0.00	Gentle	135.2.68	0.40 ± 0.89	Gentle
38.2.67	2.00 ± 1.22	Gentle	194.1.67	0.00 ± 0.00	Gentle
189.1.67	0.00 ± 0.00	Gentle	211.1.67	1.40 ± 1.14	Gentle
184.1.67	0.20 ± 0.44	Gentle	173.4.67	1.60 ± 1.81	Gentle
94.1.67	1.20 ± 2.68	Gentle	187.2.67	2.80 ± 1.09	Gentle
			229.1.67	0.60 ± 0.89	Gentle
			70.3.67	2.00 ± 2.00	Gentle
			231.1.67	81.20 ± 35.18	Aggressive
			198.1.67	5.20 ± 3.27	Aggressive
			237.1.67	100.80 ± 41.36	Aggressive
			96.1.67	65.60 ± 31.16	Aggressive
			199.2.67	55.00 ± 16.00	Aggressive

A queen produced by queen nº 194-1-67 should produce two or four types of males, which, by combining with the F_1F_2 ovules of daughters of an Italian queen would produce only gentle colonies similar to the backcrosses on the side of the Italian stock.

Colonies nºs 197, 236, 219 and 152 were the result of such crosses. When tested, they produced the following means: 0.0, 2.40 ± 1.67, 0.80 ± 0.83 and 0.40 ± 0.54 (all gentle). These data are compatible with the hypothesis formulated, even though only a small number of colonies of this type (test cross) were analyzed. No other hypothesis fits the data (Stort, 1975b).

TABLE 6. Data obtained in the Italian and Africanized backcrosses colonies in accordance with the hypothesis that the behavior defined as "number of stings in the gloves" is controlled by two pairs of genes.

	F_1 males			
Females	F_1F_2	F_1f_2	f_1F_2	f_1f_2
Italian backcrosses f_1F_2	$F_1F_1F_2F_2$ gentle; $F_1F_1F_2f_2$ gentle	$F_1f_1F_2F_2$ gentle	$F_1f_1F_2f_2$ gentle	100% gentle
Africanized backcrosses f_1f_2	$F_1f_1F_2f_2$ gentle; $F_1f_1f_2f_2$ gentle	$f_1f_1F_2f_2$ gentle	$f_1f_1f_2f_2$ aggressive	3 gentle:1 aggressive

4A - Number of stings left in the leather ball

This variable concerns the number of stings the bees left in the enemies that were moved in front of the hive entrance.

The F values of the analyses of variance for the data obtained for the Africanized and Italian backcrossed colonies were 8.19 ($p < 0.01$) and 9.33 ($p < 0.01$) respectively, suggesting that the variability is due to the different genetic constitutions of the colonies. TABLE 7 shows the data obtained in the tests of the backcrossed colonies. There was 15 gentle and 5 aggressive nuclei in the Africanized backcrosses, corresponding to a 3:1 segregation. In the Italian backcrosses there were 11 aggressive and 7 gentle colonies, corresponding to a 1:1 segregation ($\chi^2 = 0.88$, non-significant). The means for the parental colonies were 40.80 ± 12.11 for the Africanized colonies and 27.80 ± 9.41 for the Italian colonies.

The means for the three F_1 colonies tested were 38.00 ± 16.76, 63.00 ± 9.89 and 54.60 ± 42.04, indicating that dominant genes are present in the Africanized subspecies. Since there was a 3:1 aggressive to gentle segregation in the Africanized backcrosses and 1:1 aggressive to gentle segregation in the Italian backcrosses, the hypothesis was proposed that the trait "number of stings left in the leather ball" is controlled by two pairs of genes, A and B, with two alleles, A^m, A^{br}, B^m, B^{br}. The m alleles are related to gentleness and the br alleles refer to aggressiveness. The bees will be gentle only when they have the A^m/A^m; B^{br}/B^{br} genotype or when they have more m than br alleles. In other combinations, they are aggressive.

According to this hypothesis, the Italian parental queen had the A^m/A^m; B^{br}/B^{br} genotype and the parental Africanized queen had the A^{br}/A^{br};

TABLE 7. Mean results (± standard deviation)and behaviors of 18 colonies of the Italian backcrosses (*A. m. liguistica* queens mated with drones from F_1 queens) and 20 colonies of the Africanized backcrosses (*A. m. adansonii* queens mated with drones from F_1 queens)

Colonies of Italian backcrosses			Colonies of Africanized backcrosses		
Queen	No. of stings in ball	Behavior	Queen	No. of stings in ball	Behavior
8.3.78	40.20 ± 6.98	Aggressive	211.1.67	55.00 ± 14.28	Aggressive
29.4.67	45.00 ± 8.51	Aggressive	229.1.67	54.60 ± 12.62	Aggressive
145.2.68	47.80 ± 3.96	Aggressive	232.1.67	65.20 ± 18.87	Aggressive
184.1.67	69.00 ± 28.82	Aggressive	135.2.68	41.60 ± 14.05	Aggressive
47.1.67	77.20 ± 8.52	Aggressive	193.4.67	38.00 ± 10.46	Aggressive
84.1.67	60.80 ± 22.46	Aggressive	162.2.67	54.80 ± 10.47	Aggressive
45.1.67	87.60 ± 10.74	Aggressive	173.4.67	44.00 ± 7.84	Aggressive
74.1.67	58.80 ± 8.26	Aggressive	187.2.67	47.40 ± 7.48	Aggressive
94.1.67	45.60 ± 11.64	Aggressive	198.1.67	62.00 ± 5.70	Aggressive
117.1.67	63.60 ± 9.40	Aggressive	96.1.67	58.80 ± 8.76	Aggressive
124.2.67	37.60 ± 6.12	Aggressive	43.1.67	54.80 ± 9.09	Aggressive
88.2.67	35.60 ± 9.19	Gentle	199.2.67	68.20 ± 12.69	Aggressive
25.2.67	35.60 ± 13.36	Gentle	70.3.67	54.20 ± 28.93	Aggressive
91.4.68	36.20 ± 9.20	Gentle	231.1.67	76.80 ± 21.46	Aggressive
189.1.67	34.20 ± 7.98	Gentle	237.1.67	64.80 ± 6.68	Aggressive
39.2.67	32.00 ± 9.72	Gentle	210.1.67	33.80 ± 11.98	Gentle
38.2.67	29.2- ± 14.18	Gentle	69.1.67	35.20 ± 6.14	Gentle
67.1.67	25.60 ± 17.56	Gentle	194.1.67	19.40 ± 12.66	Gentle
			36.3.68	13.40 ± 8.01	Gentle
			96.3.67	25.40 ± 10.39	Gentle

B^m/B^m genotype. The F_1 queens produced by queen nº 120-1-67 had the A^m/A^{br}; B^m/B^{br} genotype and produced the following types of males: 1/4 A^mB^m; 1/4 A^mB^{br}; 1/4 $A^{br}B^m$ and 1/4 $A^{br}B^{br}$. These males were crossed with queens which produced A^mB^{br} eggs in the Italian line and $A^{br}B^m$ in the Africanized line, with 1:1 and 3:1 aggressive to gentle segregations being obtained respectively (TABLES 7 and 8). The workers originating from queen nº 194-1-67 were gentle (mean = 19.40 ± 12.66), indicating that the male with which this queen had mated was a gentle segregant originating from the F_1 queen (with A^mB^m genes). The progeny of this nº 194 queen would then have the A^{br}/A^m; B^m/B^m genotype. A queen produced by the nº 194-1-67 queen would produce two types of males, $A^{br}B^m$ and A^mB^m, in equal proportions. These drones, when fertilizing A^mB^m ovules of the daughters of the Italian queen nº 7-1-67, should produce aggressive A^mA^{br}; B^m/B^{br} colony types and gentle A^m/A^m; B^m/B^{br} colony types.

The results of these crosses (colony nºs 197, 236, 219, and 152) produced the following means when tested for aggressiveness: 28.40 ± 11.99 (gentle), 46.40 ± 12.89 (aggressive), 46.40 ± 13.31 (aggressive) and 37.00 ± 13.83 (aggressive). Even though this sample was very small (only four colonies), the two types of behavior predicted for these crosses (aggressive and gentle) were obtained, a result compatible with the hypothesis (Stort, 1975c).

5A - Distance that bees follow the observer

In this test, the black leather ball is balanced in front of the hive for 60 seconds to stimulate the workers. After this time, the observer applying the test moves away from the hive walking normally, step by step, up to a distance where no more bees are pursuing or attacking him. This variable, which was called distance to which the observer is pursued, will be analyzed in this section.

TABLE 9 shows the results obtained from the tests of the parental, F_1 colonies, Africanized backcrossed and Italian backcrossed colonies. It can be seen that the F_1 colonies showed gentle and intermediate behavior. The absence of aggressive behavior may suggest dominance of the Italian subspecies, though only three of the colonies of F_1 bees were tested.

FIGURES 6 and 7 show the histograms obtained by using all the data obtained in the 90 tests applied to the 18 Italian backcrossed colonies, and in the 100 tests applied to the 20 Africanized backcrossed colonies.

The chi-squared tests used to compare the data obtained with theoretical normal distribution data were significant for each class interval (χ^2 = 108.33 for the Italian backcrosses and χ^2= 126.75 for the Africanized backcrosses), indicating deviation from normal distribution. Thus, the two distributions were discontinuous, with the histograms being divided into three regions containing 4/8, 3/8 and 1/8 of the data, respectively. TABLE 9 shows that in both types of

TABLE 8. Segregations obtained in the Italian and Africanized backcrosses colonies in accordance with the hypothesis that the behavior defined as "number of stings in the leather ball" is controlled by two pairs of genes.

	F1 males				
Females	A^mB^m	A^mB^{br}	$A^{br}B^m$	$A^{br}B^{br}$	
Italian backcrosses f_1F_2	$A^mA^mB^mB^{br}$ gentle	$A^mA^mB^{br}B^{br}$ gentle	$A^mA^{br}B^mB^{br}$ aggressive	$A^mA^{br}B^{br}B^{br}$ aggressive	1 gentle: 1 aggressive
Africanized backcrosses f_1f_2	$A^{br}A^mB^mB^m$ gentle	$A^{br}A^mB^{br}B^m$ aggressive	$A^{br}A^{br}B^mB^m$ agressive	$A^{br}A^{br}B^{br}B^m$ aggressive	3 agressive 1 gentle

TABLE 9. Mean results (± standard deviation)and behaviors of one Africanized parental colony; one Italian Parental; 3 F_1 colonies; 18 colonies of the Italian backcrosses (*A. m. liguistica* queens mated with drones from F_1 queens) and 20 colonies of the Africanized backcrosses (*A. m. adansonii* queens mated with drones from F_1 queens).

Colony	Persecution	Behavior	Colony	Persecution	Behavior
7 (IP)	17.36 ± 12.58	Gentle	70 (AB)	94.78 ± 24.95	Aggressive
63 (AP)	89.88 ± 21.29	Aggressive	232 (AB)	83.86 ± 37.99	Aggressive
120 (F_1)	13.44 ± 5.92	Gentle	231 (AB)	162.82 ± 40.37	Aggressive
148 (F_1)	56.84 ± 38.64	Intermediate	198 (AB)	207.34 ± 47.12	Aggressive
112 (F_1)	46.20 ± 30.84	Intermediate	193 (AB)	70.00 ± 23.34	Aggressive
25 (IB)	98.70 ± 42.72	Aggressive	211 (AB)	95.90 ± 21.01	Aggressive
47 (IB)	146.58 ± 86.13	Aggressive	237 (AB)	256.34 ± 16.37	Aggressive
117 (IB)	96.04 ± 41.17	Aggressive	173 (AB)	73.78 ± 28.13	Aggressive
8 (IB)	81.77 ± 17.91	Aggressive	187 (AB)	77.84 ± 21.72	Aggressive
38 (IB)	88.20 ± 33.33	Aggressive	96 (AB)	205.24 ± 65.52	Aggressive
39 (IB)	174.58 ± 29.25	Aggressive	199 (AB)	229.18 ± 37.60	Aggressive
29 (IB)	117.60 ± 21.23	Aggressive	43 (AB)	52.92 ± 46.49	Intermediate
74 (IB)	102.00 ± 33.09	Aggressive	69 (AB)	54.05 ± 20.98	Intermediate
88 (IB)	115.92± 49.53	Aggressive	210 (AB)	20.72 ± 10.40	Gentle
45 (IB)	54.60± 15.20	Intermediate	162 (AB)	39.20 ± 11.45	Gentle
94 (IB)	51.80± 33.51	Intermediate	36 (AB)	39.20 ± 41.64	Gentle
84 (IB)	32.34± 12.58	Gentle	196 (AB)	35.00 ± 20.70	Gentle
91 (IB)	31.80± 15.75	Gentle	135 (AB)	39.20 ± 19.39	Gentle
145 (IB)	9.80± 6.38	Gentle	194 (AB)	21.14 ± 15.93	Gentle
124 (IB)	31.50± 14.05	Gentle	229 (AB)	17.92 ± 7.50	Gentle
67 (IB)	25.20± 17.56	Gentle			
189 (IB)	19.04± 9.30	Gentle			
184 (IB)	28.84± 15.92	Gentle			

IP = Italian Parental
IB = Italian Backcross

AP = Africanized Parental
AB = Africanized Backcross

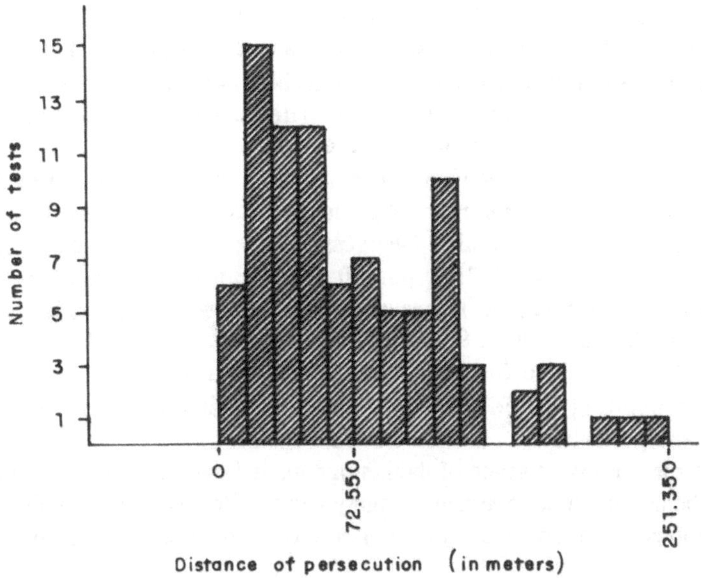

FIGURE 6. Frequency distribution of all data obtained for Italian backcross colonies. The mean is 72.550 and the interval used, corresponding to half standard deviation, is 14.900.

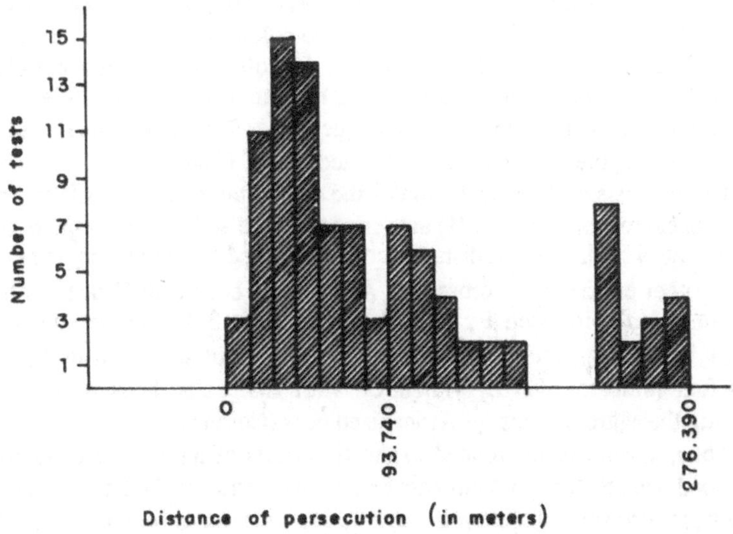

FIGURE 7. Frequency distribution of all data obtained for Africanized backcross colonies. The mean is 93.740 and the interval used, corresponding to half the standard deviation, is 14.050.

backcrosses approximately 50% of the colonies exhibited gentle and intermediate behaviors (9 Africanized backcrossed colonies and 9 Italian backcrossed colonies).

The behavior of the F_1 colonies indicates that the pursuit distance tends to be governed by dominant factors in the Italian bees, which are gentler. However, if this were a case of inheritance by simple dominance, different segregations should occur in the two types of backcrosses. Thus, a larger number of gentle colonies (or only gentle colonies) should appear in the Italian backcrosses, since all the Italian virgin queens from colony nº 7 carried the gentleness genes. The appearance of similar (1 gentle: 1 aggressive) segregations in the two types of backcrosses suggests that the pursuit trait is not controlled by simple dominance. Furthermore, the means obtained for the Africanized backcrosses and the Italian backcrosses were 93.74 $\pm$ 28.10 m and 72.55 $\pm$ 29.80 m; i.e., values close to those obtained for the Africanized parental colony nº 63 (89.88 $\pm$ 21.39), supporting the idea that the inheritance involved was not simple Mendelian inheritance.

The discontinuous shape of the histograms indicates that the inheritance of pursuit behavior is not governed by many genes. However, the hypothesis that this is polygenic inheritance with a non-additive gene effect should not be ruled out.

Colony nº 194 of the Africanized backcross showed considerable gentle behavior, indicating the presence of gentleness genes in the genotype of its workers. A queen produced by colony nº 194 was utilized to produce males which were crossed with virgin queens from the Italian parental colony that should have gentleness genes. Thus, the worker offspring of those crosses were expected to show gentle behavior. When the colonies produced by these crosses (nºs 197, 236, 219 and 152) were tested, the following means were obtained: 25.20 $\pm$ 12.15 m (gentle); 39.20 $\pm$ 21.88 m (gentle); 52.78 $\pm$ 32.57 m (intermediate), and 34.05 $\pm$ 12.55 m (gentle). Despite the small number of colonies tested, the results were as expected (Stort, 1980).

On the basis of the distribution of the data obtained for the Africanized and Italian backcrosses, Stort (1971) also suggested the action of three pairs of genes in the control of the pursuit distance trait and named them Pr_1, Pr_2 and Pr_3.

In other experiments, crosses of Africanized bees with Caucasian bees (*A. m. caucasica*) produced F_1 hybrids which were 3.3 times gentler than the Africanized parental stock with respect to "number of stings left in the enemy." This fact indicates that hybridization with Caucasian bees is important in reducing the aggressiveness of Africanized bees (Cosenza, 1972).

Thus, several genes related to the five traits of aggressive behavior were described from the experiments carried out on colonies in the field. Experiments were later performed to determine aggressive behavior in bees maintained in the laboratory. In these experiments, young workers were placed in small boxes containing water and syrup as food and the boxes were kept in a shelter under controlled conditions. A piece of cork containing paraffin oil plus the alarm pheromone, isopentyl acetate, was presented to these bees and their behavior (initial activity level, time to react and intensity of the reaction) was observed. Two bee strains of European origin, Brown-Caucasian and the gentle strain YD, were compared. F_1 and backcross progenies were obtained by artificial insemination, and analysis of the data demonstrated the existence of genetic control similar to those reported for the field experiments carried out by Stort (Collins, 1979).

DEFENSIVE BEHAVIOR AND ENVIRONMENTAL FACTORS

The importance of the environment on the behavior of bees has been demonstrated by several investigators. Schua (1952) reported that colony aggressiveness increases with temperature. Portugal-Araújo (1956) also mentioned that the heat caused by the direct incidence of sunlight on the hive influences the aggressiveness of African bees.

Data obtained by Stort (1971) with respect to the aggressiveness of Italian and Africanized backcrossed bees in hives not located under the direct incidence of sunlight showed that temperature had no effect, at least under the climatic conditions of Ribeirão Preto, state of São Paulo, Brazil. The correlation coefficients between aggressive behavior and temperature were not significant at the 5% level.

According to Lecomte (1963), the effect of wind and of storms on the aggressiveness of bees is important, but the effect of temperature is less certain. Variations in weather and cloud formations produce changes in the electric charge of each individual bee and of the colony as a whole which affect aggressive behavior (Warnke, 1976). Lecomte (1963) also reported that during the time of nectar flow European bees are less aggressive because during this time there are no bees guarding the entrance to the hive. In contrast, the aggressiveness of Africanized bees increases in the presence of large amounts of nectar (De Santis and Cornejo, 1968).

In Brazil, Brandeburgo *et al.* (1976) and Brandeburgo (1979), submitted 40 Africanized bee colonies from Ribeirão Preto, state of São Paulo, and 40 Africanized bee colonies from Recife, state of Pernambuco, to different climatic conditions for a period of more than two years. They noted that the aggressiveness of the bees, measured by the aggressiveness tests of Stort (1971, 1972, 1974), changed every time the colonies (with the same gene pool) were submitted to different environmental conditions at these two locations, in the Southeast and the Northeast of Brazil, respectively. The bees from Pernambuco were four times more aggressive than those from São Paulo. Each time the Pernambuco bees were tested in São Paulo they became gentle, and opposite responses were obtained when the São Paulo bees were tested in Pernambuco. The authors obtained significant negative correlations between aggressive behavior and changes in temperature. Thus, where the range of environmental temperature is greater (Ribeirão Preto) aggressiveness is lower. They also obtained significant positive correlations between aggressiveness and relative air humidity. Thus, the higher the relative humidity of the air (Recife), the greater the aggressive behavior. The temperature at the time of the tests also proved to have an effect; i.e., increasing temperatures caused decreased aggressiveness. Other factors, such as rainfall, and general weather conditions (rainy or dry), were not significantly correlated with aggressive behavior.

Mammo (1976), working with African bees in Ethiopia, reported that in situations where man and nature frequently destroy colonies of bees, the bees are aggressive and the least provocation is equivalent to a declaration of war. When the bees are well taken care of, however, they are reasonably gentle and easy to handle. The author observed the aggressive behavior of two groups of African bees at two experimental stations in Ethiopia (a low and warm region at 1500 m, and a high and cold region at 3000 m). He did not use aggressiveness tests but he reported that each time he moved a bee colony from the warm to the cold region, the aggressiveness of the bees decreased. When colonies were transferred from the cold to the warm region they initially appeared to adjust to the new environment, but in time they became as aggressive as the other colonies in the area.

Brandeburgo (1979) experimentally validated the observations of Mammo (1976). In a more recent study, Brandeburgo (1986) used a variation of the aggressiveness test to measure the defensive behavior of Africanized bees. The test consisted of placing a transparent box in front of the entrance to the hive, with the black leather ball balanced inside the box. The observer then recorded the time until the first attacking bee left the hive and penetrated the box, the time until the first sting was left in the ball, the time for the entire hive to become infuriated, the number of stings left in the ball and the number of bees which penetrated the transparent box. Brandeburgo tested several colonies of Africanized bees using this methodology and reached some important conclusions. He noted that defensive behavior was significantly and positively correlated with the following variables: number of field bees, number of guarding bees, colony temperature, luminosity, and wind speed, and significantly and negatively correlated with rainy weather, atmospheric pressure and precipitation. He also noted that colonies were least aggressive in April and October, which is the time of best colony development, and most aggressive in August, December and February, when food availability is lowest.

DEFENSIVE BEHAVIOR, PHEROMONES AND SENSORY STRUCTURES

The differences in defensive behavior are strongly related to the production of volatile substances (pheromones), which elicit several forms of alarm behavior. These substances are important in attracting workers to the location of the enemy. Boch *et al.* (1962) showed that isopentyl acetate (IPA) is one of the active components of venom. Blum *et al.* (1978) reported that there are at least eight more volatile compounds (alcohols and esters) in addition to IPA, which can be separated by gas chromatography from the venom of *A. mellifera* workers. Collins and Blum (1983) tested the action of these compounds in laboratory experiments, and showed that almost all of them elicit responses from the workers to a greater or lesser extent.

According to Maschwitz (1964), *Apis* workers also produce alarm pheromones in their mandibular glands. Shearer and Boch (1965) identified the alarm compound from the mandibular glands as 2-heptanone.

Kerr *et al.* (1974) showed that Africanized bees produce five times more 2-heptanone pheromone than Italian bees. Observations on F_1 and backcrossed colonies showed that colony aggressiveness increased with increasing amounts of 2-heptanone produced; i.e., the time needed by colonies to become fully aggressive decreased and the number of stings inflicted on the enemy increased. No studies have been performed in Brazil on possible differences in IPA production between Africanized and European bees, but Africanized bees are assumed to produce larger amounts of the compound.

The sensitivity to pheromone should be related to the number of olfactory sensory structures located in the antennal segments. The number of these structures has been determined in Africanized and Italian bees using a method developed by Stort (1979) which involves the following steps. Each segment of the antennal flagellum is first separated from the antenna and opened in the region opposite to the frontal one, where there are no olfactory structures, with the aid of entomological pins. Each segment is spread out on a slide containing Canada balsam and covered with a coverslip. Each slide is photographed with a photomicroscope, the film is mounted on slide frames and projected onto a paper screen which permits counting the olfactory structures whose images appear in spherical form (FIGURE 8).

Italian bees were found to have larger numbers of these structures than Africanized bees, and the data obtained for F_1 descendants suggest polygenic inheritance of the trait (Stort and Barelli, 1981).

Analysis of the data showed that the aggressive behavior traits are significantly correlated with the number of antennal olfactory structures. These correlations indicate that the larger the amount of olfactory discs, the larger the number of stings and the longer the pursuit distance of the Africanized backcrosses (Stort, 1978).

The relationship of other sensory structures, such as number of ommatidia in the compound eyes of workers, number of antennal sensory hairs and diameter of the middle ocellus, on aggressive behavior has also been studied (Stort, 1979).

Aggressive behavior depends on the interaction of pheromones and olfactory structures, and the fact that Africanized bees produce five times more 2-heptanone represents a mechanism of compensation for the lower olfactory sensitivity (smaller number of olfactory discs) they may possess.

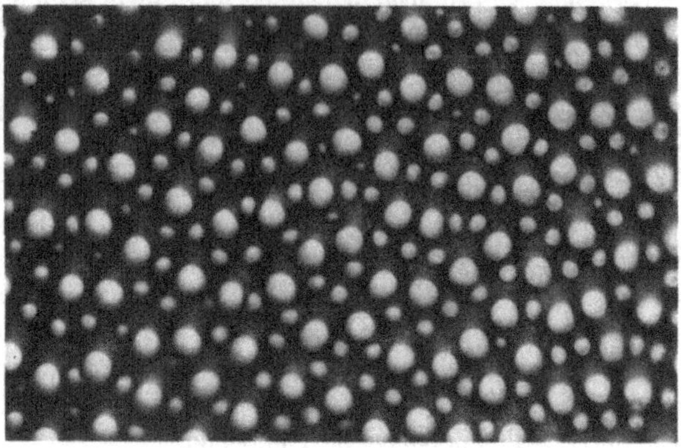

FIGURE 8. Antennal segment of Africanized bee. Bigger spherical structures are olfactory sensillae. The others structures are point of implantation of sensorial hairs.

Genetically, the defensive behavior of bees may be under complex control depending on the interaction of at least the following factors: aggressiveness genes (Stort, 1971; Gonçalves and Stort, 1978) plus genes controlling the amount of 2-heptanone, plus genes controlling isopentyl acetate production, as well as the production of other venom pheromones plus polygenes controlling the number of olfactory discs. The action of the environment should be added to this set of genes, because the genes involved in defensive behavior obviously do not function alone but interact with environmental factors to produce the final behavioral phenotype (Stort, 1979).

CONCLUSIONS

The research on the genetics of defensive behavior in experimental apiaries has shown that both aggressive and gentle descendant colonies were obtained for all five traits studied. This indicates that what is occurring in the longer term is different from what appeared to occur at the beginning of Africanization, when all colonies seemed to be aggressive.

Thus, each beekeeper has both gentle and aggressive hybrid colonies (Africanized) in his apiary, with the possibility of intermediate types also occurring. Thus, he has the option of eliminating the queens of the most aggressive colonies, if he so desires, and keeping only the gentle ones. His gentle colonies may then produce virgin queens which can be used to form other non-aggressive colonies. By this method, even if he does not know any genetics, the beekeeper may select his material and will have, within a short time, colonies of well-adapted bees which are productive and have the level of aggressiveness he prefers.

In general, this procedure has been widely implemented in Brazil. Where beekeepers do not care properly for their bees and make no selection, there are apiaries with colonies that are still quite aggressive and difficult to handle. This is the case, for example, in the region of Araraquara, state of São Paulo, where some beekeepers visit their apiaries only at the time when they collect honey.

It should also be taken into consideration that many beekeepers have imported and introduced European (especially Italian) queens into their apiaries. Although it is doubtful that these introductions have improved productivity, they should be contributing to the decrease in aggressive behavior.

The problem of controlling aggressiveness has been practically solved in Brazil. With few exceptions, beekeepers, who now use more protection than before, do not complain any longer and even prefer to raise Africanized bees.

Indications that the aggressive behavior of bees has been modified are already available. The aggressiveness of swarms collected at random in nature and tested in 1978 was much lower with respect to four of the five traits tested than the aggressiveness of colonies obtained and tested in the same region 10 years before. Colonies tested in 1976 also showed this reduction in aggressiveness of Africanized bees in the region of Ribeirão Preto (Brandeburgo, 1979).

LITERATURE CITED

Blum, M. S., Fales, H. M., Tucker, K. W., Collins, A. M. 1978. Chemistry of the sting apparatus of the worker honey bee. *J. Apic. Res.* 17:218-221.

Boch, R., Rothenbuhler, W. C. 1974. Defensive behaviour and production of alarm pheromone in honey bees. *J. Apic. Res.* 13:217-221.

Boch, R., Shearer, D. A., Stone, B. C. 1962. Identification of iso-amyl acetate as an active component in the sting pheromone of the honey bee. *Nature* 195:1018-1020.

Brandeburgo, M. A. M. 1979. Estudo da influência do clima na agressividade da abelha africanizada. Masters dissertation. Fac. Medic. Ribeirão Preto 149 p.

Brandeburgo, M. A. M. 1986. *Comportamento de defesa (Agressividade) e aprendizagem de abelhas africanizadas: análise de correlaçãoes entre variáveis biológicas e climáticas, herbabilidade e observações em colônias irmãs.* PhD Dissertation. Fac. Medic. Ribeirão Preto, Brazil. 156 p.

Brandeburgo, M. A. M., Gonçalves, L. S., Kerr, W. E. 1976. Nota sobre o estudo do efeito das condições climáticas sobre a agressividade das abelhas africanizadas. *Cienc. Cult. São Paulo* 28:276-277.

Cale, G. H., Rothenbuhler, W. C. 1979. Genetics and breeding of the honey. bee. In *The Hive and the Honey Bee*, ed. Dadant & Sons, pp. 157-84. Hamilton, Ill: Dadant & Sons.

Collins, A. M. 1979. Genetics of the response of the honey bee to an alarm chemical, isopentyl acetate. *J. Apic. Res.* 18:285-291.

Collins, A. M. 1980. Effect of age on the response to alarm pheromones by caged honey bees. *Ann. Entomol. Soc. Am.* 73:307-309.

Collins, A. M. 1982. Behavior genetics of honey bee alarm communication. In *The Biology of Social Insects, Proc. Ninth Cong. IUSSI.* ed. M. D. Breed, C. D. Michener, H. E. Evans, pp.307-311. Boulder, Colorado: Westview Press.

Collins, A. M., Blum, M. S. 1982. Bioassay of compounds derived from the honey bee sting. *J. Chem. Ecol.* 8:463-470.

Collins, A. M., Blum, M. S. 1983. Alarm responses caused by newly identified compounds derived from the honey bee sting. *J. Chem. Ecol* 9:57-65.

Collins, A. M., Kubasek, K. J. 1982. Field test of honey bee (Hymenoptera: Apidae) colony defensive behavior. *Ann. Entomol. Soc. Am.* 75:383-387.

Collins, A. M., Rinderer, T. E. 1985. Effects of empty comb on defensive behavior of honey bees. *J. Chem. Ecol.* 11:333-338.

Collins, A. M., Rinderer, T. E., Harbo, J. R., Bolten, A. B. 1982. Colony defense by Africanized and European honey bees. *Science* 218:72-74.

Collins, A. M., Rothenbuhler, W. C. 1978. Laboratory test of the response to an alarm chemical, isopentyl acetate, by *Apis mellifera*. *Ann. Entomol. Soc. Am.* 71:906-909.

Cosenza, G. W. 1972. Comportamento e productividade da abelha africana e de suas híbridas. *Serv. Pesq. Ext. Sete Lagoas, Minas Gerais.* 19:1-8.

De Santis, L., Cornejo, L.G. 1968. La abeja africana *Apis (Apis) adansonii* en America del Sur. *Revista de la Facultad de Agronomia. La Plata.* 44:17-35.

Free, J. B. 1961. The stimuli releasing the stinging response of honey bees. *Anim. Behav.* 9:193-196.

Ghent, R. L., Gary, N. E. 1962. A chemical alarm releaser in honey bee stings (*Apis mellifera* L.). *Psyche* 69:1-6.

Gonçalves, L. S., Stort, A. C. 1978. Honey bee improvement through behavioral genetics. *Ann. Rev. Entomol.* 31:197-213.

Kerr, W. E., Blum, M. S., Pisani, J. F., Stort, A. C. 1974. Correlation between amounts of 2-heptanone and iso-amyl acetate in honey bees and their aggressive behavior. *J. Apic. Res.* 13:173-176.

Kerr, W. E., Duarte, F. A. M., Oliveira, R. S. 1975. Genetic component in learning ability in bees. *Behav. Genet.* 5:331-337.

Kerr, W. E., Gonçalves, L.S., Blotta, L.F., Maciel, H.B. 1972. Biología comparada entre as abelhas italianas (*Apis mellifera ligustica*), africana (*A. m. adansonii*) e suas híbridas. In *1º Cong. Bras. Apic. (Florianópolis, 1970)*, ed. H. Wiese, pp. 151-185, Santa Catarina, Brazil.

Laidlaw, H. H. Jr. 1949. Development of precision instruments for artificial insemination of queen bees. *J. Econ. Entomol.* 42:254-261.

Laidlaw, H. H. Jr. 1958. Inseminação artificial de rainhas de abelhas. *Brasil Apic.* 4:57-59.

Lecomte, J. 1963. *Le comportment agressif des ouvrières d'Apis mellifica L.* PhD Dissertation. Univ. Paris. 116 pp.

Mackensen, O., Roberts, W. C. 1948. A manual for the artificial insemination of queen bees. *US Dep. Agric. Bur. Entomol. Plant Q.* ET-250.

Mammo, G. 1976. Practical aspects of bee management in Ethiopia (with *A. m. adansonii*). In *Apiculture in Tropical Climates*, ed. E. Crane, pp. 69-78. England: Int. Bee Res. Assoc.
Maschwitz, 1964. Alarm substance and alarm behavior in social insects. *Vitams. Horm.* 24:267-290.
Nye, W. P., Mackensen, O. 1968. Selective breeding of honey bees for alfalfa pollen: fifth generation and backcrosses. *J. Apic. Res.* 7:21-27.
Portugal-Araújo, V. de 1956. Notas bionomicas sobre *Apis mellifera adansonii* Latr. *Dusenia* 2:91-102.
Rothenbuhler, W. C. 1960. A technique for studying genetics of colony behavior in honey bees. *Am.. Bee J.* 100:176-198.
Rothenbuhler, W. C. 1964. Behavior genetics of nest cleaning in honey bees. IV. Responses of F_1 and backcross generations to disease-killed brood. *Am. Zool.* 4:111-123.
Ruttner, F. 1975. Las razas de abejas de Africa. *XX Int. Apic. Cong., Apimondia.* Grenoble, France. pp. 347-366.
Sakagami, S. F., Akahira, Y. 1960. Studies on the Japanese honey bee, *Apis cerana cerana*, Fabricus. VIII. Two opposing adaptations in the post stinging behavior of honey bees. *Evolution* 14:29-40.
Schua, L. 1952. Untersuchungen über den Einfluss meteorologischer Element auf das Verhalten der Honigbien. *Z. Vgl. Physiol.* 34:258-277.
Shearer, D. A., Boch, R. 1965. 2-Heptanone in the mandibular secretion of the honey-bee. *Nature* 206:530.
Stort, A. C. 1971. *Estudo genético da agressividade de Apis mellifera.* PhD Dissertation. Fac. Fil. Cienc. Letras, Araraquara, Brazil. 166 pp.
Stort, A. C. 1972. Metodologia para o estudo da genético da agressividade de *Apis mellifera. 1º Cong. Bras. Apic. (Florianópolis, 1970)*, ed. H. Wiese, pp. 36-51. Santa Catarina, Brazil.
Stort, A. C. 1974. Genetic study of the aggressiveness of two subspecies of *Apis mellifera* in Brazil. I. Some tests to measure aggressiveness. *J. Apic. Res.* 13:33-38.
Stort, A. C. 1975a. Genetic study of the aggressiveness of two subspecies of *Apis mellifera* in Brazil. II. Time at which the first sting reached the leather ball. *J. Apic. Res.* 14:171-175.
Stort, A. C. 1975b. Genetic study of the aggressivenss of two subspecies of *Apis mellifera* in Brazil. IV. Number of stings in the gloves of the observer. *Behav. Genet.* 5:269-274.
Stort, A. C. 1975c. Genetic study of the aggressiveness of two subspecies of *Apis mellifera* in Brazil. V. Number of stings in the leather ball. *J. Kans. Entomol. Soc.* 48:381-387.
Stort, A. C. 1976. Genetic study of the aggressiveness of two subspecies of *Apis mellifera* in Brazil. III. Time taken for the colony to become aggressive. *Ciênc. Cult. São Paulo* 28:1182-1185.
Stort, A. C. 1977. Genetic study of the aggressiveness of two subspecies of *Apis mellifera* in Brazil. VII. Correlation of the various aggressiveness characters among each other and with the genes for abdominal color. *Ciênc. Cult. São Paulo* 30:492-496.
Stort, A. C. 1978. Comportamento agressivo e estruturas sensoriais em abelhas africanizadas e italianas. *Anais Simp. Int. Apic. em Clima Quente.* Apimondia. pp. 53-55.

Stort, A. C. 1979. *Estudo genético de caracteres morfológicos e suas relacões com o comportamento de defesa de abelhas do gênero Apis*. Tese de Livre Docência. Inst. Bioc, Rio Claro, Brazil. 179pp.

Stort, A. C. 1980. Genetic study of the aggressiveness of two subspecies of *Apis mellifera* in Brazil. VI. Observer persecution behavior. *Rev. Bras. Genet.* 3:285-294.

Stort, A. C., Barelli, N. 1981. Genetic study of olfactory structures in the antennae of two *Apis mellifera* subspecies. *J. Kan. Entomol. Soc.* 54:352-358.

Warnke, V. 1976. Effects of electric charges on honeybees. *Bee World* 57:50-56.

Woyke, J., Ruttner, F., Vesely, V. 1966. Manual on artificial insemination of queen bees. Preliminary edition Dol. CSSR. 86 pp.

PART FIVE

Beekeeping in South America

18

BEEKEEPING IN BRAZIL

Lionel Segui Gonçalves,[1] Antonio Carlos Stort,[2]
and David De Jong[3]

DEVELOPMENT OF BRAZILIAN APICULTURE

The history of Brazilian apiculture can be divided into several distinct phases: (1) the meliponiculture period, before 1839; (2) the European honey bee period; and (3) the Africanized bee period, after 1957.

Initially, as in all of the Americas, there were no *Apis* in Brazil and beekeeping was based on several of the hundreds of native meliponid species. The principal species used varied with climatic region. Some beekeepers cared for a thousand or more colonies each. Honey was also commonly harvested from wild meliponid colonies, especially by Indian peoples. Meliponiculture continues in Brazil, especially in the North. Species of meliponids presently exploited include: *Melipona subnitida, M. seminigra, M. bicolor, M. marginata, M. rufiventris, M. favosa, M. quadrifasciata, M. compressipes, M. scutellaris, Tetragonisca angustula,* and *Scaptotrigona* spp. (Nogueira-Neto, 1970; Nogueira-Neto *et al.*, 1986, Kerr *et al.*, 1986/1987).

The introduction of *Apis mellifera* from Europe provided a more productive and more easily managed bee. Beekeeping began to develop in the southern states, especially in Santa Catarina, Paraná, and Rio Grande do Sul, where the primarily European settlers and their descendants maintained their tradition of keeping bees. In the northern, more tropical regions, most beekeeping with European bees was done in the cleared areas along the coast, especially near the

[1]Dr. Gonçalves is in the Depto. de Biologia, Faculdade de Filosofia, Ciências e Letras de Riberão Preto, Univ. de São Paulo, Ribeirão Preto, 14.049, SP, Brasil.

[2]Professor Stort is in the Depto. de Biologia, UNESP, Instituto de Biociências de Rio Claro, 13.500, SP, Brasil.

[3]Professor De Jong is in the Depto. de Genética, Faculdade de Medicina de Ribeirão Preto, Univ. de São Paulo, Ribeirão Preto, 14.049, SP, Brasil.

cities. Meliponids continued to dominate in the heavily forested areas, and in the caatingas (dry savanna-like regions).

The European honey bee races introduced were black German bees (*A. m. mellifera*) in about 1839, followed by Italian bees (*A. m. ligustica*) in about 1870 (Nogueira-Neto, 1972). Thus, from 1870 until 1956 the honey bees in Brazil were predominantly German, Italian, and hybrids of the two. Most colonies were relatively docile, although productivity left much to be desired.

Brazil, despite its large territory, favorable climate, and wide variety of flowering plants, never played an outstanding role in world honey production. Apiculture was quite disorganized and the methods used in honey bee husbandry were not always the best. Furthermore, honey was not a product of commercial interest even in the internal market, and therefore, there were few professional apiculturists. Many beekeepers were amateurs and had little or no knowledge of apiculture. As European bees were not aggressive, it was not uncommon to find people keeping hives or rustic boxes containing bees in their backyards, both in the countryside and in the centers of cities. Because of this lack of technological progress, knowledge, and commercial incentives, Brazil did not reach a yearly production of 7,000 metric tons of honey until 1956.

The low productivity of European bees in Brazil and the tendency of these races towards decreased brood production during the winter were facts known to Brazilian researchers. These facts led to the idea that a type of bee should be imported into Brazil which would be more productive and more adapted to the tropical climate. The high productivity of African bees under tropical conditions came to the attention of beekeepers. For example, a beekeeper in South Africa, E.A. Schnetler, was reported to have obtained yields of 70 kg of honey per colony without migratory apiculture. Thus, the Ministry of Agriculture decided that African bees should be imported.

African bees were brought to Brazil in November 1956 by the geneticist Dr. Warwick E. Kerr (Kerr, 1967). At that time the honey bees in the region of Africa which served as the source of the introduced bees were known as *A. m. adansonii*. Subsequently, Ruttner (1975) called these bees *A. m. scutellata* (See Daly, Chapter 2). Unfortunately, in 1957, a well-meaning beekeeper took off the queen excluders that had been placed over some of the entrances of the colonies of African bees, and swarms from 26 colonies left the quarantine apiary in Camaquan, São Paulo. After the dispersal of these swarms and the subsequent rapid Africanization of Brazilian apiaries (FIGURE 1), the third phase of Brazilian apiculture started.

Since 1972, we have used the term "Africanized bee" for the polyhybrid which resulted from the matings between the African bees and the European races, which had previously been used in Brazil. Some authors used the term "Brazilian bees," but this is more appropriately used for the native stingless bee species (Meliponinae) (Gonçalves, 1974a).

FIGURE 1. Chronological and geographical spread of Africanized honey bees in Brazil.

The most obvious characteristic of Africanized bees first noted in Brazil was their greater defensiveness. This fact was of concern to beekeepers who were used to the docile European bees. Many of them gave up apiculture, so that, in a way, the introduction of the new bees served to stimulate the use of improved management procedures, and to select professional beekeepers from those who reared bees only for curiosity or as amateurs. Apiculturists, who up to that time had little contact with each other, felt the need to discuss the problems raised by the arrival of the new bees. Thus, in an attempt to solve these shared problems, beekeepers and technicians from governmental institutions gathered in several regional meetings, held in such places as Piracicaba, Rio Claro and Ribeirão Preto, São Paulo State; Curitiba, Paraná State; Florianópolis, Santa Catarina State; and Taquari, Rio Grande do Sul State.

The Africanized bees were severely testing Brazilian apiculture. Beekeepers had to make special efforts to adjust to the greater defensiveness and other behavioral differences. Consequently, the leaders in the field, with the support of the Brazilian Confederation of Apiculture, organized the First Brazilian Congress of Apiculture. This Congress, with about 150 participants was held May 8-10, 1970 in Florianópolis, Santa Catarina, and was of historical importance for Brazilian apiculture. General concerns, problems and efforts to find the best way to work with the more defensive bees were discussed at the meeting. Emphasis was placed on the protective clothing to be used by beekeepers, the development of new types of smokers, the correct application of smoke, the distribution of hives in the apiary, more appropriate colony management, and correct location of the apiary. In addition, many investigators presented the results of scientific studies on the biology of Africanized bees, covering such topics as aggressive behavior, swarming, reproduction, daily activity of field bees, and hybridization with Italians and Caucasians.

It became evident that the new bees also had a series of advantages. For example, the bees displayed greater productivity, with reports of colonies that had produced more than 100 kg of honey in three months (FIGURE 2). It was also concluded that with some changes in rearing methods, greater physical protection, and alertness of beekeepers, as well as a little more collaboration on the part of governmental institutions, apiculture using Africanized bees could grow very quickly. We also observed a significant reduction in the defensiveness of Africanized bees in Ribeirão Preto when we compared the results of aggressiveness tests by Stort in 1965 and Brandeburgo in 1976 (Brandeburgo *et al.*, 1982). Nevertheless, the bees have maintained a high percentage of African characteristics (Lobo *et al.*, 1989).

However, while apiculturists and Brazilian scientists were ready to make efforts to find the best way to develop apiculture with these bees, an intensely negative campaign was being waged against Africanized bees both in Brazil and abroad. Many of the more serious problems resulted from the negative reaction of the public towards beekeepers, which was exacerbated by sensationalized accounts in local news media. The negative campaign was further fueled by the U.S. news media, and "legitimized" by a report of North American scientists (Anonymous, 1972) compiled, mostly from second- and third-hand translated accounts, during a three-week tour of Brazil. Unfortunately, this report focussed on the problems rather than the solutions being developed by Brazil's beekeepers and scientists. Much of what was discussed was not objectively treated (see Gonçalves, 1974a, 1974b, 1975).

A very curious aspect of the "Africanized bee problem" is that despite a great deal of international concern among scientists, legislators, and the public, Brazil's 33 years of relevant experience have been largely ignored. The more than 140 masters and 60 doctoral theses and over 300 scientific papers published here on the biology of these bees are virtually unread outside of Brazil. A listing of all masters and doctoral theses on Africanized bees done in the main universities and institutes of research of Brazil was prepared by our group for the XXXII International Beekeeping Congress of Apimondia in Rio de Janeiro in 1989. A book which includes information about University bee labs, English and Portuguese summaries of all theses, and an indexed listing of several thousand papers concerning research with bees done by Brazilian scientists is being

FIGURE 2. An apiary of Africanized bees in Araraquara, São Paulo. Colony 105 produced over 100 kg of honey in three months.

prepared. Copies will available from the Genetics Department in Ribeirão Preto in early 1991.

Due to the sensationalism of the press, television accounts, and catastrophe films, about 20 years ago African and Africanized bees became the insects most talked about throughout the world. Concern about the bees grew among scientists and authorities in many South and Central American countries, as well as in the United States, where the arrival of the "killer bees" was greatly feared. There is no doubt that African and Africanized bees are more defensive than European bees but this difference has been exaggerated. In Brazil, the sensationalism of the press, which occurred mainly from 1964-1970, represented an obstacle to the development of Brazilian apiculture, because non-beekeepers were afraid of having bees in their commercial orchards and *Eucalyptus* plantations.

The use of new methods for the management of Africanized bees based on acquired experience was discussed at six additional national congresses (in Sete Lagoas, MG, 1972, 200 participants; Piracicaba, SP, 1974, 400 participants; Curitiba, PR, 1976, 500 participants; Viçosa, MG, 1980, 1,500 participants; Florianópolis, SC, 1984, 1,500 participants; and Salvador, BA, 1986, 2,500 participants). These meetings finally led to the full development of Brazilian apiculture in terms of productivity. Thus, in 1977, an annual output of 10,000

TABLE 1. Honey production (in metric tons) and number of hives in Brazil.

Year	Production	Number of hives
1957	6,527	
1958	6,779	
1959	6,949	
1960	7,539	
1961	7,749	
1962	7,540	
1963	7,500	
1964	7,784	
1965	7,904	
1966	7,931	
1967	7,303	
1968	7,049	
1969	6,770	
1970	6,313	
1971	5,052	
1972	5,052	320,000
1973	4,700	348,000
1974	4,120	375,000
1975	5,490	420,000
1976	5,900	540,000
1977	10,000	840,000
1978	15,000	970,000
1979	18,000	1,080,000
1980	19,000	1,100,000
1981	20,000	1,340,000
1982*	22,000	1,400,000
1983*	24,000	1,600,000
1984	25,000	1,600,000
1985	28,000	1,850,000
1986	31,000	1,900,000
1987	30,500	1,900,000
1988	36,000	1,980,000

Source: Brazilian Institute of Statistics, Brazilian Confederation of Apiculture (1989).
* - estimated

tons of honey was reached for the first time in history, and production has continued to increase, reaching 36,000 tons in 1988 (see TABLE 1). However, the mean honey yield in 1988 was still only 18 kg per hive, a very low figure if it is considered that annual yields of 60-80 kg per hive have been commonly reported when migratory apiculture is performed with Africanized bees.

CHANGES IN BRAZILIAN BEEKEEPING TECHNOLOGY

Apiary Arrangement

Formerly, apiaries were kept near houses and domestic animals. The hives were normally kept close together, on common hive stands and in an open apiary site. Beekeepers found that Africanized bees defended themselves very vigorously and that it was necessary to move the apiaries farther away from people and confined animals. While working with these bees, they determined that it was convenient to separate the hives by about two meters whenever possible because the vibrations caused by working on one of several colonies on a common stand disturbed the others. The solution was individual hive stands. Barriers of bushes and trees around the apiary reduced the impact of the bees on neighbors.

Clothing

Before the arrival of the Africanized bees, beekeepers were generally quite casual about the kinds of clothes they wore. Most used veils, though often they were fairly flimsy. Some used leather gloves, but most did not. The pants and shirt were of varied materials and colors and little thought was given to appropriate footwear. In some regions where the bees tended to be more defensive, especially in some southern areas where the German black bees (*A. m. mellifera)* predominated before the arrival of the Africanized bees, the beekeepers were more concerned with protective clothing.

After Africanized bees appeared, some changes were necessary. The bees especially attacked dark and rough textured clothing. Modern protective clothing now includes a good light colored hat, often of straw, and a sturdy wire veil, completely light colored on the outside and black only on the inside of the front panel. A veil that is dark on the outside soon gathers many angry bees. Beekeepers also discovered that only light colored smooth textured clothing was suitable. Generally, this requirement is fulfilled by use of an oversized white zippered overall, with elastic at the sleeve and pants cuffs.

The gloves, used especially at harvest time, are made of light colored plastic or rubber. The bees are not inclined to sting them and they do not retain the alarm odors. Skilled beekeepers often check their colonies bare handed, but keep gloves in their pockets.

Boots or shoes must also be light colored and smooth textured. Anyone daring to expose dark socks soon pays for the mistake. Presently, many beekeepers use light colored hard leather boots, though the heavy white plastic boots made for butchers have been adapted quite widely as a bee industry standard.

The Smoker

It was soon apparent that the smoker made for European bees was completely inadequate for Africanized bees. The new bees were much more sensitive to disturbance and reacted violently when not smoked adequately. Many attempts were made to develop a device that would serve the purpose, culminating in a contest in 1970 at the first National Beekeeping Congress in Florinanópolis, Santa Catarina State. Though the entries included machine powered models and even a man-high giant smoker, the winner was a design which, further adapted, has become the industry standard. The smoker is large, 30 cm high by 15 cm in diameter, and has a strong bellows (Bradbear and De Jong, 1985).

Bait Hives

One result of the migratory habits and the highly developed reproductive capacity of the Africanized honey bees was the appearance of swarms in nearly any abandoned hive, honey super, old box, etc. As beekeepers began to take advantage of these "free bees" they made systematic efforts to utilize this new resource. Presently it is quite common to encounter beekeepers who capture 100-200 or more swarms per year by setting out empty hives filled only with wired frames containing strips of foundation. Such colonies often have honey to harvest within two months after bees enter the hives.

The use of bait hives brings a direct benefit to the beekeeper in the form of increased income. The beekeepers also perform a service because each swarm that they capture is one less that might have gone into someone's house, or some other place inconvenient to the public. Moreover, it is apparent that if carried out on a large scale, swarm capture considerably reduces the competition for limited food resources by wild colonies.

An interesting alternative to the use of conventional beehives as baithives has appeared in the last few years. Cardboard boxes placed under a roof, covered with plastic or painted, and filled with frames, are cheap, very attractive to the bees and widely used by the beekeepers (Soares, 1985).

BEEKEEPING REGIONS IN BRAZIL

The most productive region is the South, including the states of Paraná, Santa Catarina and Rio Grande do Sul. Santa Catarina has the highest density of hives in the country and consequently the highest honey yield per unit occupied area. Honey production in Santa Catarina is about 6,000 metric tons per year, 50% of which is consumed within the state. Apiculture is better organized and developed in this state due to the support of the Apiculture Institute of Santa Catarina (IASC). Cooperatives are widespread, and the Apiculture Association of Santa Catarina forms a Federation comprising 28 municipal or regional associations. Also, honey bees are widely used to pollinate apple orchards. The largest beekeeper in the state has 6,000 colonies and rents bees for pollination as well as producing honey.

Rio Grande do Sul used to rank highest among the honey-producing states in Brazil. However, between 1968 and 1978 local apiculture declined sharply owing to (1) the expansion of soybean culture, involving a high level of pesticide use, (2) deforestation (which has affected almost 500,000 hectares of wooded areas), and (3) slow adaptation by beekeepers to working with Africanized bees. Annual honey production, which used to be as high as 2,500 tons, fell to 800 tons before recovering to a level of 1,500 tons in 1983. This recovery was due to a strengthened spirit of cooperation and to the extension activities of the Honey Bee Park of Taquari.

In 1983, honey productivity in the state of Paraná reached its highest level in 15 years, an estimated 8,000 tons. This increase, extraordinary in relation to previous mean yearly yields of 2,500-3,000 tons, was due to the excellent weather conditions that year. Apiculture has increased significantly in this state, with migratory methods being widely used and with the use of bees for pollination in large orchards.

In southeastern Brazil, the state of São Paulo has production problems as a result of the enormous expansion of the monoculture of sugar cane. With the development of the "Pro-Alcohol Project," little of the wooded habitat remains for honey bees. The only major possibilities for forage are the large orange and *Eucalyptus* plantations. The total honey yield has increased over the last few years owing to different factors; one is the fact that beekeepers have carried Africanized bee colonies to the more northern areas of the state where tropical forest is still present (De Jong and Gonçalves, 1981). At present, much of the migratory apiculture of São Paulo has been preferentially directed towards the states of Minas Gerais and Paraná, where there is competition among beekeepers for the best apiary locations. Some beekeepers of the state of São Paulo preferred to migrate to Ceara and Piaui states, 3,000-4,000 km away in the north and northeast region of Brazil, where the flowering is much more intense.

In addition to honey, São Paulo also has produced large amounts of royal jelly. However, owing to the better price of honey on the internal market, most beekeepers now prefer to produce only honey. One of the larger outfits, Seabra apiaries, owned 1,200 colonies in the state of São Paulo in 1984, 1,700 in 1985 and 2,500 in 1986. Their mean yields have been 100-120 kg per colony per year, mostly from two flows, *Eucalyptus* and orange (Seabra Filho, 1985).

The state of São Paulo is also known for the extensive research done at the state universities: University of São Paulo (USP) and the University of the State of São Paulo (UNESP). These two institutions are responsible for more than 80% of the work published in the country in the areas of apiculture and bee biology, genetics and behavior. The graduate program at the Ribeirão Preto campus of USP alone has produced 89% of all the Brazilian theses on these subjects.

Beginning in 1985, with funding from the Food and Agriculture Organization of the United Nations (FAO), and in 1986, with funding from the Interamerican Development Bank (IDB) administered by OIRSA (Organismo Internacional Regional de Sanidad Agropecuária, an international organization which monitors and controls agriculture and livestock disease problems in Mexico, Panama and all of the Central American countries), the bee specialists at these two São Paulo universities began to receive and to train researchers and technicians from this region. Since 1987, seven Brazilian professors from the same group have been periodically teaching, training and helping to organize and evaluate an extensive Africanized bee project run by OIRSA in all of the Central American region countries. With the help of the Brazilian professors, the OIRSA Africanized bee project has published seven books and many booklets and pamphlets in Spanish. OIRSA received three medals at the 1989 Apimondia Congress, held in Rio de Janeiro for this work. Beginning in 1990, the University of São Paulo has been training students from this region at the graduate level, specifically in the areas of Genetics and Entomology, so that they may better develop their own bee programs.

The state of Minas Gerais is currently developing a project sponsored by the Department of Agriculture whereby several thousand new apiaries will be established in various regions, together with four wax-processing centers and four laboratories for honey analysis. The number of bees hives in Minas Gerais is growing explosively because of the mild climate, with neither the intense cold of the southern winter nor the intense heat of the northern summer, and because pesticides are not abused in this state.

In the state of Mato Grosso do Sul, in the west central region of Brazil, there has not been much development of apiculture. There are about 85 beekeepers who own about 2,000 hives of Africanized bees which produced about 26 tons of honey in 1986. Most beekeepers are small producers who use primitive rearing techniques and collect honey by squeezing the combs, thus obtaining a low-quality product. Only 20 of these 85 producers use centrifugal honey extractors, and these 20 are responsible for 80% of total production.

In the northeast, despite hard and unfavorable climatic conditions, the fight for the development of apiculture has persisted and deserves mention. The state of Pernambuco produces and markets extracted honey, honey pressed from the comb (called "irá") and the molasses-like honey produced by colonies placed in sugar cane fields. A novelty in this state is the use of cement hives to withstand the high humidity of mangrove swamps, making it possible to expand beekeeping into mangrove areas. The flowering in this region can be so intense that at least one beekeeper collects honey up to five times a year, with an excellent mean annual honey production of 70 kg of honey per hive. Only by using Africanized bees have beekeepers been able to utilize the semi-arid region of the northeast, which is very favorable for honey production. Recent prolonged droughts, however, have reduced honey production by more than 50% in Pernambuco.

In the state of Ceará, drought is the greatest enemy of beekeepers. The current drought has lasted uninterruptedly for six years, impairing apicultural development. The state has about 70 active beekeepers owning, on average, 100 hives each. However, even though honey yields are negatively affected by drought, the use of bees for pollination of cashew, coffee, orange, and chayote plantations has yielded excellent results.

In the state of Bahia there are about 400 beekeepers averaging 30 producing hives each; this includes some larger producers who own 300 hives each. To stimulate apiculture, a new beekeeping project has been created in this state by the state government, with support from FAO, following the Santa Catarina model, which has all the necessary infrastructure for the support of apiculture and of meliponiculture.

In the state of Sergipe, apiculture is beginning to be structured and development is being planned. The production of extracted honey in the state is estimated at 30 tons per year. The predominant type of apiculture is still non-migratory, but there is good potential for expanded yields because the state has about 30,000 hectares of citrus groves.

The state of Alagoas has practically no apiculture, with a yearly production of only 700 kg. What the state has is plenty of sugar cane fields. However, the "sertão" (dry savanna-like) region has no sugar cane and has possibilities for apicultural development. It has few flowering plants due to the scarcity of rain, but is being reforested with plants of good apicultural potential.

Piauí is a state in an interesting situation. Historically, it has practically no tradition of apiculture but it is one of the largest honey producing regions in Brazil. During the last five years, for example, the production was 2,000-5,000 tons. It is also the largest producer of wax which is marketed throughout the country. This yield is due to migratory beekeepers and "meleiros," bee-hunters who harvest honey and wax from feral colonies. Their methods are to cut down the trees where bees are nesting and to kill the bees. In the last years, the government has become seriously concerned with apiculture in this state, due to the significant number of beekeepers from the southern states who are moving to Piauí as well as to Ceara with excellent results. In Picos, Piauí there are many beekeeping companies. One of them, the Wenzel's, consistently moves its 4,000 colonies three times a year and has the highest yearly average production per hive of the country (well over 100 kg) (De Jong, 1987).

SUMMARY AND CONCLUSIONS

Since African bees were introduced into Brazil over 30 years ago, Brazilian apiculture has gone through significant changes. Primitive methods have been increasingly replaced with more technical apiculture. During the transitional period, extreme positions and attitudes arose, including sensationalism by the press. There is no doubt that the arrival of African bees turned out to be the most important element in the history of Brazilian apiculture.

Together with government technicians and university scientists, beekeepers faced the challenge of working with the new bees. After an initial period of adaptation to the very different behavior and biology of the Africanized bees, the country started to produce large quantities of honey again and may take its place as one of the largest honey producers in the world within the next few years.

Today most apiculturists are satisfied with their conditions and actually prefer Africanized bees because it is easy to get bees and because they are good producers. Indeed, the large number of people now taking up beekeeping, including many young men and women, attests to the manageability of these bees.

As scientists, we are no longer concerned with the strong defensive behavior of the bees. In fact, today most commercial beekeepers in Brazil prefer to work with Africanized bees (De Jong, 1984). The defensive behavior of these bees has decreased and beekeepers slowly realized they could exploit the high capacity for adaptation, reproduction and productivity of the Africanized bees. An important advantage is the ease with which beekeepers can obtain large numbers of swarms with simple bait hives (Soares, 1985).

In conclusion, Brazil has all the conditions to be one of the most important honey producers of South America, and of the world. Based on the progress of beekeeping observed in this country during the last ten years we predict that production will reach 50,000 tons by the year 2000.

Acknowledgements

The authors thank Warwick E. Kerr (Professor of Biology and Genetics, Brazil), Ademilson E. E. Soares (Professor of Genetics, Brazil), Roger A. Morse (Professor of Apiculture, USA), and many others un-named, for their collaboration in discussions related to this theme. Richard Nowogrodzki provided valuable comments to the manuscript. Part of the funding for the research cited came from grants by CNPq and FAPESP in Brazil and NSF (DAR 7920922, INT 8201200, and INT 8402240) and OICD/USDA in the USA.

LITERATURE CITED

Anonymous, 1972. *Final Report of the Committee on the African Honey Bee.* Washington, D. C.: Natl. Res. Counc. Natl. Acad. Sci. 95 pp.

Bradbear, N., De Jong, D. 1985. The management of Africanized honey bees. *Information for beekeepers in tropical and subtropical countries. Leaflet 2.* London: Int. Bee Res. Assoc.

Brandeburgo, M. M., Gonçalves, L. S., Kerr, W. E. 1982. Effects of Brazilian climatic conditions upon the aggressiveness of Africanized colonies of honey bees. In *Social Insects in the Tropics,* ed. P. Jaisson, pp. 255-280. Paris: Université Paris-Nord.

Confederaçao Brasileira de Apicultura (Brazilian Confederation of Apiculture). 1989. ed. H. Wiese. *Boletim CBA - Informativo de Associatismo Apicola* No. 003 - Julho 1989.

De Jong, D. 1984. Africanized bees now preferred by Brazilian beekeepers. *Am. Bee J.* 124:116-118.

De Jong, D. 1987. The Wenzel's, a family of Brazilian beekeepers. *Gleanings in Bee Cult.* 115:530-531.

De Jong, D., Gonçalves, L. S. 1981. The *Varroa* problem in Brazil. *Am. Bee J.* 121:186-89.

Gonçalves, L. S. 1974a. The introduction of the African bees (*Apis mellifera adansonii*) into Brazil and some comments on their spread in South America. *Am. Bee J.* 114:414-15, 419.

Gonçalves, L. S. 1974b. Comments on the aggressiveness of the Africanized bees in Brazil. *Am. Bee J.* 114:448-450.

Gonçalves, L. S. 1975. Do the Africanized bees of Brazil only sting? *Am. Bee J.* 115:8-10, 24.

Kerr, W. E. 1967. The history of the introduction of African bees to Brazil. *S. Afr. Bee J.* 39:3-5.

Kerr, W. E., Absy, M. L., Souza, A. C. M. 1986/1987. Especies nectaríferas e poliníferas utilizadas pela abelha *Melipona compressipes fasciculata* (Meliponinae, Apidae), no Maranhao. (Nectar and pollen yielding species used by the bee *Melipona compressipes fasciculata* (Meliponinae, Apidae), in Maranhão). *Acta Amazonica* 16/17:145-156.

Lobo, J. A., Del Lama, M. A., Mestriner, M. A. 1989. Population differentiation and racial admixture in the Africanized honey bee (*Apis mellifera*). *Evolution* 43:794-802.

Nogueira-Neto, P. 1970. *A criacao de abelhas indigenas sem ferrao.* (Beekeeping with stingless honey bees) São Paulo: Editora, Chacaras e Quintais. 365 pp. 2nd ed.

Nogueira-Neto, P. 1972. Notas sobre a historia da Apicultura Brasiliera. (Notes concerning the history of beekeeping in Brazil). In *Manual de Apicultura,* ed. J. M. F. de Camargo, pp. 17-32. São Paulo: Editora Agronômica Ceres.

Nogueira-Neto, P., Imperatriz-Fonseca, V. L., Kleinert-Giovannini, A., Viana, B., de Castro, M. S. 1986. *Biologia e Manejo das Abelhas Sem Ferrão.* (Biology and management of Stingless Bees). São Paulo:Tecnapis. 54 pp.

Ruttner, F. 1975. Las razas de abejas de Africa. (Honey bee races in Africa) *XXV International Apiculture Congress.* pp. 347-366. Grenoble, France: Apimondia.

Seabra Filho, J. R. 1985. O dia a dia de um grande apicultor. (Activities of a large scale beekeeper). *Apicul. no Brasil* 2(10):24-25.

Soares, A. E. E. 1985. Cardboard bait hives: A practical alternative to capturing swarms. *Int. Bee Res. Assoc. Newsletter for beekeepers in tropical and subtropical countries.* 6:3.

19

THE AFRICANIZED HONEY BEE IN PERU

Robert B. Kent[1]

One of the most dramatic incidents of the introduction of an insect invader in the present century has been the arrival from East Africa of the African honey bee, *Apis mellifera scutellata*, in Brazil, in the mid-1950s. The bee escaped from an experimental apiary, hybridized with other honey bee populations in Brazil, and diffused over most of South America and into Central America and the extreme south of Mexico between 1957 and 1987. During this period it hybridized with other exotic honey bee races introduced to the New World at various times since the 1800s (Kent, 1988). The Africanized bees' purported aggressiveness and danger to beekeepers and the public at large has created considerable concern in Mexico and the United States as the bee has diffused north (McDowell, 1984; Taylor, 1985). As a consequence, the two governments attempted the establishment of a biological barrier across the Isthmus of Tehuantepec (Moffett *et al.*, 1987; Rinderer *et al.*, 1987; Fierro *et al.*, 1988; Tew *et al.*, 1988).

The disruptive consequences of the arrival of this hybrid on beekeepers, beekeeping industry, and the public in countries into which it has diffused, have been reported in the popular press and even dramatized in books, movies, and television specials. Frequently these accounts are exaggerated or simply untrue (Nunamaker, 1979). In the academic and trade literature the focus of research has been on the biological characteristics of the hybrid and general discussions of the bee's impact. Yet, detailed studies documenting the impact of the hybrid bee are

---

[1]Professor Kent is at the Department of Geography, University of Akron, Akron, OH, 44325, USA.

Original article (Kent, R. B. 1989. The African honey bee in Peru: An insect invader and its impact on beekeeping. *Appl. Geog*. 9:237-257) reprinted with permission of Butterworth & Co (Publishers) Ltd. Title and some introductory material changed or ommitted for consistency with the format of the present book.

rare. Consequently, public policy decisions regarding measures to halt or retard the advance of this hybrid before it reaches the United States and decisions on how to ameliorate the effects of the bee should it reach the southern United States are being taken without adequate data on the effects of the hybrid in other countries. This is surprising given the attention the Africanized has received from bee biologists and other scientists and the oft-repeated statements of ecologists who suggest that the best methods of preparing for a biological invader are detailed field study and the collection of empirical data about the invader before it invades (Calkins, 1983; Lattin and Oman, 1983; Simberloff, 1986). For example, Roughgarden (1986) in his review of modeling the invasion process, suggests that "...it is worth stressing that the best prediction is provided by empirical precedent, and not by a model at all."

The purpose of this chapter is to evaluate the impact of the Africanized honey bee in one country where it has been present for a number of years. Two countries in South America, Peru and Argentina, have regions free of the Africanized bee as well as regions where the hybrid has been established for at least ten years, and hence both provide suitable conditions for a comparative study of this nature. Because of financial limitations, only one country could be used as the study area, in this case Peru. The paper compares beekeeping practices and production between areas where the Africanized honey bee has been present for many years, and locations, which at the time of the study in the early and mid-1980s, were free from the influence of the hybrid. The conclusions can contribute to the development of sound public policy decisions by the appropriate agencies in Mexico and the United States on how to address most effectively the problems created by the hybrid when it does arrive in those countries.

ARRIVAL, DIFFUSION, AND DISTRIBUTION

The presence of the Africanized honey bee in Peru dates from the early 1970s. The hybrid first appeared east of the Andes in the Amazon Basin in 1974 in the area of Pucallpa in the Department of Ucayali (FIGURES 1 and 2) (Taylor, 1977). Later reports placed the bee at higher elevations in the foothills of the east-facing slopes of the Andes, within two to three years (Davila *et al.*, 1980). Beekeepers in Oxapampa and the Valle de Chanchamayo, in the Departments of Pasco and Junin respectively, began to experience the effects of the hybrid in late 1976 and 1977 (Alata Condor, 1976; E. Frey, pers. comm.). At about the same time, the hybrid appeared in the Huallaga River Valley in the vicinity of Tarapoto (Department of San Martin)(J. Murakami, pers. comm.), in La Merced (Department of Junin) (M. Kaufman, pers. comm.), in Quillabamba (Department of Cuzco)(M. Spivak, pers. comm.), in Jaen (Department of Cajamarca), and in other locations in the lowlands and the Andean foothills (H. Daly, pers. comm.). Beekeepers in this area cited frequent swarms of feral honey bees and increased numbers of feral colonies as initial indicators of the presence of the hybrid. Later, beekeepers noticed increased swarming in their apiaries and greater aggressiveness by their bees. In the years between 1974 and 1978, the hybrid established itself throughout the entire region below approximately 1500 m on the eastern side of the Andes (FIGURE 1).

Despite close geographical proximity the hybrid has not successfully established itself at elevations above 1500 m in the Andes. In the Mantaro Valley, in the central region of the Peruvian Andes, one beekeeper has practiced migratory beekeeping for sometime. For five years during the late 1970s and early 1980s, he moved his hives between Satipo in the Amazonian lowlands and Jauja in the Andes to take advantage of complementary seasonal honey flows in each region (A. Caballero, pers. comm.). While the bees in his apiary demonstrate predominantly Africanized characteristics, especially strong nest defense (aggressiveness), other apiaries in the Mantaro Valley did not demonstrate Africanized characteristics even after years of exposure to these bees (V. Pizarro, pers. comm.). In the intermontane valleys at lower elevations, feral swarms of Africanized bees periodically appear, as they have in the area of Cajamarca, in the northern Peru. Their permanent establishment in the intermontane valleys of the Andes does not appear to have occurred. This is probably due to consistently cooler temperatures and occasional freezes common to the highlands. Nevertheless, recent research in Colombia suggests the Africanized bee may be capable of surviving at much higher altitudes than has generally been believed possible (Villa, 1987).

The coastal regions of Peru remained free of the Africanized bee for a number of years because of the environmental barrier the Andes presented to its diffusion. However, in the late 1970s, the hybrid appeared on the north coast in the Departments of Tumbes and Piura (Davila *et al.*, 1980). It diffused south to Chiclayo and Pacasmayo by 1981 and reached Trujillo by late 1982; and by late in 1985, the hybrid had diffused about half way down the coastal margin towards Lima. The bee spread north into Ecuador, too, and by late 1981 was established throughout most of the coastal region south of Guayaquil (Harper, 1982). Now it is the dominant honey bee race throughout coastal Ecuador.

The arrival of the Africanized bee on the north coast in 1978 or 1979 can be traced, with a fair degree of confidence, to the activities of two commercial beekeepers. In 1978 these beekeepers purchased 300 colonies in La Merced on the eastern side of the Andes in an area infested with the Africanized bee (E. Lizarraga, pers. comm.). These hives were transported from there to apiary sites in the coastal departments of Tumbes and Piura. (FIGURES 1 and 2). These beekeepers reported that upon arrival in the north, they reviewed all the colonies and requeened those demonstrating aggressive characteristics with more docile Italian bees imported from the United States. This, they assert, effectively eliminated the possibility that they could have been responsible for the introduction of the Africanized bee to the coastal areas. However, other

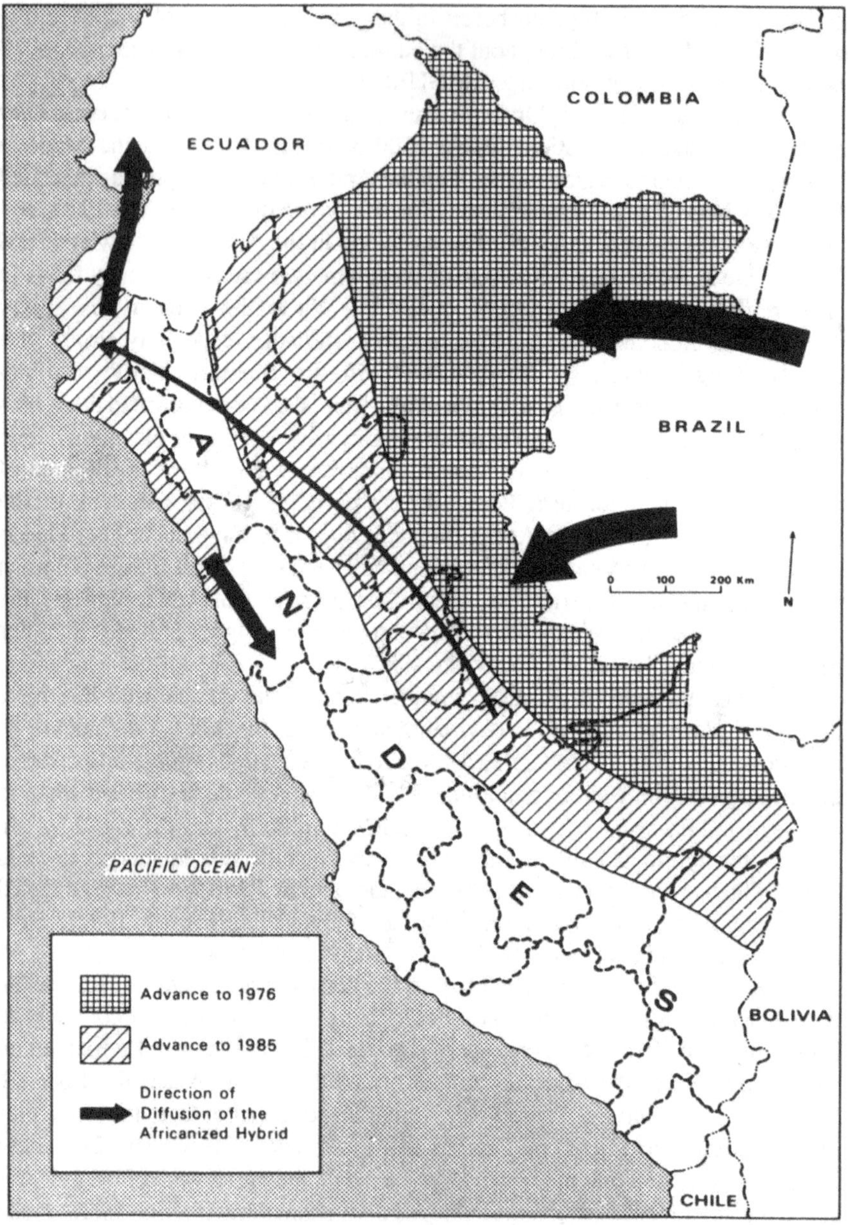

FIGURE 1. Diffusion of the Africanized honey bee in Peru.

FIGURE 2. Distribution of the Africanized honey bee in Peru and the location of case study sites.

commercial and hobby beekeepers in the area around Chulucanas, where the beekeepers established apiaries with the transferred colonies, remember increased aggressiveness and swarming by bees in their apiaries shortly after the arrival of the colonies from the La Merced (L. Ojeda, pers. comm.). Regardless of the origin or responsibility for the introduction of the Africanized bee on the north coast, the bee is now well established there and in Ecuador. If the hybrid demonstrates the same diffusion characteristics there that it has in the rest of South America, it will spread south along the coast to Lima and the irrigated valleys on the southern coast within the next few years, and then diffuse into the oases of northern Chile. The ability of the Africanized bee to abscond or swarm as far as 100 km from the original hive location allows it to diffuse relatively easily down South America's west coast desert (Seeley, 1985). It is unlikely that the hybrid will diffuse much farther south than Valpariaso or Santiago (at about 33° S) because of the cooler climates south of that latitude that make it difficult for the Africanized bee to survive.

METHODOLOGY

The analysis of the Africanized honey bee's impact in Peru and Peruvian beekeeper's response to it, is accomplished by a comparison of beekeeping practice and production between regions free of the hybrid and those infested by it. The lack of any significant historical data on Peruvian beekeeping at either the national, regional, or local level precluded the possibility of attempting to establish change within a region or case study site before and after the arrival of the hybrid. Financial and temporal constraints also eliminated the possibility of making repeated trips to Peru to evaluate beekeeping conditions in case study areas before and after the arrival of the Africanized bee. The distribution of the hybrid at the time the fieldwork for this study was initiated (1981-1982) was determined through interviews with the leading beekeepers in the country and a review of the published material on the subject available in Peru (FIGURE 2). Besides selecting case study sites in areas free of the hybrid and those invaded by it, sites were sought where beekeeping had been practiced for a considerable number of years so that conditions were generally stable and representative of conditions existing over a number of years. Furthermore, sites were selected throughout the country to ensure that local variations in beekeeping conditions and management practices would not unduly bias the results of the study. Finally, the selection of sites was constrained by two other factors. First, they had to be areas that could be readily reached by transportation available to the general public (buses, pickup trucks, or cargo trucks), since private transportation was not available in the field. Second, the case study sites had to be small towns or cities where food and lodging were available. Using these criteria, five case studies were conducted in regions where the hybrid was present, Chulucanas, Chepén/Guadalupe, Tarapoto, Oxapampa, and Quillabamba, and six case studies outside of the region affected by the hybrid, Cajamarca, San Marcos, Huaral, Camana, Concepción, and Arequipa (FIGURE 2).

In other South American countries where the hybrid bee has been present for many years, a considerable body of literature has accumulated identifying a range of variables related to beekeeping which are reported to be affected by the hybrid's presence. These fall into four major classes--management practice, honey production, apiary characteristics, and the economic status of beekeepers. Most beekeepers find management requirements more demanding and capital intensive with the Africanized bee. The need for more and better protective clothing and equipment--gloves, masks, overalls, boots, and smokers--is one of the chief reasons (Kempff Mercado, 1973; Taylor and Williamson, 1975; Wiese, 1977; Sommer, 1979). The tendency for the hybrid to swarm and abscond more frequently also necessitates more intensive management. Typically, it has been suggested that the hybrid is a more prolific honey producer and that after an initial period of adjustment when yields may decline, honey yields tend to increase notably (Cornejo, 1971; Taylor and Williamson, 1975; Gonçalves, 1978; Sommer, 1979; Stort and Gonçalves, 1979). The aggressive hive defense demonstrated by the Africanized bee is said to require changes in the location of apiaries. Before the arrival of the hybrid, beekeepers in South America generally located their hives near dwellings and farmsteads so they would be accessible and to reduce the problems of theft. Due to the hybrid's aggressive hive defense, this proximity is troublesome, and beekeepers and apiculturalists have advised that hives with hybrid bees should be located far (200 to 500 m) from homes, barns, animals, walkways, and roads (Kempff Mercado, 1973; Taylor and Williamson, 1975; Sommer, 1979; Stearman, 1981). Finally, the arrival of the Africanized bee appears to have a negative impact on the ability of many to continue keeping bees. Part-time and hobby beekeepers are most seriously affected because frequently they do not have the means to make the necessary investments in better equipment or to locate or to move apiaries to isolated sites (Wiese, 1977; Gonçalves, 1978; Presada, 1981). Consequently, these individuals may abandon beekeeping, while commercial beekeepers with greater financial resources and technical knowledge are more likely to continue keeping bees.

A questionnaire was designed to assess beekeeping conditions in each study area. It sought information on management practices, honey production, apiary characteristics, economic status, and beekeeper education. It identified specific variables, as reported from other South American countries affected by the hybrid, which are directly influenced by the hybrid bee's presence in beekeeper's apiaries, and from an earlier survey conducted in Brazil by a committee of the U.S. National Academy of Sciences (Anonymous, 1972). The questionnaire also queried beekeepers on perceived problems, changes in the behavior of their bees over the past few years, knowledge of the Africanized honey bee's presence in Peru, and the presence or absence of the Africanized honey bee in their locality. If they believed the hybrid to be present, additional information was gathered on changes in management techniques, production, apiary location, and changes in apiary size since its arrival. A total of 112 beekeepers were surveyed in the 11 study areas.

Besides the questionnaire, an apiary observation sheet was used to collect data on apiary size, location, hive type, condition, and local vegetation. The data collected included information on the distance of the apiary from human habitation and livestock pens and reports of stinging incidents of neighbors and livestock. One hundred and thirty-five apiaries were visited during the course of the fieldwork.

Because no government data are available on the number of beekeepers or apiaries at a local or even provincial level, a simple survey technique was employed in each case study area to survey beekeepers and apiary conditions. Using maps of the area surrounding each case study site, a circular area with a three kilometer radius was established. The principal roads and trails in each study area were traversed, and farmers, residents, and passers-by were asked if they knew anyone who kept honey bees or where honey bees were kept. Using this technique, it was possible to conduct complete surveys of all beekeepers and apiaries within a one week period in each study area.

The data collected concerning each of the variables are analyzed to establish if significant variations in beekeeping conditions and management practices exist between Africanized and non-Africanized regions. These differences are analyzed and tested for statistical significance using two common techniques. The chi-square test is employed for the analysis of variables measured on a nominal scale, and the analysis of variables measured on an interval scale is accomplished using a difference of means t test. For the purposes of this study, the chi-square statistic and the F-ratio are considered significant if the probability of error is equal to or less than 0.05. It is important to note that the chi-square tests reported in this paper have been conducted using the actual counts or frequencies, but to facilitate comparisons for the reader percentages have been reported in the tables. The specific variables evaluated are discussed at the beginning of each section.

REGIONAL VARIATION IN BEEKEEPING

Management Practice

Management practice examines variables which reflect both the impact of the hybrid and beekeepers' ability to respond to it. The use of additional protective equipment is a likely response of beekeepers to the presence of the Africanized bee. Variables reflecting this are the ownership and use of a smoker or another smoking device to pacify bees and of a mask and gloves to protect the

TABLE 1. Ownership and use of protective equipment

Equipment and Use	Africanized Region (%)	Non Africanized Region (%)	Total Ownership (%)
*Smoker			
Yes	73.2	56.6	65.1
Other Device	26.8	24.5	25.7
Nothing	0.0	18.9	9.2
	(n = 56)	(n = 53)	(n =109)
**Mask			
Yes	94.7	85.2	90.1
No	5.3	14.8	9.9
	(n = 57)	(n = 54)	(n =111)
***Use Mask			
Yes	92.8	84.3	88.8
No	5.4	15.7	10.3
Sometimes	1.8	0.0	0.9
	(n = 56)	(n = 51)	(n = 107)
+Gloves			
Yes	54.5	56.6	56.6
No	45.5	43.4	44.4
	(n = 55)	(n = 53)	(n = 108)
++Glove Use			
Yes	40.8	48.1	44.3
No	48.1	50.0	49.1
Sometime	11.1	1.9	6.6
	(n = 54)	(n = 52)	(n = 106)

*Chi-square = 11.77 with 2 d.f.; highly significant; p = 0.01
**Chi-square = 1.86 with 1 d.f.; not significant; p = 0.17
***Chi-square = 3.90 with 2 d.f.; not significant; p = 0.14
+Chi-square = 0.00 with 1 d.f.; not significant; p = 0.98
++Chi-square = 3.73 with 2 d.f.; not significant; p = 0.16

beekeeper from stings while the hive is being worked. Other management variables which reflect a beekeeper's ability to adapt successfully to the hybrid are hive type, (fixed vs. movable-frame) use of supers, reproduction technique, and extraction method.

Beekeepers in the Africanized region are better equipped with smokers than their colleagues in the region not infested with the hybrid (TABLE 1). Every beekeeper interviewed in the Africanized region either uses a smoker or some other smoking device, but only 80% in the non-Africanized region do so. Statistically, the variation in this practice between the two regions is highly significant ($p<0.01$) and suggests bees in the Africanized region are probably more aggressive. The fact that nearly 20% of the respondents in the areas free of the hybrid do not use smokers is surprising, since some smoking of even the most docile race of honey bees is recommended by apiculturalists. No differences were observed in the size of the smokers used in each region, and no jumbo smokers, reportedly common in Brazil and Bolivia, were seen.

The ownership and use of other protective equipment, masks and gloves, presents a slightly different picture. Mask ownership and use are more prevalent in the Africanized area than in the area free of the hybrid, about 95% and 85% for both ownership and use in each area respectively (TABLE 1). Gloves are owned by considerably fewer beekeepers than masks, and nearly equal proportions of beekeepers in each region own them, 55% in the Africanized area and 57% in hybrid-free area. Glove use is even less common, and unexpectedly, their use is more frequent in areas unaffected by the hybrid than in those areas where the hybrid is found, 48 and 41% respectively (TABLE 1). In no case, however, are the variations between the two regions significant statistically. What does appear curious is that so many beekeepers in both regions do not wear masks when working with bees. As with smokers, the use of a mask, even with docile bees is highly recommended.

Distinct practices are evident in hive type, use of supers, reproductive techniques, and extraction methods between the Africanized and the non-Africanized regions. About one-half of the beekeepers use standard movable-frame hives (modern technology) in the Africanized region, but only one quarter do so in the area free of the hybrid (TABLE 2). Similar trends are evident in the use of supers, additional hive boxes placed on top of the broad chamber to give the bees more space to store honey. In this case, 65% use them in the Africanized region versus 38% in the non-Africanized region. More sophistication is also evident in techniques employed for colony reproduction. In the Africanized region, 65% use hive division, an efficient reproductive technique, whereas only 35% utilize this technique in the non-Africanized region. Finally, honey extraction using modern equipment, a centrifuge, is almost twice as common in the Africanized region as it is in the region free of the hybrid--45 and 26% respectively (TABLE 2). In all four instances, the variables demonstrate statistically significant variation between the two regions. This

TABLE 2. Management practices

Management Practice	Africanized Region (%)	Non-Africanized Region (%)	Total (%)
Hive Type*			
Standard	47.4	24.5	36.4
Non-Standard	10.5	35.9	22.7
Fixed Comb	26.3	26.4	26.4
Mixed, Standard and Non-Standard	15.8	13.2	14.5
	(N = 57)	(N = 53)	(N = 110)
Use of Supers**			
Yes	65.8	38.1	51.2
No	34.2	61.9	48.8
	(N = 38)	(N =42)	(N =80)
Reproductive Technique***			
Hive Division	64.8	34.7	50.5
Swarm Capture	24.1	42.9	33.0
Mixture of Both	11.1	22.4	16.5
	(N = 54)	(N = 49)	(N = 108)
Extraction Technique****			
Centrifuge Extraction	45.5	25.9	35.8
Comb Cutting	50.9	74.1	62.4
Other	3.6	0.0	0.0
	(N = 55)	(N = 54)	(N = 109)

*Chi-square = 11.81 with 3 dif., p = 0.01
**Chi square = 5.07 with 1 dif., p = 0.02
***Chi-square = 9.36 with 2 dif., p = 0.01
****Chi-square = 7.21 with 2 dif., p = 0.03

tends to indicate that beekeepers in the Africanized region are better equipped and more progressive in their management practices than their colleagues in the non-Africanized region. These characteristics also suggest that those in the Africanized regions are better able to adapt to the hybrid and exploit it effectively. Standard movable-frame hives and the use of supers allow the beekeeper more control over the colony and facilitate the requeening of aggressive colonies, prevention of swarming and absconding, and better honey yields. The use of centrifuge extractors and colony reproduction using hive division, common in the Africanized region, also are instrumental for increasing honey yields.

Honey Production and Colony Numbers

Average annual honey yields and the number of hives owned per beekeeper are examined as indicators of the hybrid's impact. But it is important to note that honey production, and to some degree the number of hives a beekeeper owns, also are strongly affected by a number of other factors having little or nothing to do with the presence or absence of the Africanized honey bee. These factors include management practices, agricultural conditions, competition from other apiaries, local flora, and climatic conditions. In a study of this character it is impossible to control completely for all these complications. As a consequence, it is difficult to draw definite conclusions about variations in honey production. The primary problems in this study are the variety of hive types used which directly affects honey production, their differing concentrations in each region, and the comparatively small size of the sample.

Regardless of the basis upon which honey production between the two regions is compared, average colony production is consistently greater in the Africanized region. An initial comparison between the two regions, using reported colony production for all hive types, by all beekeepers, ($n = 72$) reveals an average production of 30.5 kg and 16.8 kg in the Africanized and non-Africanized regions respectively, ($t = 2.85$, $p = 0.006$). Even when efforts are made to control for production differences between different hive types, consistent differences persist. Using all types of movable frame hives as the basis of comparison ($n = 52$), production averages 34.5 kg and 21.4 kg in the Africanized and non-Africanized region respectively ($t = 2.28$, $p = 0.027$). Focusing only on production estimates for standard movable frame hives ($n = 28$) similar differences appear, with production averaging 37.5 kg and 26.1 kg in the Africanized and non-Africanized regions respectively ($t = 1.36$, $p = 0.187$). These production differences are statistically significant in all but the last case.

The average number of colonies owned by beekeepers in each region varies considerably, from 46 in the Africanized region to 14 in the non-Africanized region. Statistically, however, these differences are not significant ($t = 1.71$, $p = 0.092$).

Apiary Characteristics

Apiary characteristics can reflect both the impact of the hybrid as well as beekeepers' ability or need to respond to it. Apiary location, frequency of stinging incidents, and apiary tenure are assessed in this section. The distance apiaries are located from human dwellings reflects, to some degree, the problems or the perception of possible problems beekeepers or neighbors expect the bees to cause. The actual occurrence of stinging attacks on livestock and neighbors, should reveal if bees in the areas infested with the Africanized bee are more aggressive and if they cause more problems for beekeepers. Apiary tenure affects the permanence of apiary locations and can cause beekeepers to relocate their hives in response to complaints about aggressive bees. Beekeepers who rent or borrow their apiary sites are more vulnerable to displacement if their bees are troublesome to landowners or their neighbors.

Apiaries in both Africanized and non-Africanized regions are consistently located near human dwellings. Understandably, beekeepers want to locate their apiaries near their homes or the homes of others since it makes management easier and reduces the possibility of theft or vandalism. Many apiaries in Peru are located immediately adjacent to the beekeeper's house in a patio or garden. Overall, about one-half of the apiaries in Peru are situated within 10 m of a dwelling and an additional one-quarter are located between 11 and 20 m. Modest differences are evident between apiary locations in the two regions. Apiaries in the Africanized regions are situated an average of 16 m from dwellings, while those in non-Africanized regions are located an average of about 13 m. The differences between the two regions, however, are not statistically significant ($t = 1.36$, $p = 0.178$).

The tendency for apiaries to be located so close to homes and livestock appears partly to be the result of the limited number of problems and complaints beekeepers have with their bees. In fact, problems with bees stinging neighbors and/or livestock were reported in a total of only about 15% of the apiaries visited (TABLE 3). Stinging incidents demonstrate some minor differences between the two regions, being slightly more common in the Africanized areas. Problems with bees stinging neighbors are the least prevalent and exhibit the least variation; stinging incidents occur in 15% of the Africanized apiaries and 11% of the non-Africanized apiaries (TABLE 3). The differences are not statistically significant. This is not so, however, with respect to stinging incidents involving animals. These are more common in apiaries in the Africanized region, about 24% versus only 7% in the area free of the hybrid. The chi-square statistic, although small, 4.02, is statistically significant. The greater frequency of stinging incidents in the Africanized region supports the impression that the hybrid bees are more aggressive. The fact that the stinging incidents involving humans are minor and not statistically significant, probably indicates that the

TABLE 3. Stinging incidents or problems with humans or livestock reported in apiaries.

Stinging Problems or Incidents	Africanized Region (%)	Non-Africanized Region (%)	Total (%)
*With Humans			
Yes	18.6	11.1	14.4
No	81.4	88.9	85.6
	(n = 44)	(n = 54)	(n = 97)
**With Livestock			
Yes	23.8	7.3	14.4
No	76.2	92.7	85.6
	(n = 42)	(n = 55)	(n = 97)

*Chi-square = 0.57 with 1 d.f.; not significant; p = 0.45
**Chi-square = 4.02 with 1 d.f.; moderately significant; p = 0.05

public in the Africanized region is aware of the bee's aggressiveness and consequently avoids the area around apiaries. Livestock, on the other hand, can make no such conscious efforts.

Apiary sites are located predominantly on land owned by the beekeeper, his family, or property to which they have usufruct rights. About 90% of all apiary sites are owned by the beekeeper and 10% are borrowed from friends or relatives in both regions. In less than 1% of all cases do beekeepers have formal financial agreements with property owners. This is slightly misleading, since even beekeepers who borrow sites from friends or relatives are expected to give a small portion of their crop (10 to 20 kg) to the owner in consideration for the use of the site. Differences are evident in apiary site tenure between the two regions, yet, it is interesting to note that the frequency of borrowed, rented, or leased sites in the Africanized region is greater--12%--than in the region free of the hybrid -- about 7%. If the aggressiveness of the hybrid bee is a problem, one might expect the situation to be reversed, reflecting increased difficulties in getting and keeping apiary sites with the troublesome bees. Nevertheless, statistically the differences are not significant (Chi-square = 1.46 with 2 d.f., p = 0.48).

Economic Status

Two variables are examined in this section--economic status and beekeepers' future plans. The economic status of the beekeepers can indicate both the impact of the Africanized hybrid and the ability of the beekeepers to respond to it.

Many observers feel the hybrid drives hobbyists and part-timers out of beekeeping and that in a comparative sense, commercial operators are favored because they can adapt to the hybrid bees more readily. So, the proportion of beekeepers in each group may reflect on the impact of the bee on beekeepers in the Africanized region, as well as on the ability of beekeepers in the non-Africanized areas to adapt successfully to the bee in the future. Beekeepers' plans for future expansion or reduction of their apiaries reveal something of their perceptions of the possibilities for successful and expanded operations in the face of the possible problems presented by the Africanized bee.

Beekeeping is primarily a hobby for most Peruvian beekeepers, as it seems to be in other Latin American countries, the United States, and Canada. Nearly three-quarters of all beekeepers consider themselves hobbyists, and only about one-quarter report beekeeping as an economic activity. In this latter group over three-quarters reported their income from beekeeping to be less than 25%. A few individuals categorize their beekeeping as both a hobby and an economic activity and a small number of institutional owners exist (TABLE 4).

Sharp contrasts emerge in beekeepers' economic status between Africanized and non-Africanized regions. In the area free of the hybrid, 95% of the beekeepers are hobbyists, while in the Africanized area hobbyists account for only 47% of all beekeepers (TABLE 4). This difference is highly significant statistically. It is probable that hobbyists in that region have abandoned beekeeping in large numbers in response to the problems presented by the presence of the Africanized bee. This experience would be consistent to that reported in Brazil after the arrival of the hybrid there (Cornejo, 1971; Gonçalves, 1978).

Peruvian beekeepers are optimists. Over three-quarters of those interviewed planned to increase the number of hives they owned in the coming year. Moderate numbers planned to maintain the same number of hives (15%) or were uncertain about the future (7%). Less than 1% wanted to decrease the size of their holding. Beekeepers' plans for expanding or reducing the size of their apiaries exhibit few differences between the two regions. In fact, the only notable difference is that in the Africanized region a small number of beekeepers (2%) plan to decrease their hive numbers, while none have similar plans in the region free of the hybrid. It is doubtful that this represents any important divergence between the two regions and the difference is not statistically significant (Chi-square = 1.56 with 3 dif., p = 0.67).

Beekeeping Experience and Education

Beekeeping experience and education help indicate the level of sophistication beekeepers bring to their task and their ability to solve their problems or identify those who can. The variables evaluated here focus primarily on practical experience, informal learning, and formal educational

TABLE 4. Economic status of beekeepers.

Economic Status	Africanized (%)	Non-Africanized (%)	Total (%)
Hobby	47.3	93.9	69.3
Business	47.3	4.3	26.9
Business and Hobby	3.6	0.0	1.9
Institutional	1.8	1.8	1.9
	(n = 55)	(n = 49)	(n = 104)

Chi-square = 26.54 with 3 d.f.; highly significant; p = 0.01

activity related to beekeeping. These include the number of years of keeping bees, the number of other practitioners known to the beekeeper, coursework or training, and ownership of publications about beekeeping.

With regard to practical beekeeping work, most Peruvian beekeepers have significant experience. Three-fourths have worked with honey bees for four years or more. However, the length of experience varies between the two regions. Beekeepers in the Africanized bee region have more experience, an average of 16 years, than their colleagues in the region free of the hybrid, 12 years. But the differences are not statistically significant ($t = 1.43$, $p = 0.155$). However, if the data are grouped into ranks, an interesting contrast appears. Thirty-two percent of the beekeepers in the Africanized bee region are recent entrants to beekeeping (one to three years), but only 22% are in the region free of the hybrid. This is noteworthy because it is hardly a trend which would be expected given the aggressive characteristics of the bee. Beekeepers in the Africanized region seem more committed to beekeeping and nearly 20% have kept bees for over 31 years, and this contrasts to less than 2% in the region free of the hybrid. It may be that superior ecological conditions for beekeeping in the Africanized region and the potential for excellent yields encourage beekeepers to continue to keep bees despite the management problems presented by the hybrid bee.

Contact with other beekeepers represents a useful informal mechanism for beekeepers to learn from each other's experience, to obtain equipment, or to share its use. At the national level, over 85% of all beekeepers have knowledge or contact with at least one other practitioner. Comparing the two regions results in notable regional variation. Beekeepers from Africanized bee areas are acquainted with more fellow practitioners than respondents in uninfested areas, an average of 1.8 and 1.2 respectively. Statistically the differences are highly significant, and indicate that a better network for sharing experience and knowledge exists among beekeepers in the Africanized bee region ($t = 2.40$, $p = .017$). This has undoubtedly helped them a great deal with the hybrid bee, but probably grew out of an earlier commercialization of beekeeping in this region and not from the hybrid's presence.

Books, pamphlets, manuals, and other printed materials about apiculture represent an auxiliary source of information for the improvement of beekeepers' knowledge and management skills. At the national level, slightly more than one-half of all beekeepers have access to such materials, but this varies somewhat regionally. Beekeepers from the Africanized bee region are more likely to own books, manuals, or pamphlets on beekeeping than their counterparts in the area free of the hybrid, 67% against 53% respectively. And, they are likely to own more publications per person (2.5 to 2.1) but, the differences are not significant statistically ($t = 0.29$, $p = 0.772$). Again, it appears that practitioners in the Africanized bee areas are better informed about beekeeping and thus, in a better position to accommodate themselves to the negative attributes of the hybrid bee.

Few of those interviewed, 15%, had any kind of formal course or training in beekeeping. About one-half of this group received university training, whereas the remainder is divided between those trained in high school, by extension personnel, or through correspondence courses. Some form of training is more common among those who work with the Africanized bees than those who do not, 24% and 10% respectively. However, this does appear to be consistent with the pattern of other variables where beekeepers in the Africanized bee region characteristically are more knowledgeable and better trained than their colleagues in areas free of the hybrid. However, the differences between the groups are not statistically significant (Chi-square = 2.4 with 1 d.f., $p = 0.12$).

Beekeepers' Problems

Numerous problems beset Peruvian beekeepers, and over 90% of all respondents identified one or more as being significant. The nature of their problems varies, but they can be classified as primary and secondary. Primary problems are those cited by more than 10% of all the respondents. These, in order of their importance, are the lack of technical assistance, unfavorable weather, insecticide use, poor honey flora, and inadequate sources of beekeeping supplies and equipment (TABLE 5). Secondary problems, reported by less than 10% of the respondents are pests and parasites, lack of credit, Africanized bees, overcrowding of apiaries, disease, and aggressiveness. A myriad of other specialized and idiosyncratic problems are cited by slightly over one-third of the beekeepers.

The number of beekeepers with problems and the number of problems they experience vary little between the Africanized and non-Africanized bee regions.

TABLE 5. Beekeepers most frequently reported problems*

Problems	Africanized Bee Region	Non-Africanized Bee Region	Total
Percent Reporting Problems	78.9 (n = 57)	83.6 (n = 55)	81.3 (n = 112)
Average Number of Problems Reported per Beekeeper	1.8	1.4	1.6
Number of Problems Cited	80	66	146
Primary Problems	(%)	(%)	(%)
Lack of Technical Assistance	24.4	32.6	28.6
Weather	20.0	17.4	18.7
Insecticide Use	8.9	21.7	15.4
Poor Honey Flora	6.7	19.6	13.2
Inadequate Supplies Including Wax Foundation	13.3	8.7	11.0
Secondary Problems			
Pests and/or Parasites	8.9	10.9	9.9
Lack of Credit	15.6	0.0	7.7
Africanized Bees	15.6	0.0	7.7
Overcrowding of Apiaries	8.9	0.0	7.7
Disease	4.4	2.2	3.3
Aggressiveness	4.4	2.2	3.3
Other Problems	46.4	26.1	37.4

* Percentages and totals in the table are calculated on the number of respondents in each category. Since multiple responses are possible column totals add to more than 100 %

Unexpectedly, slightly fewer individuals in the Africanized bee region report problems than those in areas free of the hybrid, 79 and 84% respectively. However, they do cite more problems per capita than their colleagues, an average of 1.8 to 1.4 (TABLE 5).

Differences are evident in the kinds of problems experienced by beekeepers in the two regions. Beekeepers in the region free of the Africanized bee have considerably more complaints about lack of technical assistance, the excessive use of insecticides, and poor honey flora. But, roughly the same number in each region have trouble with bad weather and inadequate sources of beekeeping supplies. On the other hand, beekeepers in the Africanized bee region voice more dissatisfaction with regard to secondary problems. Between 10% and 15% of the beekeepers in the region infested with the Africanized bee have complaints about the lack of credit, the Africanized bee, and overcrowding of apiaries. None of their counterparts in unaffected areas cited any of these problems. Differences reported in the other subcategories are generally inconsequential.

The differences in the problems beekeepers recognize in the two regions seem to reflect variations in the environment, local agricultural conditions, and the commercial status of beekeeping. In the areas free of the Africanized bee, where large commercial agriculture is more common and natural vegetation sparse, beekeepers cite problems with insecticides and poor honey flora. Beekeepers in that region also have less technical expertise and hence frequently cite their need of technical assistance. In the Africanized bee region, which is more advanced in an apicultural sense, beekeepers feel inadequate credit and the overcrowding of apiaries, as well as the presence of the hybrid bee to be their principal problems.

Changes in Bee Behavior

About 40% of all beekeepers interviewed in the study felt some behavioral changes had taken place in their colonies during the last two to three years. Aggressiveness, swarming, and other miscellaneous changes were the most frequently reported. By almost a two to one margin more individuals noted such changes in Africanized bee region than in the region free of the hybrid (TABLE 6). Increased aggressiveness was cited by over 85% of the respondents in the Africanized bee areas, but only by 44% in areas not infested by the hybrid bee. Increased swarming was experienced by equal numbers in both regions, about one-quarter. Fully one-half of those from the hybrid free region noted other miscellaneous behavioral changes, whereas only 11% responded similarly in the Africanized bee regions (TABLE 6). Both increased and decreased honey production were cited by small percentages of the respondents in each region, and less aggressiveness and quicker flight activity were cited by similar numbers only in the Africanized bee areas. The increased perception of aggressiveness by beekeepers in the Africanized bee areas is consistent with the bee's vigorous nest

TABLE 6. Changes in honey bee behavior

Behavior Change	Africanized Region (%)	Non-Africanized Region (%)	Total (%)
Reporting Change	49.1 (n = 57)	29.1 (n = 55)	39.3 (n = 112)
More Aggressive	85.7	43.8	70.5
More Swarming	25.0	25.0	25.0
More Productive	3.6	12.5	6.8
Less Productive	7.2	6.3	6.8
Less Aggressive	3.6	0.0	2.3
Fly Faster	3.6	0.0	2.3
Other	10.7	50.0	25.0

* Percentages and totals in the table are calculated on the number of respondents in each category. Since multiple responses are possible column totals add to more than 100 %

defense. The fact that the same percentage in both regions noted increased swarming is rather puzzling, since the hybrid purportedly swarms excessively.

Management Adjustments

Most respondents in the Africanized bee region had heard of the presence of the Africanized or "killer" bees in Peru. Yet, only about one-half had noticed changes in their bees' behavior, and surprisingly only one-quarter acknowledged the presence of the Africanized bee in their apiaries. Many individuals thought the more aggressive behavior of their bees had been caused by their lack of attention to their colonies. That is, they believed the bees did not recognize them or no longer were used to humans, and that by working the colonies more regularly their aggressiveness would diminish. But this is not true.

The beekeepers who acknowledged the hybrid's presence in their apiaries had made some adjustments in their management practices. The number of management changes made by individuals averaged more than two and included a variety of responses (TABLE 7). Most popular were the use of protective

TABLE 7. Management changes reported by beekeepers in response to the Africanized bee*.

Management Changes	Percent Reporting	Number Reporting Change
Use More Clothes	35.7	5
Requeening	35.7	5
Use More Smoke	28.6	4
Burn Out Aggressive Colonies	28.6	4
Wear Gloves	21.4	3
Change Apiary Locations	21.4	3
Other	21.4	3
Make Less Noise	14.3	2
Work Bees at Specific Times	7.1	1

* Percentages and totals in the table are calculated on the number of respondents in each category. Since multiple responses are possible column totals add to more than 100%.

clothing and the requeening of aggressive colonies. These were followed closely by the use of more smoke and the destruction of aggressive colonies and swarms by burning them out. Less frequent changes included the use of gloves, changing apiary locations, making less noise, and working bees at special times of the day (TABLE 7). It is indicative of the hybrid that all these management adjustments relate to its aggressiveness. These are similar to those made in Brazil and other countries previously affected by the hybrid. A surprising fact is that requeening, a technically advanced and expensive practice, is employed by such a large proportion of the Peruvian respondents. Also notable is the limited number of beekeepers in Peru who relocate their apiaries as a result of the hybrid's presence.

Respondents also reported (multiple responses per individual) on the advantages and disadvantages of working with the hybrid bee. Concerning the advantages of the hybrid, there was almost complete agreement; they work harder and produce more (66.7% in both cases). Some suggest they are more resistant to disease and pest problems, too (20.0%). The hybrid's greatest single disadvantage, according to the respondents, is its aggressiveness, and nearly 80% cite this as a problem. Swarming follows as a distant second (20%), with the production of excessive propolis (13.3%), lower honey production (13.3%), and problems with neighbors (6.7%) following, as secondary problems.

CONCLUSIONS

Beekeeping practice and management techniques in Peru demonstrate consistent differences between the region affected by the Africanized honey bee and the region free of the hybrid. Beekeeping practice in the Africanized region is characterized by the more frequent ownership and use of modern equipment--smokers, masks, gloves, standard movable frame hives, hive supers, and centrifuge extractors, whereas beekeepers in the region free of the hybrid typically use simpler and less sophisticated and efficient equipment. Management techniques and goals also demonstrate more sophistication among those beekeepers in the region affected by the Africanized bee. Typically, they have had more experience keeping bees, their per-capita honey yields are greater, they are better informed about modern beekeeping practices, they have more contacts with other beekeepers, and they are more commercially oriented than their colleagues in the region free of the hybrid honey bee.

Surprisingly, only about half of the beekeepers in the Africanized region reported noticing any change in the behavior of their bees over the five year period before the interviews--the same time period during which many of these beekeeper's apiaries were infested with the Africanized honey bee. Of these beekeepers, the vast majority did report increased aggressiveness as the most significant change, a finding consistent with experience in other countries affected by the Africanized hybrid. This study does demonstrate stinging incidents involving humans and livestock are slightly more common in the Africanized region, but statistically these differences are not significant.

Beekeepers report making a number of management changes in order to adapt to the greater aggressiveness demonstrated by the Africanized bee, including using more smoke, wearing more protective clothing, requeening, and moving apiaries to more remote locations. Some of these changes are confirmed in the comparative analysis of management practices between the two regions, yet others are not. The siting of apiaries is a case in point. In the region free of the hybrid, apiaries are located about 13 m from human dwellings, whereas in the Africanized bee region they are located 16 m from human dwellings. The difference between in these distances is not significant statistically or in practical terms.

Beekeeper's responses to questions concerning their problems suggest that the Africanized bee, while bothersome, is not a major problem. Among beekeepers in the Africanized region, a lack of technical assistance and poor weather rank ahead of the Africanized bee as their most significant problems. Clearly, beekeepers in this region do not see the hybrid as their principal problem.

The picture that emerges from this analysis of the Africanized honey bee in Peru shows that beekeepers do not view the hybrid as a major problem. Many beekeepers did not report even noticing any changes in their colony's behavior over the last five years, the period during which many apiaries were infested. Furthermore, apiaries in the Africanized region are located fairly close to human dwellings and animal pens just as they are in the region free of the hybrid. This does not conform with the hybrid's reputation as unduly aggressive nor with expert advice to beekeepers to locate hives of Africanized bees far from humans and animals. In Venezuela and in other countries, for example, beekeepers have been advised to locate colonies 200 to 300 m from people or livestock (Moffett and Maki, 1988). Management practices and equipment in the Africanized bee region are considerably more sophisticated and advanced than those in the region free of the hybrid, and in almost all cases provide beekeepers in the Africanized bee region with the capability of dealing successfully with the negative characteristics of the hybrid bee.

The question arises as to the extent that this increased sophistication represents a response to the hybrid's negative characteristics, the earlier development of technical and managerial sophistication in the region now occupied by the Africanized bee, or the abandonment of beekeeping by part-time and hobby beekeepers who are typically less sophisticated in their equipment and management techniques. A lack of historical data on beekeeping in Peru makes this question difficult to answer. It is possible that the greater technical and managerial sophistication in the Africanized area, or at least parts of it, are the result of settlement by Italian and German colonist during the late 1800s and early 1900s. These colonists brought honey bees, beekeeping equipment, and apicultural knowledge with them from Europe, and indeed some of the most successful commercial beekeepers today are their descendants. Another possible explanation is that the region occupied by the Africanized bee has some of the best environmental conditions for beekeeping, and consequently this has simply been the most attractive area for commercially-oriented beekeepers. Finally, it is also possible that once the Africanized bee arrived in that region, hobby beekeepers abandoned beekeeping in large numbers, leaving only the larger commercially oriented beekeepers in operation. The technical and managerial sophistication exhibited in the region probably represents a combination of all of these factors (Kent, 1986).

As beekeepers, the beekeeping industry, and the public in general prepare for the arrival of the Africanized bee in Mexico and the southern United States, the experience with the Africanized bee in Peru can provide some insight for policymakers. The Peruvian case clearly demonstrates that although the Africanized bee maybe problematic, for beekeepers with sound technical and managerial knowledge of beekeeping, the presence of the Africanized bee is not debilitating to the successful practice of beekeeping. It is probable that the more sophisticated beekeepers typical of Mexico and United States will be able to adapt to the arrival of the hybrid without serious problems. This conclusion is similar to that drawn by Camazine and Morse (1988) who suggest that in the United States the negative impacts of the Africanized bee when it arrives will be relatively minor. These conclusions are supported by Taylor (1985) who notes that managing the invasion of the Africanized honey bee in the United States is clearly possible. The most severe effects according to his analysis will probably be on the queen-rearing and package bee industries concentrated in the southern states and on the practice of migratory beekeeping. It is important to note that in the Peruvian case, beekeepers in the region infested with the Africanized bee adapted to the presence of the bee without any kind of governmental or private assistance. While this might not be the most desirable state of affairs, it seems that with limited extension assistance beekeepers could adapt to the arrival of the hybrid more easily. This suggests that instead of devoting scare funds to the development of a barrier to block or retard the diffusion of the hybrid through the Isthmus of Tehuantepec (Tew *et al.*, 1988), the United States and Mexican governments should dedicate these funds to extension education for beekeepers to help them and the public prepare for the inevitable arrival of the hybrid in central and northern Mexico and the southern United States.

Acknowledgements

Numerous Peruvian beekeepers took time from their labors and their leisure to respond to the author's questionnaire and inquires. Their help is most gratefully acknowledged. The Dellplain Program in Latin American Geography at Syracuse University, directed by Dr. David J. Robinson, assisted in providing logistical support for this research in several of the case study sites in Peru.

LITERATURE CITED

Alata Condor, J. 1976. Las abejas africanas y/o sus hibridos y su posible presencia en el Valle de Chanchamayo. *Informe Especial No. 3. Ministerio de Alimentacion. Direccion General de Investigacion.* Lima, Peru.

Anonymous. 1972. *Final Report of the Committee on the African Honey bee.* Washington, D.C.: Natl. Res. Counc. Natl. Acad. Sci. 95 pp.

Calkins, C. O. 1983. Research on exotic insects. In *Exotic Plant Pests and North American Agriculture*, ed. C. L. Wilson, C. L. Graham, pp. 321-359. New York: Academic Press.

Camazine, S., Morse, R. A. 1988 The Africanized honey bee. *Am. Sci.* 76:465-471.

Cornejo, L. G. 1971. Management technology of the African bee. *Am. Bee J.* 111:262-263.

Davila, M., Ortiz, M. S., Huiza, I. R. de 1980. Presencia de la abeja africanizada en el Peru. *Rev. Peruana de Entomol.* 23:125-127.

Fierro, M. M., Munoz, M. J., Lopez, A., Sumuano, X., Salcedo, H., Roblero , G. 1988. Detection and control of the Africanized bee in coastal Chiapas, Mexico. *Am. Bee J.* 128:272-275.

Gonçalves, L. S. 1978. Impacto causado por las abejas africanizadas en la America del Sur. *Gaceta del Colmenar* 40:474-483.

Harper, J. D. 1982. Africanized bee has spread to Ecuador. *Speedy Bee* 11:7-8.

Kempff Mercado, N. 1973. The African bees: Contribution to their knowledge. *Apiacta* 8:121-126.

Kent, R. B. 1986. Beekeeping regions, technical assistance, and development policy in Peru. *Yearbook, Conference of Latin Americanist Geographers* 12:22-33

Kent, R. B. 1988. The introduction and diffusion of the African honey bee in South America. *Assoc. of Pacific Coast Geographers, Yearbook* 50:21-43.

Lattin, J. D., Oman, P. 1983. Where are the exotic insect threats?. In *Exotic Plant Pests and North American Agriculture,* ed. C. L. Wilson, C. L. Graham, pp. 93-137. New York: Academic Press.

McDowell, R. 1984. The Africanized Honey Bee in the United States: What Will Happen to the U.S. Beekeeping Industry? *Agric. Econ. Rep. No. 519.* Washington, D.C.: Government Printing Office.

Moffett, J. O., Maki, D. L. 1988. Venezuela and the Africanized bee. *Am. Bee J.* 128:827-830.

Moffett, J. O., Maki, D., Andre, T., Flierro, M. M. 1987. The Africanized honey bee in Chiapas, Mexico. *Am. Bee J.* 127:517-519, 525.

Nunamaker, R. A. 1979. Newspaper accounts of Africanized bees are designed to frighten people - Being stung by the press. *Am. Bee J.* 119:587-588, 591-592,646-647,657.

Presada, W. A. 1981. Rebirth of Brazil's bee industry. *Am. Bee J.* 121:630.

Rinderer, T. E., Wright, J.E., Shimanuki, H., Parker, F., Erickson, E., Wilson, W. T. 1987. The proposed honey bee regulated zone in Mexico. *Am. Bee J.* 127:160-164.

Roughgarden, J. 1986. Predicting invasions and rates of spread. In *Ecology of Biological Invasions of North America and Hawaii,* ed. H. A. Mooney, J. A. Drake, pp.179-188 New York: Springer-Verlag.

Seeley, T. D. 1985. *Honey bee Ecology: A Study of Adaptation in Social Life.* Princeton: Princeton University Press.

Simberloff, D. 1986 Introduced insects: A biogeographic and systematic perspective. In *Ecology of Biological Invasions of North America and Hawaii,* ed. H. A. Mooney, J. A. Drake, pp. 1-26. New York: Springer-Verlag.

Sommer, P. G. 1979. Comportamento das abelhas africanizadas no Parana diante das observacoes feitas no periodo de 1960-1970. In *Apiculture in Hot Climates,* pp. 77-79. Bucharest: Apimondia.

Stearman, A. M. 1981. Working the Africans in eastern Bolivia. *Am. Bee J.* 121:28-35,43-44.

Stort, A. C., Gonçalves, L. S. 1979. A abelha africanizada e a situaçao atual da apicultura no Brasil. *Ciencia e Cultura* 31:32-43.

Taylor, O. R. 1977. The past and possible future spread of Africanized honey bees in the Americas. *Bee World* 58:19-30.

Taylor, O. R. 1985. African bees: Potential impact in the United States. *Bull. Entomol. Soc. Am.* 31:15-24.
Taylor, O. R. Williamson, G. B. 1975. Current status of the Africanized honey bee in northern South America. *Am. Bee J.* 115:92-92,98.
Tew, J. E., Bare, C. H., Villa, J. D. 1988. The honey bee regulated zone in Mexico. *Am. Bee J.* 128:673-675.
Villa, J. D. 1987. Africanized and European colony conditions at different elevations in Colombia. *Am. Bee J.* 127:53-57.
Wiese, H. 1977. Apiculture with Africanized bees in Brazil. *Am. Bee J.* 117:166-168,170.

20

BEEKEEPING IN VENEZUELA

Richard L. Hellmich II and Thomas E. Rinderer[1]

Stingless bees, *Melipona* and *Trigona* spp., were the only source of honey and wax in Venezuela until European honey bees, *Apis mellifera* L., were introduced, perhaps as early as the sixteenth century. The Iberian honey bee native to Spain, *A. m. iberica* (Goetze), was the first subspecies imported, followed by the black European, *A. m. mellifera* L., Italian, *A. m. ligustica* (Spinola), Carniolan, *A. m. carnica* (Pollmann), and Caucasian, *A. m. caucasica* (Gorbatschev) subspecies. For many years beekeeping was practiced primarily by hobbyists. Then, early in the 1960s, these European-evolved honey bees, primarily the Italians, formed the basis of a honey producing industry (M. Calvo-Díaz, pers. comm.). This industry peaked in 1976 at which time there were nine major commercial beekeepers. These commercial beekeepers and most of the other beekeepers were located in an agricultural belt that extends from Caracas southwest to San Cristobal (FIGURE 1). In total they had approximately 50,000 colonies. Although many of these bees were kept in primitive hives, they still produced 530 metric tons of honey; 323 tons were exported to Germany, England and the United States (R. Gómez-Rodríguez, pers. comm.).

Africanized bees entered Venezuela from Brazil and Guyana in 1975 as they colonized forest areas in the Amazon Territory and the State of Bolivar (Gómez-Rodríguez, 1986). They were first detected near Santa Elena de Uairén in April of 1976 (Taylor and Levin, 1978). In the next five years they advanced across the country at approximately 320-400 km/yr and were found in the beekeeping region of northeastern Colombia by 1981. During this time Venezuelan honey production decreased drastically. Average colony yields of several commercial beekeepers decreased from 75-125 kg to 25-30 kg or even less. In 1981, Venezuelans produced only 78 metric tons of honey and no longer had surplus

---

[1]Drs. Hellmich and Rinderer are with the USDA--ARS, Honey Bee Breeding, Genetics and Physiology Research Laboratory, 1157 Ben Hur Rd., Baton Rouge, Louisiana, 70820, USA.

FIGURE 1. Map of Venezuela.

honey to export (R. Gómez-Rodríguez, pers. comm.). This near collapse of the Venezuelan bee industry was due solely to the Africanized bee.

Many traits of the Africanized bee make them undesirable for beekeeping. Heading the list is their unpredictable and sometimes excessive defensive behavior (Collins *et al.*, 1982). Approximately 400 Venezuelans have been killed by excessive stinging from the bees. Most of these deaths, however, occurred during the first few years after the bees arrived, and have decreased considerably since then. For example, nearly 100 deaths were reported in 1978, compared with 12 in 1988 (R. Gómez-Rodríguez, pers. comm.). Consequently, such behavior requires that beekeepers isolate apiaries and change many of their management procedures. Additionally, this bee's tendency to abscond, swarm, and invade other colonies adds to the problems they cause the beekeeper. Necessary management changes have increased production costs and apiary sites, since they must be isolated, are more difficult to find and are more vulnerable to vandalism and theft.

Despite these problems, only two of the original nine commercial beekeepers went out of business (M. Calvo Díaz, pers. comm.; W. I. Vogel, pers. comm.). Because of increased management difficulties, though, these beekeepers have had to reduce their operations by 50-75%. Prior to Africanization a few beekeepers managed 1,000 colonies or more; but in 1981 no beekeeper managed over 500 colonies. When the beekeepers had European bees, they often placed 50 or more colonies in an apiary and harvested respectable amounts of honey from each colony (R. Gómez-Rodríguez, pers. comm.). Now increased competition from feral Africanized colonies appears to have reduced nectar availability. This, plus problems with defensive colonies, requires that beekeepers place no more than 30 colonies in an apiary.

Africanization has had an even more serious impact on part-time beekeepers, as 90% or more were forced to quit (W. I. Vogel, pers. comm.). People with rustic equipment in particular have found the bees unmanageable, and most backyard beekeepers, disillusioned by stinging instances, have realized that keeping bees is no longer an enjoyable hobby.

Interestingly, many people who abandoned beekeeping, plus many others, have reverted to the ancient methods of the honey hunters. They locate feral colonies during the day and harvest honey during the night. Increased numbers of feral colonies have made this a common activity.

Both honey hunters and beekeepers are motivated to produce honey by high honey prices.[2] Government import restrictions on foreign honey have helped to maintain high domestic prices for honey. Most beekeepers agree that without such support they would be forced to quit.

Nevertheless, Venezuelan beekeeping has improved somewhat since its worst year, 1981. Although annual colony yields are considerably less than they were prior to Africanization, they have begun to increase again. Presently a commercial beekeeper usually can average 45-50 kg from a colony each year. National honey production also has increased. In 1985, 480 metric tons of honey was produced (R. Gómez-Rodríguez, pers. comm.). Such improvements suggest that at least some of the beekeepers have learned to work in an Africanized environment.

VENEZUELA'S MAJOR BEEKEEPING REGION

Most of Venezuela's agriculture occurs in the dry tropical forest life zone; this region covers 38% of the country and accounts for about 85% of the national honey production (Gómez-Rodríguez, 1986). The vegetation of this zone includes virgin forest, secondary forest and savanna grassland. Extensive plains and some of the foothills of the Andes Mountains make up the major land formations; altitude ranges from sea level to between 400 to 1000 m. The region is characterized by a dry season that lasts from four to six months during which the major nectar flows occur. Typically, this season starts toward the end of October and lasts through April, but the length varies regionally and yearly.

---

[2]As of August 1989, a kg of honey is selling for 120 Bolivares (Bs) retail and Bs 60 wholesale. Exchange is approximately $1 = Bs 35.

The rainy season, a dearth period for bees, occurs during the remaining part of the year.

The first major nectar flow occurs in the savanna and some of the secondary forest areas, typically from the first part of November through December. Nectar sources are primarily annual species such as: tara, *Oyedea verbesinoides* DC; cruceta, *Thevetia peruviana* (Pers.) Schum.; mastranto, *Hyptis sauveolens* (L.) Poit.; and bejuquillo, *Ipomoea nil* (Roth.). Another significant nectar flow occurs in all three areas of vegetation, usually from mid-to-late February though March. The important nectar sources during this flow are primarily perennial species such as: chapparo, *Byrsonima crassifolia* (L.) HBK.; matarratón, *Gliricidia sepium* (Jacq.) Steud.; araguanéy, *Tabebuia chrysantha* (Jacq.) Nich.; apamate, *Tabebuia pentaphylla* (L.) Hemsl.; and bucare, *Erythrina glauca* (Willd.). Beekeepers in the states of Portuguesa, Barinas, and Cojedes produce honey from both major nectar flows, whereas beekeepers in the states of Carabobo, Aragua, and parts of Miranda, produce honey only during the second nectar flow. For more information concerning Venezuelan nectar sources consult: Thiman and Aymard (1982), Gómez-Rodríguez (1986), and López-Palacios (1986).

SEASONAL MANAGEMENT

Dry Season

Before the first nectar flow, each colony is inspected for population strength (September or October). Strong colonies are not inspected further, but weak colonies are checked for brood quality and the presence of a queen. If requeening is necessary the beekeepers either give the colony a new queen or allow the colony to rear a queen from young brood taken from another colony. Generally all colonies are fed sugar either as syrup (50% vol) or in a fondant form that is four parts sugar mixed with one part honey. Pollen or pollen supplement, if available, is added to this mixture (~5% vol). Such feeding ensures proper colony development in the absence of reliable pollen and nectar sources during this time of year. The beekeeper's objective is to have hives with approximately 3-4 kg of bees before the nectar flow starts.

Most beekeepers place only one empty super on a colony at a time. Africanized bees, when given two or more empty supers at once, have a tendency to scatter honey stores and often swarm before enough honey is capped for harvesting. Restricted comb space forces the bees to consolidate their honey, and requires that beekeepers visit apiaries every 10-14 days during strong nectar flows. During these visits beekeepers replace a honey-filled super with an empty one and in some colonies replace honey-filled combs with empty combs. Prior to Africanization, such intensive management was not required because swarming was not a serious problem. During this time, beekeepers generally

FIGURE 2. Beekeepers removing honey supers.

gave their colonies two or three empty supers, then harvested honey every three or four weeks. Consequently, beekeeping with Africanized bees compared to beekeeping with European bees requires two or even three times more visits to the apiary during strong nectar flows.

When harvesting honey many Venezuelan beekeepers remove Africanized bees from honey supers by smoking the supers with large amounts of smoke, then by brushing or shaking out any remaining bees. A major disadvantage of this method, especially when bees are agitated, is that as many bees often enter the supers as are shaken out. Thus, beekeepers prefer to use bee blowers when they have access to them (FIGURE 2). With either of these methods, though, most beekeepers put honey supers into a closed vehicle. This lessens the threat of a robbing frenzy and also reduces stinging occurrences en route to the honey house. An increasing number of beekeepers remove honey at night because it is an effective way to reduce stinging encounters. Fume boards, which also tend to reduce stinging problems, are being used by a growing number of beekeepers (W. I. Vogel, pers. comm.).

Capturing swarms in bait hives (particularly in the early dry season) is a procedure that is being used by many beekeepers to increase colony numbers. Africanized colonies, at least in the tropics, produce more swarms than European colonies (Otis, 1982). This, therefore, explains the observed increase in the incidence of swarming in Venezuela since Africanization. In one case, over a hundred swarms were caught in bait hives from October to January in a ten hectare mango orchard near Acarigua (unpubl. data). When such swarms are caught early in the dry season, they often produce surplus honey. Capturing swarms not only increases the beekeepers' colony numbers, but also reduces competition for nectar sources by decreasing feral colony populations. Beekeepers usually do not requeen such colonies. However, this eventually could cause problems as the beekeepers unintentionally select for bees that are more apt to swarm.

Rainy Season

Honey bee colonies are not as active during the rainy season because, in most areas, there are no major nectar flows. Colony populations decrease to one-half to one kg levels during July and August and do not build up appreciably until October. Often apiaries are relocated before they are threatened by flooding or before access roads are made impassable by heavy rains. High honey prices motivate many beekeepers to harvest nearly all the honey from their colonies at the end of the dry season. Thus, colonies must be fed on a routine schedule throughout the rainy season until nectar is once again available. Feeding colonies, building and repairing equipment, and clearing weeds at the bee sites are normal beekeeper activities during this season.

In May, the beekeepers often divide strong colonies in an effort to increase colony numbers and to reduce absconding. Africanized colonies, particularly larger colonies with depleted honey stores, have a tendency to abscond. In September some beekeepers again divide strong colonies as a means of increasing colony numbers, but also to reduce swarming.

Prior to Africanization many of the commercial beekeepers moved their colonies during the rainy season to dry areas near the coast. Colonies foraged enough to maintain themselves, and honey was occasionally harvested. Since Africanization, migratory beekeeping has been practiced by fewer beekeepers (M. Calvo-Díaz, pers. comm.). The defensive and absconding behavior of Africanized bees has discouraged beekeepers from making such moves. Additionally, increased competition by feral Africanized colonies for nectar sources appears to have reduced nectar availability, and thus the value of such areas to beekeepers.

RECOMMENDATIONS FOR BEEKEEPERS IN AFRICANIZED AREAS

Beekeepers in Venezuela, like many beekeepers around the world, seldom agree on beekeeping methods. Nevertheless, the following are common-sense recommendations that the authors feel represent the opinions of the majority of

Venezuelan beekeepers. These recommendations are based on talks with several Venezuelan beekeepers and the experience of members of the USDA-ARS laboratory in Baton Rouge, Louisiana, some of whom have been working with the Africanized honey bee in Venezuela since 1979.

Apiary Locations

Apiaries with Africanized bees should be located no closer than 200 m, preferably 300 m, from people or livestock. Bees commonly follow a walking person 300 m or more and can be transported several kilometers inside vehicles. Consequently, access roads also require some isolation and should not pass near houses or areas where livestock are confined. If beekeepers anticipate driving past such areas while leaving an apiary, they should stop one or more times to brush and smoke bees out of their vehicle. These procedures, plus simply driving faster than the bees can fly, around 20-25 km/hr, reduces or even eliminates stinging encounters. Stinging encounters also appear to be reduced when dense and relatively high vegetation separates the apiary or access road from potential problem areas.

Beekeeper Safety

Safety should not be compromised in an Africanized apiary. The most important recommendations are to wear a reliable bee suit, never work Africanized colonies alone, carry an adequate smoker, and, if possible, bring an emergency sting kit. Additionally, beekeepers should always be prepared to reassemble open colonies and leave the apiary if the defensive response of the bees at any time becomes unmanageable.

A reliable bee-tight suit is the beekeeper's best defense against dangerous stinging encounters. A suit includes overalls with some type of leggings, gloves, helmet and veil. Many of the commercially available suits are adequate. However, stings through overalls and gloves are common, even with the most reliable suits. Some beekeepers reduce such stings by padding the shoulder area and by wearing sting-proof leather gloves. A few beekeepers wear two pairs of overalls when they expect the bees to be excessively defensive. A nylon suit is a common choice for the second pair because its slickness reduces the bee's ability to hold to the surface and then sting. This two-suit combination, though, can be uncomfortably hot. Under such conditions drinking water should always be brought along in order to lessen the risk of dehydration and heat exhaustion.

Most colonies in Venezuela are placed on hive stands to protect them from predatory ants. Most beekeepers recommend that colonies should be placed on individual stands; and that these stands should be spaced five m or more apart. Such an arrangement tends to reduce stinging incidents. As if Africanized bees were not enough to worry about, Venezuelan beekeepers also must be wary of poisonous snakes, primarily *Bothrops* spp.

Colony Inspections and Requeening

Generally, two people are required to inspect a colony because smoking Africanized bees is often a full-time job. When beekeepers are prepared to enter an apiary, that is, when the suits are bee tight and the smokers are lit, they should avoid walking near colony entrances. If possible they should approach a colony from the back or side so that the guard bees are not alerted. When ready to work a colony, the entrance should first be smoked lightly (three or four puffs), then a similar amount of smoke should be applied under the inner cover. Then it is necessary to wait 30-60 seconds. During this time beekeepers often smoke entrances of colonies that are within five meters.

When working an Africanized colony, beekeepers have learned that it is necessary to move supers and frames deliberately in order to avoid jerky movements and crushing bees. These procedures, plus directing a constant flow of smoke over the area that is being worked help to keep the colonies manageable. Beekeepers also try to keep the part of the colony not being worked covered in order to curtail defensive behavior and robbing. Despite all these precautions, inspection procedures are often difficult because Africanized bees tend to fly or run off frames and often form large festoons.

Festooning and the tendency to run make queen finding difficult, as queens often move to the bottom board or one of the side boards. On the other hand, some Venezuelan beekeepers actually take advantage of a queen's aptness to run when disturbed to find her. They smoke the entrance of the colony (10-15 puffs), then, after four of five minutes, remove the inner cover. Frequently (50-75% of the time) the queen is found on or just under the inner cover. Some beekeepers assign one person to find and cage queens in several colonies with this smoking technique, and then follow that person with a two-man crew that works colonies in which queens have been found.

Introducing queens, especially European queens, into Africanized colonies can be problematic. Often new queens are not accepted or are quickly superseded. Such problems appear to be minimized when queens are introduced into colonies that have only young bees and emerging brood. This is accomplished by putting a new queen into a second hive which has most of the bees and all the brood, and then by moving this hive to a different location within the apiary. Bees of flight age return to the original hive leaving only young bees and emerging brood in the relocated colony. The colony with the Africanized queen and the older bees generally is used to produce brood for future divisions.

Maintaining European or Hybrid Bees in an Africanized Area

Obtaining pure-bred European queens is a common problem for Venezuelan beekeepers. Such queens are not produced in Venezuela and often importing queens is too expensive. Fewer than a thousand honey bee queens are produced commercially in Venezuela each year, but many beekeepers rear their own (M. Calvo-Díaz, pers. comm.). European queens from the U.S. are usually used as breeders to produce queens that are naturally mated; these queens produce mostly hybrid progeny. Hybrid progeny, in general, display defensive behavior which is intermediate to European and Africanized colonies (Collins *et al.*, 1988). Many beekeepers prefer hybrid colonies because hybrids often produce more honey than colonies that are European or colonies that are more Africanized (M. Calvo-Díaz, pers. comm.; W. I. Vogel, pers. comm.). Many Venezuelan beekeepers are trying hybrids from different strains in an attempt to find less defensive and better honey-producing bees. Hybrids produced from the Carniolan subspecies are popular among several of the beekeepers.

Success in producing mated queens is lower when Africanized bees are used to populate mating colonies than when European bees are used (Hellmich *et al.*, 1986). The efficiency of Africanized mating colonies is decreased by absconding and population dwindling. These problems are greatest in five-liter nuclei, the type of mating colony most commonly used by commercial queen producers. Efficiency of Africanized mating units is improved when bee populations and hive volume are increased and when brood is added. Thus, queen production with Africanized bees is possible if large mating colonies are used. A few beekeepers have had success mating queens in large (32 l) mating colonies.

Colony queens have to be clearly marked so that they can be distinguished from supersedure or foreign queens. A foreign queen may enter a colony accompanied by a small cluster of bees. These so called "invader swarms" appear to be more successful when they enter queenless colonies or colonies with failing queens. Little else is known about their biology except that most invading queens are mated and begin laying immediately. The best way to retain a chosen queen is to inspect the colony every two or three weeks. Such frequent inspections are not practical for large-scale beekeepers, but are essential if maintaining a certain stock is important. A more practical approach that some Venezuelan beekeepers have tried involves requeening all colonies in an apiary once a year. This approach reduces management procedures but does not completely eliminate Africanization.

Certainly the most successful beekeepers are those who requeen colonies on a regular basis with queens produced from breeding stock or, more simply, from queens produced from their best colonies. However, there still appears to be a need for increased commercial queen production as queens from the U.S. become more expensive. Quality queens, selected from colonies with favorable European and Africanized traits, could be produced by mating the queens in an area in which a high percentage of the drone population comes from colonies of desirable stock. Populations of drones can be controlled by saturating mating areas with desirable drones, decreasing feral colony populations, or both (Hellmich *et al.*, 1988). Also, mating colonies could be located in areas where feral colony densities are low. Venezuela has many diversified habitats, some of which probably support few, if any, feral honey bee colonies.

OTHER ASPECTS OF VENEZUELAN BEEKEEPING

Pollination Management

Most crops in Venezuela prior to Africanization were pollinated by native bees, primarily stingless bees. Honey bees were not used in an organized system. The impact Africanized bees have had on native bees and how much they contribute to pollination is likely to be significant (Roubik, 1978, 1979, 1980, and Chapter 13).

Information based on recent foraging studies suggests that Africanized bees pollinate crops as well as European bees (Danka, 1987). But problems develop more frequently in colonies of Africanized bees than in colonies of European bees when they are moved repeatedly to different crops. Problems arose from population losses and excessive stinging (Danka *et al.*, 1987). Such difficulties give Venezuelan farmers and beekeepers little incentive to manage Africanized bees for pollinating crops.

Since 1983, the government has encouraged farmers to use honey bees for pollinating sunflowers, *Helianthus annuus*. However, farmers are advised to use European bees to reduce stinging incidents to field workers (R. Gómez-Rodríguez, pers. comm.).

Honey Bee Enemies and Diseases

•*Ants*. Predatory ants, primarily from the subfamily Dorylinae, destroy more honey bee colonies in Venezuela than all pests and diseases combined. The inability of European bees to establish large numbers of feral colonies prior to Africanization is probably attributable, at least partly, to these ants. In a single night tens of thousands of ants can destroy a large colony. Often all they leave are wax combs, patches of capped brood and a pile of severed wings and legs. Beekeepers protect colonies from such invasions by placing them on hive stands. These stands have a barrier, usually motor oil around the legs, to prevent ants from reaching the colonies.

•*Wax moths*. Venezuelan beekeepers have problems with both the greater (*Galleria mellonella*) and the lesser (*Achroia grisella*) wax moths. The former causes the most damage. Many beekeepers avoid infestations by keeping supers on the colonies even during nectar dearths. A growing number of beekeepers store supers after they spray combs with a *Bacillus thuringiensis* solution. The problem with wax moths appears to have increased since Africanization (W. I. Vogel, pers. comm.). This may be attributed to more feral nests that have been abandoned.

•*Stingless bees.* The species *Trigona trinidadensis*, often robs honey bee colonies, especially weak colonies, during the rainy season. Another species, the fire bee, *Trigona* (*Oxytrigona*) *mellicolor* even uses chemical warfare while conducting its robbing raids, as it repels the defending honey bees with a secretion from its cephalic gland (Rinderer *et al.*, 1988)

•*Parasitic Mites.* The honey bee tracheal mite, *Acarapis woodi*, was reported to be present in Venezuela in 1957 (Gómez-Rodríguez, 1986). In a survey conducted in 1984 about one percent of colonies had low infestations of this mite which was not considered an economically important pest (Gómez-Rodríguez, 1986). Fortunately, the more serious mite, *Varroa jacobsonii*, is not found. Import restrictions on packages and queens appear to have been successful in keeping this pest out of the country.

•*Diseases.* Honey bee brood diseases are not a major problem in Venezuela. Sacbrood virus has been reported (Gómez-Rodríguez, 1986), and European foulbrood, *Streptococcus pluton*, has been observed occasionally (pers. obs.). Several diseases common in the temperate areas, such as American Foulbrood, *Bacillus larvae*, and chalkbrood, *Ascosphaera apis*, do not occur. Nosema disease, *Nosema apis*, (Stejskal, 1972), amoebal disease, *Malpighamoeba mellificae*, (Stejskal, 1966) and gregarinosis, from the larger order of Gregarinida (Stejskal, 1955), are reported adult diseases; although, the pathogenicity of gregarines is questioned by Steinhaus (1967).

Hive Products

Honey in Venezuela is used primarily for medicinal purposes and for the baking and candy industries. Very little is used in food preparation except during the Easter holiday when traditional "bunuellos" (yuca friters), a popular desert, are made. Most of the honey produced by the commercial beekeepers is sold at wholesale prices to national supermarkets. It is not uncommon, however, to see people along the road selling honey to motorists. Pollen and royal jelly are also popular hive products and commonly are used for medicinal purposes. Wax is used in the manufacture of cosmetics and decorative candles.

Beekeeping in the Amazon Territory

A project with the Sanemas Indians in the Amazon Territory was started in 1981 as part of an effort to introduce these people to Venezuela's civilization. In 1985 this project plus two similar ones in the area produced 8.5 metric tons of "puuna pudu" (bee honey) (J. R. Barragan and H. Gonzalez, pers. comm.).

SUMMARY

Most beekeepers in Venezuela were not prepared for the Africanized honey bee, so the initial effects of Africanization on the industry were severe. After nearly a decade of experience, though, many beekeepers are learning to manage this bee. Prospects for beekeepers will probably continue to improve as long as high honey prices remain an economic incentive. A new generation of beekeepers appears to be emerging, one that is better educated and better prepared. University courses, government-sponsored workshops, and national conferences have helped to inform beekeepers and non-beekeepers of potential problems and solutions. Yet the bee's unpredictable defensive behavior will pose a lasting threat to the general public. Additionally, Venezuelan beekeepers, even those with European stock, probably will seldom, if ever, see the 75-125 kg annual honey yields per colony which were common prior to Africanization. Increased competition for nectar sources by feral colonies appears to have made this a permanent problem in Venezuela.

Acknowledgements

In cooperation with the Louisiana Agricultural Experiment Station. Mention of a proprietary product does not constitute an endorsement by the USDA. The authors wish to thank Waldemar I. Vogel, Ricardo Gómez-Rodríguez, Marcelino Calvo Díaz, Alcides Escalona, Judith Principal de D'Aubetterre, James Bozdech-Hellmich, J. Anthony Stelzer, Anita M. Collins, John R. Harbo, José D. Villa, and Robert G. Danka for valuable discussion. The authors also wish to thank J. Anthony Stelzer and Lorraine Davis for technical assistance.

LITERATURE CITED

Collins, A. M., Rinderer, T. E., Harbo, J. R., Bolten, A. B. 1982. Colony defense by Africanized and European honey bees. *Science* 218:72-74.

Collins, A. M., Rinderer, T. E., Tucker, K. W. 1988. Colony defence of two honey bee types and their hybrids 1. Naturally mated queens. *J. Apic. Res.* 27:130-140.

Danka, R. G. 1987. *An assessment of Africanized honey bees (Apis mellifera L.) as crop pollinators*. PhD Dissertation. Louisiana State Univ, Baton Rouge.

Danka, R. G., Rinderer, T. E., Collins, A. M., Hellmich R. L. 1987. Africanized honey bee (Hymenoptera: Apidae) responses to pollination management. *J. Econ. Entomol.* 80:621-624.

Gómez-Rodríguez, R. 1986. *Manejo de la Abeja Africanizada.* Dirreccion General Desarrollo Ganadero, Caracas, Venezuela. 280 pp.

Hellmich, R. L., Rinderer, T. E., Collins, A. M., Danka, R. G. 1986. Comparison of Africanized and European queen-mating colonies in Venezuela. *Apidologie* 17:217-226.

Hellmich, R. L., Collins, A. M., Danka, R. G., Rinderer, T. E. 1988. Influencing matings of European honey bee queens in areas with Africanized honey bees (Hymenoptera: Apidae) *J. Econ. Entomol.* 81:796-799.

López-Palacios, S. 1986. *Catálogo para una Flora Apícola Venezolana.* Consejo de Desarrollo Científico y Humanístico de la Universidad de los Andes, Merida, Venezuela.

Otis, G. W. 1982. Population biology of the Africanized honey bee. In *Social Insects in the Tropics*, Vol. I, ed. P. Jaisson, pp. 209-219. Paris: Université Paris Nord.

Rinderer, T. E., Blum, M. S., Fales, H. M., Bian, Z., Jones, T. H., Buco, S. M., Danka, R. G., Lancaster, V. A., Howard, D. F. 1988. Nest plundering allomones of the Fire Bee *Trigona (Oxytrigona) mellicolor. J. Chem Ecol.* 14:495-501.

Roubik, D. W. 1978. Competitive interactions between neotropical pollinators and Africanized honey bees. *Science* 201:1030-1032

Roubik, D. W. 1979. Africanized honey bees, stingless bees, and the structure of tropical plant-pollinator communities. *Proc. IVth Int. Symp on Pollination, Md. Agric. Exp. Stn. Spec. Misc. Publ.* 1:403-417.

Roubik, D. W. 1980. Foraging behavior of competing Africanized honey bees and stingless bees. *Ecology* 61:836-845.

Steinhaus, E. A. 1967. *Principles of Insect Pathology.* New York: Hafner Publishing.

Stejskal, M. 1955. Gregarines found in the honey bee *Apis mellifera* Linnaeus in Venezuela. *J. Protozool.* 2:185-188.

Stejskal, M. 1966. Amibiasis. *Am. Bee J.* 106:292-293.

Stejskal, M. 1972. Akklimatisierung de Carnica in den Tropen (Acclimatization of Carniolan bee in the tropics). *Allg. Deut. Imkerztg.* 6:319-322.

Taylor, O. R., Levin, M. D., 1978. Observations on Africanized honey bees reported to South and Central American government agencies. *Bull. Entomol. Soc. Am..* 24:412-414

Thiman, R., Aymard, G. 1982. *Flora Apícola de la Mesa de Cavaca.* UNELLEZ, Guanare, Venezuela.

AUTHOR INDEX

For Product Safety Concerns and Information please contact our EU representative GPSR@taylorandfrancis.com Taylor & Francis Verlag GmbH, Kaufingerstraße 24, 80331 München, Germany

Batch number: 10725935

Printed by Printforce, the Netherlands